I0893261

CHINA'S TRANSITION TO MODERNITY

CHINA'S TRANSITION TO MODERNITY

The New Classical Vision of Dai Zhen

MINGHUI HU

UNIVERSITY *of* WASHINGTON PRESS | SEATTLE & LONDON

© 2015 by the University of Washington Press
Printed and bound in the United States of America
Composed in Minion Pro, typeface designed by Robert Slimbach
18 17 16 15 5 4 3 2 1

All rights reserved. No part of this publication may be reproduced or transmitted in any form or by any means, electronic or mechanical, including photocopy, recording, or any information storage or retrieval system, without permission in writing from the publisher.

University of Washington Press
www.washington.edu/uwpress

Library of Congress Cataloging-in-Publication Data
Hu, Minghui.
China's transition to modernity : the new classical vision of Dai Zhen / Minghui Hu.
p. cm.
Includes bibliographical references and index.
ISBN 978-0-295-99476-5 (hardcover : alk. paper)
1. Dai, Zhen, 1724–1777. 2. Philosophy, Chinese—1644–1912. 3. China—Intellectual life—1644–1912. 4. China—History—Qianlong, 1736–1795. I. Title.
B5234.T324H83 2015
181'.11—dc23
2014050175

The paper used in this publication is acid-free and meets the minimum requirements of American National Standard for Information Sciences—Permanence of Paper for Printed Library Materials, ANSI Z39.48–1984. ∞

To Chiung-chi and our daughter, Clio

CONTENTS

ACKNOWLEDGMENTS

My interests in both Chinese history and the history of science commenced at UCLA. Benjamin Elman and Theodore Porter consistently and methodically guided me through difficult and uncertain times; Joyce Appleby, Perry Anderson, Ted Huters, and Sharon Traweek all left important imprints in my intellectual formation. Kenneth Pomeranz kindly offered me my first academic position at the University of California, Irvine, in 2002. Prasenjit Duara selected me to be his postdoctoral fellow at the University of Chicago from 2003 to 2005. Their support in launching my career is truly appreciated. Many institutions nudged me along my journey as a professional historian: the Needham Research Institute, the Institute of Modern History at Academia Sinica, the Ricci Institute at the University of San Francisco, Tsinghua University, Korea University, the Vatican Library, the British Library, the Library of Congress, Bibliothèque nationale de France, the Fu Sinian Library at Academia Sinica, the Institute of Qing History at Renmin University, the Palace Museum in Beijing, and the Palace Museum in Taipei. The following foundations provided me with financial support that eventually led to the completion of this book: the Chiang Ching-kuo Foundation for International Scholarly Exchange, the Andrew W. Mellon Foundation, the Ricci Institute at the University of San Francisco, the Center for the Pacific Rim Studies at the University of California, the academic senate committee on research and the Institute of Humanities Research at the University of California, Santa Cruz, and Academia Sinica.

I started this book soon after I joined the faculty in the Department of History at the University of California, Santa Cruz. My East Asian colleagues at Santa Cruz, including Gail Hershatter, Emily Honig, Noriko Aso, Alan Christy, Christopher Connery, and Lisa Rofel gave me a nourishing and stimulating intellectual home. Many friends and colleagues read and commented on the entire manuscript or parts of it: R. Kent Guy, Laura Hostetler, J. G. A. Pocock, Sam Gilbert, Isabelle Landry-Deron, Chu Pingyi, Lu Miaw-fen, Chang So-an, Jiangping Chen, Catherine Jami, Guy Allito, Edward Shaughnessy, Toby Huff, Françoise Waquet, Wang Hui, Charles Hedrick, and Nathaniel Deutsch. I owe special gratitude to Gail, Kent, and So-an, who are the unwavering supporters of this book. My special thanks to Dong Jiangzhong, who translated and helped publish earlier forms of chapters five and seven in Chinese. My editor, Lorri Hagman, continued to support me with integrity and honesty. She is the best editor one could hope for. The production team at the University of Washington Press is also outstanding, and I would like to thank Jacqueline Volin for her attention to detail.

A note about the cover image: Dai Zhen's portrait is included in the collection *The Portraits of Qing Scholars* (Qingdai xuezhe xiangzhuan), volume 1. The first of the two volumes originated from the private collection of the painter and scholar Ye Yanlan (1823-98), one of many students who attended the Hall of Sea Learning (Xuehai Tang), the private Confucian academy founded by Ruan Yuan in Canton in 1820. Ye Yanlan studied under the leading instructor Chen Li (1810-82), who advanced the intellectual agenda of evidential scholarship movement. Ye passed the metropolitan exam in 1856 and managed to navigate through the Taiping civil war to the end of the Sino-Japanese War in 1895. He claimed that he had collected many authentic portraits of Qing scholars from reliable sources. He Yikai's recent monograph *A Study of the Portraits of Qing Scholars* identifies and confirms these sources.

After inheriting the family collection, Ye Gongzuo (1881-1968, Ye Yanlan's grandson), edited and published most of his grandfather's lifelong work in 1928, to immediate acclaim from intellectual celebrities. Ye Gongzuo experienced the fall of the Qing dynasty and the political ups and downs of the new republic; he was appointed to several minor positions on cultural affairs during the republican years. Throughout his career, he continued to collect and copy portraits of other Qing scholars and published the second volume in 1953, which he submitted to Mao Zedong. Mao expressed interest in the collection, asking him: "May I read the first volume as well?"

Perhaps as a result of Ye Yanlan's evidential training with the accuracy and reliability of the authentic portraits as well as his fascination with photography later in his life, *The Portraits of Qing Scholars* is endowed with an eerie sense of photorealism. Dai Zhen's portrait also conforms to his friend Hong Bang's description of Dai's physical appearance.

What else can I say to Chiung-chi and our daughter, Clio? Raising Clio with Chiung-chi has been just as exhilarating and frustrating as writing this book. I have enjoyed every minute of our life together, even during intractable challenges.

CHINA'S TRANSITION TO MODERNITY

CHAPTER 1

THE MAN AND HIS TIMES

Individuality was subversive in imperial China. Dai Zhen's 1776 book *An Evidential Analysis of the Meaning of Terms in the Mencius* (Mengzi ziyi shuzheng) would come to be recognized as a fully developed and systematically articulated vision of the new individuality in China, but when it was first published, it failed to arouse much interest—just the occasional condescending remark about a lowly scholar without an examination certificate speaking about challenging philosophical principles. It was not until the twentieth century that historians began to identify the text as the fusion of early modern individualism and objective methodology, a blow against the Neo-Confucianism that had dominated the intellectual and cultural scene in China for centuries.[1]

In early modern China, individualism was not typically articulated in philosophical or political discourses. It was buried in commentaries on the classical scriptures. By conceiving of the individual as a social unit rather than a thinking or moral subject, Dai Zhen (1724–77) began to systematically articulate the social, economic, and biological needs of each individual in society. This intellectual agenda was cleverly disguised in his studies of ancient texts and appeared to conform to the acceptable practices in a rather long commentary tradition of the Confucian canon.[2]

Any articulation of individuality was, however, meant to challenge the existing imperial orthodoxy, political hierarchy, or social order. Why would Dai make this risky move? Situated in Huizhou, the commercial hub of eighteenth-century China, Dai did not belong to any corporate, military, or merchant lineage. He did not succeed in the civil service examination, and he did not become influential politically. In short, Dai did not enjoy any initial privilege accruing to local, commercial, and national elites in his times. While sophisticated drinking games and poetry composition were expected of polite society in the imperial capital, to which Dai relocated, his manner was ever serious, awkward, and stoic. Throughout his life, he never composed a poem in private or for social etiquette. As a misfit in literati society, Dai would not have socialized with the many "power elites" who controlled the intellectual agenda of his times. Dai took as his starting point the intellectual endeavors of Yan Yuan (1635–1704) and Li Gong (1659–1733), two critics of the Neo-Confucian orthodox view of self and society. Neo-Confucianism was a system of moral, social, and metaphysical philosophy first articulated in the twelfth century and later enshrined in the newly consolidated Ming dynasty elaborately at the beginning of the fifteenth century. The Neo-Confucian orthodoxy did not evolve without significant challenges in the following three centuries. Yan Yuan and Li Gong, in particular, provided an effective deconstruction of the core of Neo-Confucianism that made it possible for Dai to build a positive picture of how individual members of a society should and could weave their social, economic, or biological needs into a collective formation.[3] This was a pragmatic, but not radical or subversive, intellectual move. Dai's humble background surely affected how he viewed the structure of intellectual elites of which he became a part later in his career.

Dai Zhen has long been portrayed as a forerunner of modern individualism in China. The celebrated intellectual Hu Shi (1891–1962) was the first to detect this new approach to Confucianism in Dai Zhen's writings. In *The Philosophy of Dai Zhen* (Dai Dongyuan de zhexue), published in 1927, Hu presented *An Evidential Analysis of the Meaning of Terms in the Mencius* as an intellectual tour de force that should have altered the course of intellectual history in China. Hu called it a "new philosophy" from which Dai could simultaneously attack metaphysics and pseudoscience. According to Hu, Dai's methodology later became an indigenous embryo for an uncompromising and potentially revolutionary humanism. Dai not only broke down the distinction between textual and nontextual methodologies in the eighteenth century but also anticipated the nineteenth-century debates on the distinction between essences and tools in classical learning. It was Dai's revolutionary stance, according to Hu, that kept his critics and even his followers from grasping his true message. In other words, Hu Shi reinvented Dai as a prophet of intellectual modernity in China.[4] Mizoguchi Yūzō, an intellectual historian with keen comparative insights, dubbed Dai's systematic formulation of individualism the "germination of modernity" (*kindai no hōga*), and he felt that it anticipated the emergence in the nineteenth and twentieth centuries of a scientific worldview based on the idea of "public principles" (*gong li*), which turned the normative structure of heavenly principles into a universal system of human values.[5]

Despite his laudable contribution, Dai remained a follower of Confucius after he rejected the imperial orthodoxy and Neo-Confucianism of his times to develop his classical vision. In what sense was Dai still a Confucian? The question is loaded with difficult political and intellectual implications. In political terms, few scholars in imperial China would have denied Confucius's role as the mediator and transmitter of the ancient system of thought and value (*dao*), since that is in fact how Confucius characterized himself. When Li Zhi (1527–1602) cast doubt on the great teacher's record, he was arrested for sabotaging the imperial orthodoxy of the Ming dynasty and committed suicide while in prison at the age of seventy-six. Unlike Li Zhi, Dai repeatedly attempted to pass the highest hurdle in the civil service examination system and sought employment as a staff member in the Qianlong emperor's court (r. 1735–96), proof that political subversion was not his goal. Was he, then, a genuine believer in the Confucian canon? The answer should most definitely be affirmative. His commitment was nevertheless complicated in two senses: he turned into a fierce critic of the Neo-Confucian tradition in his mature years, as indicated by his *An Evidential Analysis of the Meaning of Terms in the Mencius*, and he disagreed with Hui Dong's authoritative approach based on the Han dynasty exegetes, evident in his studies of the Palace of Light.

If he was neither a Neo-Confucian defined by the Qing imperial orthodoxy nor an original Confucian defined by the exegetical legacy of the Han dynasty, what kind of Confucian was he? Dai's methodology, based on early modern European cosmography and spherical trigonometry, was the main tool for his reconstruction of the classical world. The combination of his technical methodology and classical vision shows that he was neither Neo-Confucian nor a conservative who advocated the return of the original Confucianism. He was, rather, a visionary thinker with distinction.

Dai and his admirers did not always share the same commitment to the Confucian canon; they did not always agree on how to interpret specific passages; and they probably did not agree on how to use empirical evidence for interpreting the classical texts. But they had a common utopian vision of the classical world and believed that their innovative methodology was the only way to realize that vision. The methodology they all shared accommodated early modern European sciences and adopted a more tolerant and inclusive view of "Western learning." Their shared vision was to search for a cosmological and political order, and they all believed this vision was more splendorous and magnanimous than that of their Neo-Confucian predecessors. In other words, they developed a blueprint of the cosmological and political order of the classical world that they believed their Neo-Confucian predecessors had systematically and purposively distorted and ignored.

Despite their shared classical vision and innovative methodology, Dai and his admirers remained Confucians in the sense that they believed that the classical world, the essential touchstone for moral and political action, was accessible only via the ancient records mediated by Confucius. But they were quite certain that ancient records had been changed and damaged over time—the very reason why every effort had to be made to reconstruct them. They did not hesitate to seek in unexpected places the means needed to rebuild antiquity. On the contrary, that these "new" Confucians still operated within the interpretative paradigm of the Confucian canon should not preclude our appreciation of their innovative methodology and vision. Dai was, above all, the trendsetter and innovator in both.

Dai Zhen began life as a nobody. But he nevertheless managed to seize the attention of Beijing's cognoscenti in 1755, and for over a decade he was the intellectual toast of the Lower Yangzi Delta. This came to a halt in 1777, with his move to Beijing after appointment to the imperial library project called The Complete Collection of Four Treasuries (Siku Quanshu), a post he occupied until his death. Dai rode the high tide of classical scholarship, but how his legacy was perceived has changed over time. Before 1870 he was principally regarded as an outstanding evidential scholar and mathematical astronomer; it was not until the beginning of the twentieth century that he was elevated to serve as a critical thinker who freed China from its feudal past.

Dai considered Neo-Confucian orthodoxy to be insensitive and reified tools for political oppression. This should be understood, however, in the context of expressions such as "self and society" and "the private and the public," which could be formulated only with the rise of new ideas of individuality. Dai, his followers, and his critics were all keenly aware of the political implications of his description of the relations among individuals as the basis of society. The study of individuality in China has shed new light on individualism as a modern ideology, and comparison of the histories of self and society in early modern China and Japan complement the study of these concepts in the framework of European liberalism.[6] Dai's new social philosophy was made possible by the methods he developed to explore the objective existence of the classical world.[7]

More importantly, Dai's political vision was derived from his methodology, from mathematical astronomy to analytical phonology.[8] All told, he produced twenty-two publications consisting of 2,129 pages (104 *juan*).[9] Most of what he wrote can only be described as highly technical. Yet contemporary historians and philosophers have focused mainly on Dai's political-moral philosophy, which accounts for less than 5 percent of his writings. The topics to which he devoted so much labor—mathematical astronomy, historical geography, analytical phonology, and historical etymology—have rarely been integrated into any investigation of his thought.[10]

Hu Shi was, once again, the first modern thinker to recognize Dai's methodology. "Classical scholars of the Qing dynasty," wrote Hu, "stressed how to *ground* their discourse. There were two distinct approaches to this. The first group relied on evidence, which could, according to Gu Yanwu (1613–1682), involve either direct evidence [*ben zheng*] or indirect evidence [*pang zheng*]. The second group relied on authority, meaning the use of ancient precedent to support their arguments. Hui Dong (1697–1758) and his followers belonged to the second group, and they worshiped the exegetes of the Han dynasty."[11] Hu prized the achievements of the first group, asserting that without the development of robust techniques for evaluating evidence there would have been no true classical scholarship in the Qing dynasty. The second group—Hui and his followers—had blindly worshiped the authority of classical scholars in the Han dynasty, and insisted that all classical knowledge should follow and expand the ancient authority of Han classical scholarship.[12] The approach advocated by Hui and his followers can be best encapsulated by historian Kai-wing Chow's phrase: a purist hermeneutics of classical scholarship.[13]

In Hu's view, Dai made two revolutionary contributions to evidentiary methodology: he broadened the discipline's scope by combining textual and nontextual evidence, and he maintained a zealous commitment to rebuilding the material base of the classical world. More importantly, Dai's methodology became the groundwork for a new political philosophy. Dai's followers and admirers usually invoked the honorific *tongru* (the complex associations of this term will be discussed later in this chapter) in referring to each other, a custom that Hu connected to the methodology of this new Confucianism.[14]

For example, Ruan Yuan (1764–1849), a high-standing official and a follower of Dai, called Dai a *tongru* and praised his colleagues Jiao Xun (1763–1820) and Lin Tingkan (1757–1809) with the same honorific because both Jiao and Lin brilliantly elaborated Dai's methodology and argued eloquently against a "purism" hermeneutics of classical scholarship.[15] Disinclined to attack Hui Dong's academic orientation directly, they chose to deconstruct the idea of *kaoju* (lit., rectification through investigation) and the scholarship associated with it. Jiao put it bluntly:

> It strikes me as bizarre that some classical scholars have invented a category called *kaoju*. Grouping thinkers in terms of chronology, they say that the scholars of the Tang dynasty are superior to those of the Song, and those of the Han are superior to those of the Tang. . . . Every word written by Xu Shen and Zheng Xuan [of the Han dynasty] is cherished like a rare and precious jade; no *kaoju* scholar would ever doubt their interpretations. I cannot help thinking that if this fashion carries on, *kaoju* scholars will eventually look either stupid or hypocritical. They will render ancient words unpronounceable and incomprehensible; they will also make the ancient institutions look trivial, burdensome, and utterly inefficient. They resemble those cowards and imbeciles obliged to rely on a powerful family rather than looking after themselves. These cowards and imbeciles brutalize and deceive the foolish and unworthy in the name of the powerful family, without ever being permitted to enter that family's mansion and

> so [are] denied all knowledge of its rooms and hallways, the decoration on its walls, doors, and windows.[16]

For Dai, Jiao, and Lin, methods that were logically sound and based on evidentiary principles should have been sufficient by themselves to yield a convincing picture of ancient China. There ought to have been no need to rely on ancient authority. The implications of such an epistemological stance, which emancipated scholars from the reliance on a tradition of commentary on classical texts that had lasted millennia, were radical. It was as if these new Confucians had lost interest in the endless and convoluting loops of textual interweaving. They had little patience for those who considered any commentary on classical texts sacrosanct, elevating the Han commentaries to a position of infallibility, immune from intellectual challenges.

Dai's methodology had its historical bounds. Some historians go as far as recasting Dai's agenda as a form of "humanist positivism" because he prioritized evidence over ancient authority.[17] But Dai did not value evidence alone. Putting together the various clues gleaned from evidentiary analysis required a utopian vision, itself constitutive of his interpretations. If we apply to Dai, Jiao, and Lin the language of modern positivism, we anachronistically misconceive that vision. Viewed from another angle, the history of positivism in the twentieth century, particularly in the form of logical positivism, has been a history of antirealism.[18] But Dai was always a utopian realist. Dai's moral and social philosophy could become conceivable only if the totality of the classical world demonstrated a cohesive whole of cosmological and political order. Any evidentiary research on a specific aspect of the classical world presumed, in his mind, to enhance and increase understanding of the classical world as a single whole, a Confucian "cosmopolis."[19] Because of his ambitious, or even unrealistic, goal in the reconstruction of the classical reality, Dai was repeatedly denounced by high-standing officials and criticized by his esteemed colleagues. Even more striking was Dai's application of mathematical astronomy from post-Renaissance Europe to interpret the classical Confucian texts, connecting him paradoxically to the struggle between Chinese scholars and European missionaries. This connection necessitates a different line of historical research, following the traces from evidentiary methodology to mathematical astronomy.

This book is about the contexts and development of Dai Zhen's new vision of cosmological and political order. To counter previous characterizations that made him a prophet of intellectual modernity, we shall neither view him through the lens of Neo-Confucianism nor focus exclusively on his most celebrated philosophical treatise. Instead, we should begin tracing his involvement with European missionaries and their mathematical astronomy. We should rethink the relevant circumstances and premises from which Dai began to develop his methodology and vision.

LOCALITY AND SCHOLARSHIP

Dai Zhen was identified as the founder of the so-called Wan school in both Chinese and Japanese scholarly literature. Historians in China have conventionally grouped scholars by their native city, county, or province—such as Yangzhou, Huizhou, Wan (Anhui), Wu (Suzhou), and Tongcheng. This habit reflects how literati spoke of themselves. But we must critically examine scholarly self-portraits since the fabrication of schools often reveals less than it conceals. When a literatus—say, Zhang Xuecheng (1738-1801)—described the distinct local culture of the area "east of Zhe" (Zhedong) in contrast to the area "west of Zhe" (Zhexi), he was certainly not defining a "school" in our modern sense; he was speaking to his fellow literati in accordance with specific intellectual, political, and social purposes and contexts.[20] This is something quite different from the historian Zhang Shunhui identifying scholarship in Suzhou with specialization (*zhuan*), scholarship in Huizhou with refinement and precision (*jing*), and scholarship in Yangzhou with comprehensiveness and interconnectedness (*tong*).[21] These two types of mapping should be carefully separated in our analysis.

In 1924 Liang Qichao (1873–1929), a prominent reformer and intellectual, wrote a pioneering essay on the "geographical distribution of scholarship in early modern China," asking what it meant to attribute an idea or a line of inquiry to a specific group in a given locality. He posed a series of questions:

- Why did the Lower Yangzi Delta virtually monopolize academic production?
- Why did evidential scholarship thrive in the Lower Yangzi Delta

while Neo-Confucianism remained the standard intellectual commitment on the North China Plain?
- Why did the number of excellent scholars from the North China Plain decline between the Early Qing and the High Qing?
- Why did the number of fine scholars from the middle Yangzi increase dramatically from the Early Qing to the Late Qing?
- Why did some transportation hubs never become centers of scholarship?
- Why did neighboring regions have markedly different scholarly approaches?
- Why did certain regions produce more sophisticated and differentiated scholarship than others?
- Why did some regions have a long and continuous tradition of scholarship?[22]

Although he did not attempt to answer all of these questions, he did challenge the uncritical association between place and scholarship. Few heeded him, and throughout most of the twentieth century nationalist historiography reigned supreme along with labels like "the Tongcheng school" and "the Yangzhou school." The so-called regional or local approaches, often promoted as antidotes to the excesses of nationalism, have mostly been sidelined throughout the twentieth century in China.[23]

Some case studies could illuminate the examined association between place and scholarship. The influential political and intellectual movement known as New Text Confucianism originated from a specific locality, as has been shown in a study examining the continuity of scholarly interests within the Zhuang and Liu lineages of Changzhou that links the reemergence of New Text Confucianism in Changzhou, after fifteen hundred years of obscurity, to the factional struggle that broke out during the latter part of the Qianlong emperor's reign. One of the main players, Zhuang Cunyu (1717–88), encountered so much frustration at the hands of the powerful faction led by Heshen that he resigned from his post and abandoned the court. He had hoped that a new reading of the classics would permit him to revive Donglin-style literati activism.[24] In contrast, another study analyzes the pursuit of native-place literature in twentieth-century China as a search for cultural authenticity in the midst of fierce ideological battles. Presenting the locality as an authentic homeland meant absorbing a variety of discourses connected to frontier communities, nature, the notion of a timeless essence, and ancestral aesthetics.[25]

These explorations of unexamined and vague assumptions surrounding the discussions of "schools" offer a variety of means to associate locality with intellectual endeavors. Similarly, merely labeling Dai as a founder of the "Wan school" has done more harm than good to our attempts to understand him. Rejection of the conventional label "Wan school," a sketch of relevant circumstances from which Dai developed his vision, requires that we search for new historical connections between European missionaries and the classical exegesis of the Confucian canon. The past scholarship regarded such connections as merely sporadic, local and rustic. The power elites of the imperial bureaucracy never dignified European missionaries with a serious intellectual response, let alone considering it as part of the classical exegesis of the Confucian canon. This book, by contrast, considers two such connections: the struggle between Chinese scholars and European missionaries and their debate over the competing visions and origins of civilizations, as they established many interlocutors with whom Dai had implicitly or explicitly been in dialogue throughout his career.

STRUGGLE BETWEEN SCHOLARS AND MISSIONARIES

The global legacy of the Society of Jesus (founded in 1540, dissolved in 1773, reconstituted in 1814) has been extensively documented and analyzed.[26] The most significant challenge the Jesuits faced in eighteenth-century China was the so-called Rites Controversy, when the willingness of the missionaries to tolerate local practices was tested by traditions such as the worship of dead ancestors and of the sage Confucius—idolatry prohibited by the Roman Catholic Church.[27] While the controversy has fascinated those interested in intercultural clashes, figuring out the role played by the Jesuits can be challenging, since their own accounts vary and they appear to have presented different faces to the different people they met.[28] Historians interested in the Jesuit mission to China have tended to rely on one of the three following perspectives:[29]

Accommodation. Some historians have shown how Jesuits and the Chinese scholar-officials they encountered "made room for" each other in their respective epistemological frameworks. These historians have tended to assume that both parties were somehow benign historical agents who would negotiate and compromise in order to serve higher purposes.

Incommensurability. The historians devoted to this approach have focused on communication failures and the conflicts that ensued. They assume that the two cultures' epistemes were so different that they could not possibly communicate on the same terms; cross-cultural communication was by definition miscommunication. Research has focused on linguistics, communication strategies, and the cultural contexts that complicated translation.

The rise of the West. The dominant mode of analysis grew out of comparative historical sociology. Historians who subscribe to this Weberian paradigm refuse to assign the Chinese literati or the Jesuit missionaries a moral character. Instead, they emphasize how and why the parties succeeded or failed in their attempts to develop rational institutions. Rationality is the main concern.

Our concern here is with the *consequences* of the Jesuit legacy. What kind of technical and quantitative knowledge was brought to China, and how was it perceived and used? What kinds of claims were made about transcendental or supernatural beings? What kinds of visions and origins of civilizations were articulated? How were they perceived and used? Two questions of paramount concern: How was it that the foreigners possessed technical superiority? How did the origins of the European and Chinese civilizations overlap or diverge? This book does not provide an overarching narrative about the Jesuit sciences in early modern China, but looks at the Jesuit legacy from the Chinese scholars' perspectives, explaining why they cared about it.[30]

The Jesuit legacy has recently been the subject of both positive and negative assessments.[31] Many have seen its rejection as a tragic missed opportunity—a view echoed by politicians, policy-makers, and intellectuals over most of the twentieth century. Xi Zezong (1927–2008), for instance, a distinguished astronomer who later turned to the history of science, offered a different and interesting reflection on the Jesuit legacy. In 2000 he published an article entitled "On the Shortcomings of the Kangxi Emperor's Scientific Policies" (Lun Kangxi kexue zhengce de shiwu). When Xi delivered a lecture on the topic in Singapore, the *Straits Times* covered the story under the headline "New Culprit for China's Scientific Slide."[32]

The "culprit" was neither a band of crafty Jesuit missionaries nor the xenophobic classical scholars who rejected Western sciences. The problem, as Xi's title revealed, lay in a series of policy mistakes made under the Kangxi emperor's watch. The historian was issuing a warning at the dawn of a century that, many claim, will belong to China. Soon to become the world's most important economy, the country is set to invest heavily in science, technology, and medicine for many decades to come.[33] Xi seemed to be drawing a parallel between policies of the imperial past and the potential for huge mistakes in the present. His historical judgment was clear and blunt: the indigenous development of science, technology, and medicine culminated between 1573 and 1644. In fact, the accomplishments of the day compared favorably with those of the West. The Manchu conquest brought a sudden halt to this progress, until political and economic stability, along with the Jesuit translation into Chinese of relevant European works, set the stage for a slow revival. According to Xi, that revival was balked and China experienced a crucial loss of momentum at the inception of the global scientific competition. This missed opportunity was due entirely to the mistakes of the Kangxi emperor.[34]

What were the mistakes? Xi listed five. The Kangxi emperor did not (1) put the best talent in charge of the Qin Tian Jian (the imperial institution in charge of the timing of state rituals and mathematical astronomy), (2) promote a research institute at court, (3) recognize the true utility of scientific instruments, (4) fully appreciate Western theories, and (5) reject the fairy tale that ancient China was the ultimate source of Jesuit mathematical astronomy.[35] Each criticism is, in a sense, anachronistic, and the "what if" quality fails to grapple with the complexities of temporality; yet, they serve as helpful correctives to the received wisdom. After all, the Jesuit legacy was not exactly modern or progressive, and Chinese scholars pursued it vigorously.

Moreover, on their arrival at the Qing court the Jesuits became part of an intellectual hierarchy built up of many coexisting ideas. Speaking of the "first Western impact" casts the West as a more advanced and sophisticated civilization, a generalization in keeping with a long discarded worldview. It is more fruitful to speak of the Jesuits as intermediaries who simultaneously affirmed both the Catholic and the Chinese hierarchy. Even if we were to attempt a quantitative comparison of hosts and visitors, we would soon find that the astronomy imported by the Europeans was only a bit more accurate than its Chinese counterpart. Only a fraction of Copernican science reached the Qing court, which essentially subsisted on outdated cosmology and cosmography designed by the Danish aristocrat Tycho Brahe (1546-1601) until 1911.[36] Historians of science still debate the relative merits of Tycho Brahe's cosmos and China's indigenous astronomy, but there was never any question of the utter difficulty in reconciling them. The classical scholars of the eighteenth century, especially Dai Zhen, applied their own tradition—the mathematical astronomy accumulated in Chinese imperial courts over two thousand years—to mastering Jesuit mathematical astronomy, which they viewed as a branch of their own cultural heritage.

When Jesuits presented themselves as mathematical astronomers, they gained entry to a vast and complicated realm of classical learning and other practical knowledge.[37] Xi reduced this extremely complex interaction to a matter of science policy. Because he made modern rationalism his standard and telos, he fell prey to an ahistorical misunderstanding: anything but modernizers, the Jesuits were militantly opposed to the Reformation and Copernicanism. By many measures, Dai Zhen and his admirers were more open-minded and critically inclined than the Jesuits. Why then did they greet the Jesuit science with open skepticism? Dai Zhen and his admirers saw mathematical astronomy as a tool for deciphering the cosmological order, something that could not be entrusted to anyone outside the state apparatus—it had to be firmly coupled with the Confucian canon. For the imperial state, foreign evangelical agents had to be placed under strict surveillance. The need for political and legal constraints on Jesuit sciences became increasingly obvious to the Qing court and literati society as the Manchu empire consolidated control at the beginning of the eighteenth century. The outcome of the bitter struggle between classical scholars and European missionaries shifted as the political landscape continued to be restructured in the eighteenth century. The classical scholars in Qing China gradually gained an upper hand over the interpretation of Jesuit legacy, and Dai Zhen stood to benefit from it.

THE *TONGRU* AS VISIONARY SCHOLARS

China scholars usually render the term *tongru* as "polymath." That translation, proposed by *The Cambridge History of Chinese Literature* and supported by several dictionaries, does capture *tongru* as a generic noun.[38] Su Shi (1037–1101), for example, was in many ways a *tongru*, a "broadly learned scholar" deeply versed in hydraulic engineering, art, poetry, and belles lettres.[39] But Dai and his followers would have dismissed him as a dabbler in this and that (*zaxue*) rather than an erudite with a profound vision of the classical world.

Few historians would dispute that Dai was a visionary thinker, but not many would agree upon what that vision was. Dai's utopian thought hinged upon the word *tong*, which is usually rendered as "comprehensive" in English, while *ru* means "Confucian" or "literati." Some etymology may help. According to both the Chinese-Japanese dictionary *Dai Kan-Wa jiten* and the Chinese dictionary *Hanyu da cidian*, two monumental works comparable to the *Oxford English Dictionary*, a *tongru* is a person of great and varied learning, a definition that makes the received translation, "polymath," seem unproblematic. As one reads on, however, both dictionaries provide quotations that raise doubts. It is beyond dispute that in most official documents written between 200 and 1900, the word *tongru* connoted an elevated social status and broad learning. But many Qing scholars exhibited a more specific and nuanced understanding of the phrase. The first quotation in both dictionaries reads as follows: "Wild animals do not make good sacrificial animals, just as hybrid learning does not make one a *tongru*."[40] This passage, probably the first occurrence of *tongru* in any surviving record, comes from chapter 11, "Foundations of Governing" (Zhi ben), in the ancient military text *Wei Liaozi*, which dates far back to the Warring States period (500–221 BCE). This is probably the earliest text in which the term *tongru* appears. *Wei Liaozi* reads like a collection of fragments, hence trying to contextualize this quotation is useless. Still, we learn that a *tongru* is not just someone "of great or varied learning" or "a person acquainted with many fields of study" but someone whose erudition is refined, something superior to vulgar hybrid learning. A *tongru*, in other words, is civilized and mature, superior to unsystematically amassed knowledge.

The dictionaries' next few quotations list a few individuals respected as *tongru* who lived in the Eastern Han dynasty (25–220 CE). For instance, both Du Lin (d. 47), who discovered the Old Text version of *The Book of Documents* (Shang shu), and Jia Kui (174–230), who left us the enigmatic *Jia Kui on Mathematical Astronomy* (Jia Kui lunli), the highest achievement of its kind in ancient China, were referred to as *tongru* in *The Standard History of the Later Han Dynasty* (Hou Han shu). At that time the term likely referred to someone who went beyond specializing in just one of the Confucian classics.

What did *tongru* mean in the eighteenth century? A search of China's standard dynastic histories and in *A Draft History of the Qing* (Qing shi gao) for meaningful uses of the honorific reveals that during the eighteenth century *tongru* was an honorific applied to Dai and his followers centered on Ruan Yuan's patronage, or Ruan Yuan's circle.[41] It lost that meaning after the devastation of the Taiping civil war in the nineteenth century.

The final quotation cited in *Hanyu da cidian*, from a late-Qing anarchist named Liu Shipei (1884–1915), confirms not only that *tongru* meant something more than "polymath" in both the ancient period and the eighteenth century, but also the subsequent change in meaning it underwent. Liu claimed, "During the reigns of the Qianlong and Jiaqing emperors [1730s–1820s, roughly what is commonly called the High Qing], many *tongru* emerged."[42] This was certainly a reference to a specific group of scholars, but it said nothing about their agenda. Liu was not a naïve and careless scholar who thoughtlessly praised his predecessors; he saw himself as perpetuating the Old Text branch of classical studies championed by eighteenth-century Yangzhou scholars. Here the honorific meant someone more specific than polymath, implied something more profound than scholarly erudition, and appears to have pointed to something besides success in the civil service examination.

The piece from which Liu's quotation was taken, "On the Evolution of the Study of Writing in the Recent Past" (Lun jinshi wenxue zhi bianqian), discusses the emergence of a new group of philosophical and philological scholars in the seventeenth, eighteenth, and nineteenth centuries. This group had privileged learning as the pursuit of a higher truth and dismissed the cultivation of elaborate writing as trivial. Liu had belonged to that group before he embraced modern anarchism. His was a somewhat negative assessment of *tongru* because they turned away from belles lettres and focused on evidentiary analysis. The most brilliant writers of the day, said Liu, considered philosophy and philology to be lacking in authenticity and spontaneity (*xingling*), and the label *tongru* was part of a dismissal of their unadorned writing style.[43]

A second example fleshes out the picture. In 1902 Zhang Zhidong (1837–1909) advocated the creation of a short-lived institution called the Tongru Research Institute (Tongru Yuan). It was a unit for postgraduate studies attached to the Imperial University (Jingshi Daxuetang), the first state-sponsored academic institution that offered a modern curriculum modeled after those of European universities and a forerunner of today's Beijing University. Zhang's proposal was filled with references to *tongru* as men who had achieved the highest level of academic scholarship, more or less equivalent to those who had distinguished themselves by succeeding in the metropolitan and palace examinations. Zhang's use of *tongru* to describe those who mastered both Chinese and Western learning was clearly a novelty, but it reflected the legacy of the eighteenth century. The word *tongru* no longer belonged to bureaucratic eulogies. To home in on the moment when the word began to take on new associations, we need to go back another century or two.

In the eighteenth century, the word *tongru* appeared routinely in the literary anthologies published by Dai's followers and admirers. The following scholars were all praised as *tongru* in the prefaces to their collected writings.

Jiao Xun (1763–1820) set out his ideas about the technical methods that could shed light on classical knowledge in *Collected Writings from [the Studio of] Engraved Bamboo* (Diaogulu wenji), which was published posthumously. His ideas were presented in a wide variety of genres, such as rhymed prose (*fu*), poetry (*shi*), encomium (*song*), critique (*bian*), disquisition (*lun*), explanation (*shuo*), exposition (*jie*), exegesis (*shi*), and inquiry (*kao*).

Lin Tingkan (1757–1809), in *Collected Writings from the Hall of Collating Rituals* (Jiaolitang wenji), commented on the technical methods used in as varied a spectrum of genres as Jiao had presented in *Diaogulu wenji.*

Wang Chang (1724–1806) and Sun Xingyan (1753–1818), in their respective *Collection from the Hall of Warm Springtime* (Chunrongtang ji) and *Collection from the Hall of Inquiring Words* (Wenzitang ji), discussed and debated many issues prompted by the use of the sort of technical methods championed by Dai.

Any of these four scholars should provide us a witness account of Dai's vision. Like Dai, Lin Tingkan attempted to transcend the long-standing dispute between the proponents and the foes of so-called Western learning, which largely referred to the Jesuit legacy during the eighteenth century. By addressing this foreign body of knowledge, Lin set the ancients and the moderns in opposition—a familiar trope of his time. Typically the idealized scholars of the distant past—deeply versed in all manner of practical knowledge—served as foils for contemporary scholars, whose unfounded moral speculations were readily debunked. Those who never mastered analytical tools, Lin argued, were easily impressed—even amazed—by the Jesuits' achievements in mathematics, cosmography, and cartography, though their heretical implications repulsed more than a few.[44]

Lin mentioned Xu Guangqi (1562-1633), a Catholic convert who proposed reform on the imperial calendar at the end of the Ming dynasty, and Yang Guangxian (1597-1669), a defender of imperial orthodoxy against Catholic faith. Xu embraced the Catholic faith, studied mathematical astronomy, and admired a global map drafted by the pioneering Jesuit Matteo Ricci (1552-1660). In contrast, Yang, an indignant defender of imperial orthodoxy, viciously persecuted the Jesuits at the Qing court from 1664 to 1667. In Lin's view, neither had taken the correct path. To embrace the Jesuit sciences without having thoroughly plumbed them was as foolish as rejecting them out of moral indignation. Instead, Lin urged his fellow scholars to rely on their meticulous glosses and their close reading of classical scriptures, which would show them that these ideas from the West were neither entirely new nor utterly heretical.

The solution, Lin declared, lay in the study of both classical Chinese *and* Western knowledge, assessing each based upon its merit. Aspects of Western science squared with ideas found in *The Book of Documents* and *The Gnomon of the Zhou Polity: An Arithmetical Classic* (Zhou bi suanjing). The great scholars of ancient times could not have been ignorant of mathematical astronomy, Lin suggested. Granted, old texts did not provide the detailed exposition of mathematical astronomy the Jesuits offered—that was precisely why later scholars had to study Western learning. A study of both classical and Western knowledge would permit Confucians to reconstruct ancient cosmology and the institutions valued by the ancients. Lin warned his fellow scholars that as yet undiscovered ancient records might turn out to resonate with the ideas from the West.[45]

The second example comes from the patron saint of these new Confucians. Ruan Yuan, an admirer of Dai, directed the production of *The*

Biographies of Astronomers and Mathematicians, a collection comprising 243 entries that included many scholars mentioned in the standard dynastic histories, as well as thirty-seven Europeans introduced by the Jesuits.[46] The contributions of all these astronomers, Chinese and foreign, made up a grand progression that culminated in the Qianlong emperor's reign. The year in which this biographical project commenced (1799) was also the year the emperor died. For Ruan, the juxtaposition symbolized the end of an important era—for astronomers and for scholars in general. Many had contributed to the task of formulating a specifically Chinese astronomy based on a quantitative approach: "Over the past two thousand years, there have been seventy systems of mathematical astronomy. These go far beyond what any one man could accomplish by himself. In their astronomical measurements and adaptation of mathematical techniques, some were more accurate than others. However, each contributed something unique. As a result of these many contributions, mathematical astronomy continued to expand and should be preserved as exemplary [*dianyao*] for the future."[47]

Ruan believed that mathematical astronomy was an integral part of the ancient Zhou polity, an indispensable body of knowledge for all scholars. For two millennia, it had been abused, misused, or ignored by fraudulent, superficial, or feeble-minded scholars; only during the splendid reign of the Qianlong emperor had it been reintegrated with the imperial state, reviving the Zhou status quo.[48]

By publishing *The Biographies of Astronomers and Mathematicians*, Ruan invented—he would have preferred "rediscovered"—a professional category: *chouren*. There had been *chouren* in the past. The philologist and mathematician Tan Tai (birth and death dates unknown, *juren* 1786) traced the character *chou* to ancient agriculture, in which context it meant both "charting fields" and "counting rods." In fact, he traced it even farther back, to a hereditary occupation in the ancient Zhou polity: *chouren* were laborers. But the word was also linked to the official hereditary mathematical astronomers (*tianguan*), and later it expanded to include any hereditary group with scholarly pretensions, such as those with expertise in music, poetry, and rituals.[49] "Records of Ancient Technology" lists many professions, including masons, craftsmen, and palace builders; according to Ruan, mathematical astronomers should have been listed among those hereditary specialists.[50]

Ruan's rehabilitation of *chouren* as a respectable ancient occupation was part of his overall effort to promote technical methodology, advocated by Dai. Mathematical astronomy, he believed, was not merely a useful art; it was rare and intricate like ancient palaces and ritual vessels in the classical world.

Evidently the label "Confucian" was no longer restricted to the study of Confucius. Some of the paths to the classical world demanded a much broader knowledge. Ruan himself explained in the preface to *The Biographies of Astronomers and Mathematicians*, paraphrasing the ancient thinker Yang Xiong: "This work is designed to illustrate how the wonder of mathematics can explain our world, with all of its obscurities and nuances. Mathematics, a system of moral principle, is sufficient to guide us in both the cosmic and the human realms. For the Confucian, mathematics is the search for correctness in actual facts, not the immoral tool craftsmen and charlatans use to gain wealth. Those who are determined to connect [knowledge of] the celestial, human, and terrestrial realms would do well to study it in detail."[51]

Dai and his followers were those who recognized the equal importance of humanist vision and technical methodology for understanding the classical world. As Yang himself had proclaimed, "A Confucian [*ru*] is one who can connect [*tong*] knowledge of the celestial, human, and terrestrial realms; a technician [*ji*] is one who can only connect [*tong*] knowledge of the celestial and the terrestrial realms but not the human."[52] The emphasis here was on *tong* or making connections. Yang's adage suggested a comprehensive and symmetrical understanding of both the sciences and the humanities of the classical world, or, in Yang's language, knowledge of the celestial, human, and terrestrial realms. The unity of systematic knowledge ascribed to ancient China was considered one of its greatest merits.

For Ruan Yuan, these *tongru* were not just scholars whose broad expertise (*bo xue*) encompassed a significant number of different subject areas. Neither were they interested in merging several disciplines into a form of hybrid learning (*zaxue*). Instead, they believed in the undivided unity of the classical world, which could be understood on both technical and humanistic terms.

The model of *tongru* was Dai Zhen, whose name appeared in many compilations of scholarly lives by 1800. Ruan included Dai in *The Biographies of Astronomers and Mathematicians*, citing his accomplishments in mathematical astronomy, historical geography, analytical phonology, and historical etymology. For Dai, these four disciplines were the foundation for classical knowledge—he composed treatises on mathematical astronomy with its classical implications in mind. This was not merely an exegetical tool, but a part of the body of classical knowledge. After Dai, it was no longer possible to speak lightly of mathematical astronomy. Ruan concluded that his contribution was as significant as that of Mei Wending, who received a lengthy and laudatory entry in *The Biographies of Astronomers and Mathematicians*.[53] Mei's achievement, as Ruan saw it, was to incorporate the most advanced mathematical astronomy into classical knowledge.

At the beginning of the nineteenth century, the utopian ideal of *tongru* fell afoul of a resurgent emphasis on political realism. Utopian realists of the classical world seemed to have gradually faded out of the political and intellectual scene. Liu Wenqi (1789–1865), the scion of a prominent Yangzhou family, declared that the classical achievements of his family had been significantly influenced by Dai Zhen.[54] In 1877 he published a study of historical geography in the canonical Confucian classics. Wang Shiduo (1802–89), an eccentric scholar, wrote the preface:

> The ancients wrote books primarily to record events, and to make manifest the principles of how things were arranged and whether they succeeded or failed. Sometimes these records were too simplistic, and it is up to posterity to follow the ancient categories and seek proofs. However, since some proofs cannot be arrived at directly, they must be reached indirectly. For that which cannot be indirectly reached, [one seeks] contextual corroboration, going over it in exhaustive detail. One must study the [Zhou] polity of several millennia ago as if it still existed today. This is what I mean by *tongru*.[55]

What Wang had in mind was probably not the same approach Ruan had conceptualized. Some of the core elements of that earlier agenda remained recognizable in post-Taiping China, but the sort of mathematical astronomy that Dai had championed more than a century before now seemed outdated and irrelevant. By 1924, when Liang Qichao organized a conference in honor of Dai's two-hundredth birthday, he praised him as a great thinker and a pioneer of "the scientific method" and "humanistic philosophy."[56] Reformers and revolutionaries alike began to see Dai not as a Confucian, but as a genius who stood at the mouth of the Chinese river that flowed, to borrow Joseph Needham's image, "into the ocean of modern science."[57]

This book represents a new perspective on both intellectual history and the history of science in China. Historians of Jesuit science at the Qing court generally impose a chronology that terminates with the ejection of the missionaries from the court. They pay little or no attention to the legacy those Europeans left behind and how it affected the eighteenth and nineteenth centuries.[58] Intellectual historians, in contrast, often study Dai in terms of the dichotomy between the empiricist paradigm of Zhu Xi (1130–1200) and the idealist paradigm of Wang Yangming (1472–1529) (or swap in "intellectualism" versus "antiintellectualism"). They pay little or no attention to Dai's technical accomplishments, instead subscribing, for instance, to a three-stage development of Confucianism from classical Confucianism to the Neo-Confucianism that arose in the Song and Ming dynasties to the rebirth of Neo-Confucianism in the twentieth century. To these conservative historians, Dai has been seen as a deviation from Neo-Confucianism's true mission, with a "paradigm shift" taking place during the seventeenth and eighteenth centuries, as leading scholars broke with Neo-Confucian tradition.[59]

In a broader context, the historical sociologist Max Weber (1864–1920) undertook a breathtaking comparative history of religions and cultures globally and often spoke affectionately about "occidental rationality" as a standard bearer of European civilization. His theory of cultural development and comparative history can be encapsulated as "the rise of Western rationalism."[60] Is it possible to portray Dai Zhen in the rise of Western rationalism? What was Dai's place in global and comparative history? What was his distinct vision and methodology in the early modern world? These are the questions that drive the following exploration of Dai's life and thought, and the context in which it emerged and developed.

CHAPTER 2

HOW JESUIT SCIENCE CONQUERED THE KANGXI COURT

In 1664, a body of Manchu regents acted on a mildly annoying accusation that had passed through the central bureaucracy of the newly established Qing court. The accusation was drafted by a disenfranchised political elite named Yang Guangxian (1597-1669), who had already attempted to launch his career several times by pointing fingers at corrupt officials in the former Ming dynasty. Yang accused the Jesuit missionaries and their converts as conspirators who aimed to undermine or even overthrow the new Manchu Qing dynasty. To everyone's surprise, the regents decided to take the accusation seriously and consequently purged the entire Jesuit faction at court.

These events—perhaps the best-known persecution of the Jesuits in early modern China—informed Chinese scholars' assessment of the Jesuit legacy over the following century. United behind the indignant, low-ranking bureaucrat Yang Guangxian, a group of literati and court officials accused the Jesuits of using their technical skills as a cover for malevolent intentions. Most Jesuit missionaries were exiled to Macao, although some of their converts were executed, an admonition for others.[1] Historians have characterized this sequence of events in several ways, such as a calendar dispute, cultural incommensurability, or religious persecution. Most of them describe Yang as manipulative and scientifically incompetent, but few would deny that he prepared his case with consummate care or that his arguments resonated.[2] His legal reasoning convinced many literati and the case had long-lasting political and cultural consequences.

One of them was for mathematical astronomy to assume an importance in imperial politics as it claimed a place in the front row of techniques employed for reconstructing classical knowledge. During the eighteenth and nineteenth centuries, Jesuit mathematical astronomy became a bureaucratic standard for imperial astronomy. Meanwhile, Yang's collection of anti-Catholic writings, titled *I Cannot Do Otherwise* (Bu de yi), had dropped out of scholarly discussions by the end of the eighteenth century. The contrast would seem to suggest that Yang's impact on literati society had irrevocably dissipated. An anecdote from the end of the eighteenth century, however, provides a rather different idea of how classical scholars viewed Yang's screed.

One day in the winter of 1799, a book merchant carried a copy of *I Cannot Do Otherwise*, by then rare, to the wealthy bibliophile and bibliographer Huang Pilie (1763–1825). Huang found the title unusual, so he paid a piece of silver for the book. A piece of silver would have weighed between 148 and 185 grams, and in those days it could have purchased as much as 142 kilograms of rice, or roughly a six-month food supply for a small peasant household. Many wealthy landlords and collectors, however, would have paid far more for a copy of, say, a rare Song edition. Huang Pilie had hundreds of such editions, so in his library *I Cannot Do Otherwise* would have been picayune. However, Li Rui (1768–1817), a classical scholar and mathematician who was, like Huang, a native of Suzhou, told the collector that his friend and former mentor Qian Daxin (1728-1804) had never laid his eyes on the book and would very much like to read it. Huang then came to believe that he had made a good investment. He hired a craftsman to frame his copy of *I Cannot Do Otherwise* and asked Qian to write a postscript for it.

When Huang got his book back, he was quite taken by the postscript, in which Qian cited his colleague Dai Zhen: "Europeans used to pay a lofty price for this book and burned them in large numbers. They intended to stop its circulation entirely."[3] Qian then went on to remark that Yang had failed in eliminating the Jesuits "because he was not a competent mathematical astronomer. Also, he had no competent mathematical astronomers among his staff to assist him."[4] What did Qian mean by questioning Yang's competence? It is hardly unlikely, after all, that the book merchant had lauded Yang as a heroic defender of the Confucian canon and sworn foe of the loathsome Christian heretics. As a book collector, preserving this book was both a good investment and good politics. There was, however, this thorny business of mathematical astronomy. Huang's next move was to ask Li, known as both a competent mathematician and classical scholar, to explain mathematical astronomy to him. At that moment Li was working for Ruan Yuan, who was about to assume the governorship of Zhejiang, to direct the compilation of *The Biographies of Astronomers and Mathematicians* (Chouren zhuan). The anecdote about the book merchant's curiosity and Qian's postscript made its way into the brief entry for Yang Guangxian. The thrust of the biographic sketch was that Yang had not been a fine astronomer but a great defender of imperial orthodoxy.[5]

The anecdote links Yang Guangxian to the two main protagonists of this book: Dai Zhen and Qian Daxin. The two men quite unexpectedly met in Beijing in 1754, at a time when both had to cope with the duality of the Jesuit legacy—the challenge to the Confucian canon and the technical utility to the imperial state—which was symbolized by the conundrum of Yang's life.

The story of Dai Zhen begins at the moment in 1664 when Yang denounced the Jesuits as seditious foreigners; it ends at the moment when Dai's legacy provided the intellectual framework for *The Biographies of Astronomers and Mathematicians* around 1800. Yang's anti-Jesuit writing set in motion four key parameters that evolved into the backdrop against which Dai Zhen could argue for his case and develop his new vision in the eighteenth century. These parameters operate along the axes of imperial state and its rituals, cosmology and cosmography, astronomical accuracy, and the political and legal constraints on the missionary operation in China.

State Rituals. The imperial authority had long been wrapped with elaborate rituals to demonstrate as spectacularly as possible the mandate of Heaven. Mathematical astronomy had historically played a substantial role in timing the rituals. In some critical cases, such as solar eclipses, the imperial state had to predict when each celestial anomaly would happen in the future and prepare a specific ceremony in response to it. The European missionaries hijacked the role of mathematical astronomers in this process and, in many scholars' view, dangerously threatened the integrity of the ritual.

Cosmography. European missionaries decoupled cosmography from cosmology. They strategically avoided the exegetical challenge of deciphering the *Book of Changes* (Yi jing), which virtually all classical scholars in Qing China believed to be the only legitimate source of cosmology. The missionaries presented cosmography as the ad hoc conceptual device to improve astronomical accuracy, with the aid of telescopes and astronomical instruments. They disassociated cosmology from their presentation and argued only for the indispensible utility of their sciences.

Astronomical Accuracy. Catholic missionaries were indeed better equipped with their observational astronomy and spherical trigonometry than were their Chinese counterparts in terms of depiction and prediction of the movement of celestial bodies. They were also familiar with the difficulty of correlating solar and lunar cycles in order to implement the imperial calendar in Qing China. Between a rejuvenated Catholic Church, which militantly opposed Reformation, and the expanding Qing dynasty, which desperately required a functioning and better calendar than their Ming predecessor, the historical junction allowed the missionaries to transfer their observational astronomy and spherical trigonometry to Qing China in a timely manner. The missionaries' bricolage in choosing observational data and computational techniques, however, came from both the Protestant and Catholic worlds and gave them the technical edge to achieve astronomical accuracy at a higher level.

Political Constraints. Political circumstance changed, for a brief window, from a cosmopolitan style of ruling to a stern "Manchu" style that appeared more suspicious of Chinese subjects and foreign missionaries. Yang Guangxian took advantage of this opportunity and made a legal case against the missionaries. His complaint successfully went through both the Ministry of Rites and the Ministry of Punishment. Consequently, the missionaries were exiled to Canton and their converts executed. Yang's persecution permanently established the ambivalent loyalty of the European missionaries to the Manchu emperor as the imperial authority gradually detected their unwavering vow to the pope. The intellectual elites in China liked to portray the European missionaries not only as suspicious foreigners but also as spineless opportunists.

All four parameters were set in motion after Yang died in 1669. Yang's posthumous anti-Catholic writings became the straw man many European missionaries attacked with a heightened intensity. The counterattack from the classical scholars came much later.

THE SIGNIFICANCE OF MATHEMATICAL ASTRONOMY

Astronomy was no bit player in Chinese history.[6] Long embedded in the apparatus of power and known as heavenly learning (*tianxue*), it was a large and complex body of technical knowledge that was carefully guarded by the imperial state and difficult to master.[7] It pertained to three traditional practices: judicial astrology (forecasting the future by calculating the position of celestial bodies), hemerology (identifying auspicious and inauspicious calendar dates), and the siting of necropolises for the imperial family. All three relied on techniques that belonged to a unified cosmology.

So important were these practices that, during much of Chinese history, the standard punishment for applying them for a subversive purpose—by predicting, say, the death of the emperor or by forecasting the collapse of the current dynasty—was death.[8] Not surprisingly, mathematical astronomers were kept on a short leash within the central bureaucracy. In periods of displacement—during dynastic transitions, during the coexistence of rival dynasties—the surveillance of astronomers became a top priority. As a result of such official anxieties, only a privileged few were given access to this body of knowledge.

Like the Mongols who ruled China in the thirteenth and fourteenth centuries, the Manchus who conquered the Ming dynasty in 1644 were foreigners. Unlike the previous regime, which had benefited from its identification with a resilient and long-lasting literati tradition, the Manchus were determined outsiders who had to cope with assumptions of Chinese superiority. The newly established Qing dynasty adopted Jesuit mathematical astronomy from the outset, and the foreign missionaries who inserted themselves into the court as technical consultants soon became a political faction intent on converting the sovereign.

When these foreign missionaries showed that they possessed unique technical skills, the Manchu conquerors did not hesitate to employ them, confident that a keener grasp of such things as calendar making would legitimate the Manchu claim to the mandate of Heaven. The Qing commitment to European astronomy did not last long, however, and the missionaries shuffled in and out of imperial favor. The Jesuits put up a strong fight on a number of fronts, and if they did not ever do very well in the religious, cultural, or political areas, they did maintain a foothold at the court thanks to their scientific know-how.

The story of Jesuit success is similar to that of their Muslim predecessors, another group of outsiders who came from Baghdad in the thirteenth century and acted as technical experts. After the Ming dynasty expelled the Mongols in 1368, the newly established court recruited two groups of Muslim astronomers (twenty-five people in all) and confiscated more than two hundred documents from the astronomical observatory in Beijing for the Ming court. All of the Muslim and Chinese astronomers and astrologers who had served under the previous regime signed on to work for the Ming. More Muslim astronomers arrived, working at the Ming court in Nanjing until the Astrocalendrical Bureau (Qintian Jian) moved with the court back to Beijing in 1403.[9]

When the Jesuits arrived in China in the seventeenth century, they were quickly cast as technical experts in competition with the Muslims.[10] A dynastic house that knows nothing of celestial events cannot long retain the mandate of Heaven: each time the faltering Ming attempted to repair its astronomy, the better to predict solar and lunar eclipses, the invading Qing army gained momentum. The Ming dynastic house's failure to predict and decipher the will of Heaven stood in sharp contrast to the Manchu conquerors' swift adoption of Jesuit mathematical astronomy.

When the first Manchu emperor, Hong Taiji (r. 1626–43), deployed his large army of Manchu, Mongol, and Han soldiers along China's northeastern border in 1636, he declared that he had received a new mandate from Heaven.[11] The next year, well before his forces had seized the Ming capital, Hong promulgated an imperial calendar. But Manchu elites had little knowledge of astronomy, and the calendar was a rather shoddy affair.

After the Manchu coalition forces pushed out the rebels who had overwhelmed the impotent Ming just days earlier and entered Beijing on June 5, 1644, they issued an edict ordering all residents of the area closest to the imperial palaces—known as the inner city—to vacate their residences and move south to an adjacent area—the outer city—so that soldiers and their families could take up residence around the court, creating a vast armed garrison. An estimated 300,000 members of the alliance, known as Banner people because of the system of social registration employed by the Manchus, repopulated the inner city. Shifting the populations was a slow and painful process that was not accomplished until the mid-1650s.[12] Such partitioning also took place in other cities, part of a vast and systematic urban segregation in major metropolitan areas.[13]

Shoring up the Manchus' declaration of political legitimacy (their claim to the mandate of Heaven) from an unexpected quarter was a group of foreign missionaries. After the chaotic fall of the Ming dynasty, the Jesuit missionaries who had made Beijing their base for attempts to convert the ruler of China were left vulnerable. On June 15, nine days after the Manchus had swept away all resistance in Beijing, they responded to the imperial edict that ordered them to move to the outer city. The German Jesuit Johann Adam Schall von Bell (1592–1666) submitted a memorial to the young Shunzhi emperor (r. 1644–61):

> I purchased a mansion that I have used as my church in Beijing. I have been living and praying in this church for more than twenty years. The imperial calendar of the previous dynasty went terribly wrong in 1629, and I was ordered to preside over its reform. I compiled more than 140 chapters on mathematical astronomy that can help predict solar and lunar eclipses and the movements of the five planets. I had planned to submit all of this to the imperial court, but it was my good fortune that the great Qing dynasty saw that the people were suffering injustices and expelled the malevolent rebels to restore order to our realm. Thanks to the greater glory of God the Qing dynasty came to our rescue, saving every person in the realm.[14]

Although Schall never explicitly requested that his church be treated as an exception to the edict of eviction, there can be no doubt that his memorial was part of an effort to protect the interests of the Jesuits and to advance their mission. As Martino Martini (1614–61), one of Schall's fellow Jesuits, testified, the Manchu conquest, which he conceived as a world-historical event, was often swift and ruthless—the Jesuits did all they could to survive.[15] Schall and Martini, among others, shifted their allegiance from the Ming rulers to the Manchus rather quickly; others, like Michael Boym (1612–59), remained loyal to a displaced Ming court, called the Southern Ming, headed by the Yongli emperor (r. 1646–62).[16]

In his memorial, Schall offered to come to the new emperor's assistance with a number of powerful intellectual tools. The Jesuit offered to assist the Qing: "I have more than three thousand books from the West, among them useful books on nature and principle, mathematical astronomy, agriculture, and hydraulics. In addition to those books, I also have had many woodblocks cut to print copies of a revised imperial calendar that I had prepared for the last Ming emperor."[17] Publishers had long relied on woodblocks for printing books. Storage of the blocks ensured that copies of books could be printed on demand, with no need to set type afresh for each page. Schall boasted that not only had he transported many books from Europe to China, he had also had them translated into Chinese and carved onto woodblocks, ready for printing and dissemination.

Schall's petition was accepted, setting a paradigm for other Jesuits to follow. The Manchu prince Dorgon (1612–50), regent for the young Shunzhi emperor, liked what Schall proposed and quickly granted his wish. That act marked the beginning of the acceptance of Jesuit astronomy in the Qing court, a process that would culminate with the exit of the Chinese and Muslim technicians who staffed the Astrocalendrical Bureau. Those displaced and disgruntled staff members later joined the anti-Jesuit forces at court to purge the Jesuit faction.[18]

In addition to segregating the larger cities of the realm, the Manchu conquerors drove home their domination over the local population through a series of fiats. Even as the emperor commanded all Han men to adopt the Manchu hairstyle (shaving the forehead and leaving a long queue in back), he subscribed to what must have been perceived as timeless mandate from Heaven to rule China: the technical competence to revamp the imperial calendar demonstrated that the Qing possessed the mandate of Heaven. These symbolic coercions were enforced from 1645 to the fall of the Qing dynasty in 1911.

Every Qing subject shared the belief that Heaven endowed the ruling house with the right to rule China. If astronomers could read the will of Heaven in the movement of the seven celestial bodies—the Sun, the Moon, Mercury, Venus, Mars, Jupiter, and Saturn—then the emperor had best monopolize astronomy. To assist him, he needed a variety of experts. Mathematical astronomers offered descriptions and, more importantly, predictions of the position and movements of the celestial bodies.[19] Astrologers interpreted the political implications of the position and movements of the celestial bodies, a set of practices that evolved into the art of judicial astrology, also called hemerology. (Note that astrology was also used in state and family sacrifices, burials, and rituals, and to forecast the outcome of military engagements.)

In 1645 Adam Schall and his assistants produced their first "temporal model calendar" (*shixian li*). They acknowledged their reliance on *Mathematical Astronomy According to the New Western Methods* (Xiyang xinfa lishu), a collection, then ten years old, of Western treatises translated into Chinese and compiled by Xu Guangqi and a team of Jesuits.[20] Schall's calendar certainly attracted a great deal of attention. In *The Collected Statutes of the Qing Dynasty* (Daqing huidian), its unveiling was documented in technical detail and its formal presentation was accompanied by the performance of elaborate state rituals.[21]

The calendar of 1645 comprised seven tables tracing the locations of the celestial bodies over the coming year, day by day. Such tables were called ephemerides. Schall and his team had devised a step-by-step sequence of computations that permitted them to plot planetary, lunar, and solar locations.

Following the template of the imperial Ming calendar, the temporal model calendar comprised three sets of ephemerides and almanacs. The first set, *Ephemerides of the Seven Celestial Bodies* (Qizheng li), and the second set, *Ephemerides of Eclipses and the Obscuring of Celestial Bodies* (Lingfan li), were both confidential documents to be read by only the imperial family. The third set, civil ephemerides and almanacs called simply *The Temporal Model Calendar* (Shixian li), was published for the bureaucracy, the gentry, and leading merchant families, as well as neighboring tributary states, such as Choson Korea and the Mongol Zhungars.

Bestowing the imperial calendar was itself an important state ritual. Indeed, in China, as in most early modern states, rituals were powerful symbolic actions. They conveyed political messages that were simultaneously vague and simple, ambiguous and multivalent. The spectacular, meticulously crafted rituals were often the sole means of delivering such messages.[22] The Qing rulers advanced their political goals by participating in three major ritual arenas: Confucian practices, through which the emperors asserted control over the Han heartland; shamanism and Tibetan Buddhism, through which the emperors linked the peoples of Northeast Asia and Central Asia to the Qing empire; and private rituals, in which the imperial household acted as a patriarchal lineage group practicing a mixture of shamanic, Daoist, Chinese Buddhist, Tibetan Buddhist, and popular religions.[23] The rituals that accompanied the presentation of the imperial calendar were primarily Confucian, but they included other elements to draw various minority groups and tributary states into the imperial embrace. Complex and multivalent, the rituals provide great insight into the character of imperial rulership.[24]

The Qing imperial household took seriously the responsibility of performing state rituals and was attentive to their legitimating implications.[25] Schall established a precedent when he offered his new calendar to the throne on the second day of the tenth lunar month in the second year of the Shunzhi emperor's reign (November 19, 1645); the calendrical rituals continued to be performed annually on that date until the Qing dynasty fell in 1911.

Before the ceremony commenced, all civil and military personnel associated with the court were to don formal attire and convene outside the Meridian Gate (Wu Men, the southern entrance to the Forbidden City). A table covered with yellow silk was set up in the middle of the Hall of Supreme Harmony (Taihe Dian), which rested on three stacked marble terraces at the very center of the Forbidden City. This was the site, too, of other important state rituals, such as the celebration of the winter solstice and the announcement of the results of the palace examination. A second table was set up in front of the empress's palace, and a third and fourth stood outside of the Meridian Gate—all dressed in yellow silk. Eight more tables—these decorated with red silk—were set on either side of the Meridian Gate.[26]

As the ritual began, music surged and a parade of officials left the offices of the Astrocalendrical Bureau, passed through the Eastern Gate of Eternal Peace (Dong Chang'an Men), and came to a halt at the Meridian Gate. As they wended their way along that route, the parades distributed ephemerides and almanacs to the yellow and red desks, leaving only the emperor's and empress's tables bare. That omission was soon rectified, as two high-ranking officials from the Ministry of Rites approached the Hall of Supreme Harmony, while two others proceeded toward the empress's palace. Having placed the books on those two yellow tables, they knelt down and kowtowed three times, then backed out. At that moment, the ritual announcer loudly pronounced, "Kneel!" then read the following imperial edict: "In the second year of the Shunzhi emperor's reign [1645], this temporal model calendar is bestowed on you. You shall now disseminate it to the entire realm!" All of the officials knelt, pressed their foreheads to the stone floor nine times, rose, then knelt again, repeating the action thrice. That concluded the ritual.[27]

The task of making the astronomical calendars that were so dramatically unveiled was entrusted to the Astrocalendrical Bureau, whose three subsections were the division of mathematical astronomy (*li ke*), the division of judicial astrology (*tianwen ke*), and the timekeeping division (*louke ke*).[28] The bureau also published annual civil almanacs provided to different bureaucratic units and tributary states. In 1735 more than 24,300 copies of civil calendars were distributed, the equivalent of 900 copies for each province. In addition, 100 copies were sent to each Mongolian tribal leader and to the ruler of Korea.[29] For the commanders of the Eight Banners, civil calendars were prepared in Manchu, Mongolian, and Chinese.[30]

Mathematical astronomy was not merely the timekeeper for state rituals. Together, the imperial ephemerides and almanacs monopolized by the imperial household and the commercial almanacs circulated in gentry-merchant society provided a single consistent cosmology that persisted through the Qing dynasty. Mathematical astronomy performed a key function in developing what we might call the "narrative temporality" in the historical consciousness of Qing literati.[31]

NEW COSMOGRAPHY: 1610–1700

The Jesuit contribution to Chinese astronomy commenced at the beginning of the seventeenth century.[32] The great pioneer was Matteo Ricci, who had studied mathematics and *scientia* with Christopher Clavius at Collegio Romano from 1572 to 1577 and later at Portugal's University of Coimbra in 1577 and 1578. By 1607 he had collaborated with the scholars Xu Guangqi and Li Zhizao (1565–1630), both of whom had converted to Catholicism, to render into Chinese some texts about Euclidean geometry and Ptolemaic astronomy—most significantly, Ptolemy's *Almagest*. Xu, Li, and Ricci also appropriated portions of three Latin texts from Nicolaus Copernicus (1473–1543) and Tycho Brahe (1546–1601).[33] The Ming court showed little interest. In 1610, after the Astrocalendrical Bureau failed to predict a solar eclipse, the minister of rites decided that the bureau needed new blood. He presented a list of candidates to the emperor for approval, including Xu and Li, whom he hoped might reform the state's astronomical system. Again the Ming leaders failed to act.

From 1620 to 1627, a clique surrounding the powerful eunuch Wei

Zhongxian (1568–1627) waged a vicious struggle against the literati faction known as the Donglin party. This virtually paralyzed the Ming bureaucracy for nearly a decade.[34] Nothing was done to improve the competence of the Astrocalendrical Bureau until the Chongzhen emperor (r. 1628–44), the Ming's last ruler, came to power. He knew of Xu, who had studied mathematical astronomy and irrigation technology with learned Jesuits from 1607 to 1610, and in 1629 the Chongzhen emperor appointed him a senior minister in the Ministry of Rites. Without delay, Xu set out to reform the astronomical system.

Xu's activism took off at an important historical juncture: it was on the eve of the fall of the Ming dynasty and the dawn of the Copernican revolution. In 1621 Adam Schall and his fellow Jesuit Giacomo Rho (1593–1638) brought with them to China a new cosmography. The work of the Danish aristocrat Tycho Brahe was a compromise between Ptolemaic geocentricism, which stated that all celestial bodies revolved around the Earth (fig. 2.1), and Copernican heliocentrism, which stated that all local celestial bodies revolved around the Sun. In Tycho's compromise, called geoheliocentricity, the Earth remains the center around which the Sun, the Moon, Mars, Jupiter, and Saturn revolve, but Venus and Mercury revolve around the Sun.[35] Tycho's model served as the underlying cosmology for *Mathematical Astronomy of the Chongzhen Reign* (Chongzhen lishu), compiled by Schall and Rho and renamed *A New Western Method* (Xiyang xinfa) in 1645.[36] That work, never published in a definitive version and perpetually under revision, was in fact a somewhat heterogeneous collection of materials from different eras.[37]

As early as 1628, Li Zhizao and his Jesuit collaborators compiled *A Preliminary Collection of Celestial Studies* (Tianxue chuhan); it included nineteen works on arithmetic, geometry, and Ptolemaic astronomy. The promulgation of *A New Western Method* in 1645 concluded a century-long translation of early modern mathematical astronomy from the Latin West to Qing China. From *A Preliminary Collection* (published 1629) to *A New Western Method*, the Jesuit treatises were warmly received at court, and Tychonic cosmology and cosmography became the new official methods. Whereas the Tychonic cosmos faded quickly from the European scene, it lasted for centuries in China, serving as the foundation for imperial astronomy and the baseline for any debates in the court and beyond.[38] It was Tycho's model (fig. 2.2) that permitted the Jesuits to supplant the

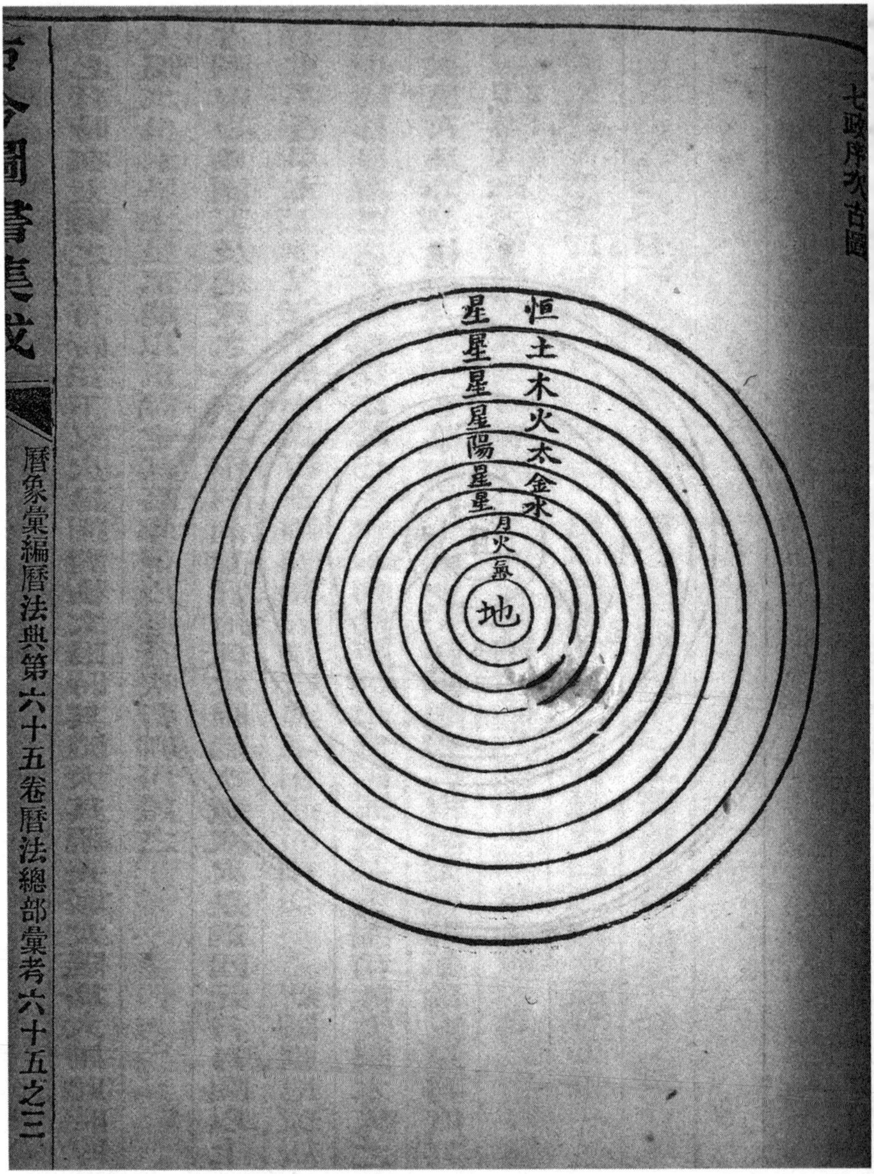

FIGURE 2.1. Ptolemaic cosmography: "Ancient illustration of the order of seven luminaries/polities." The concentric orbs from inner to outer are Earth, Ether, Light, Moon, Mercury, Venus, Sun, Mars, Jupiter, Saturn, and Stellar Sphere. From *Gujin tushu jicheng*, 65–3.

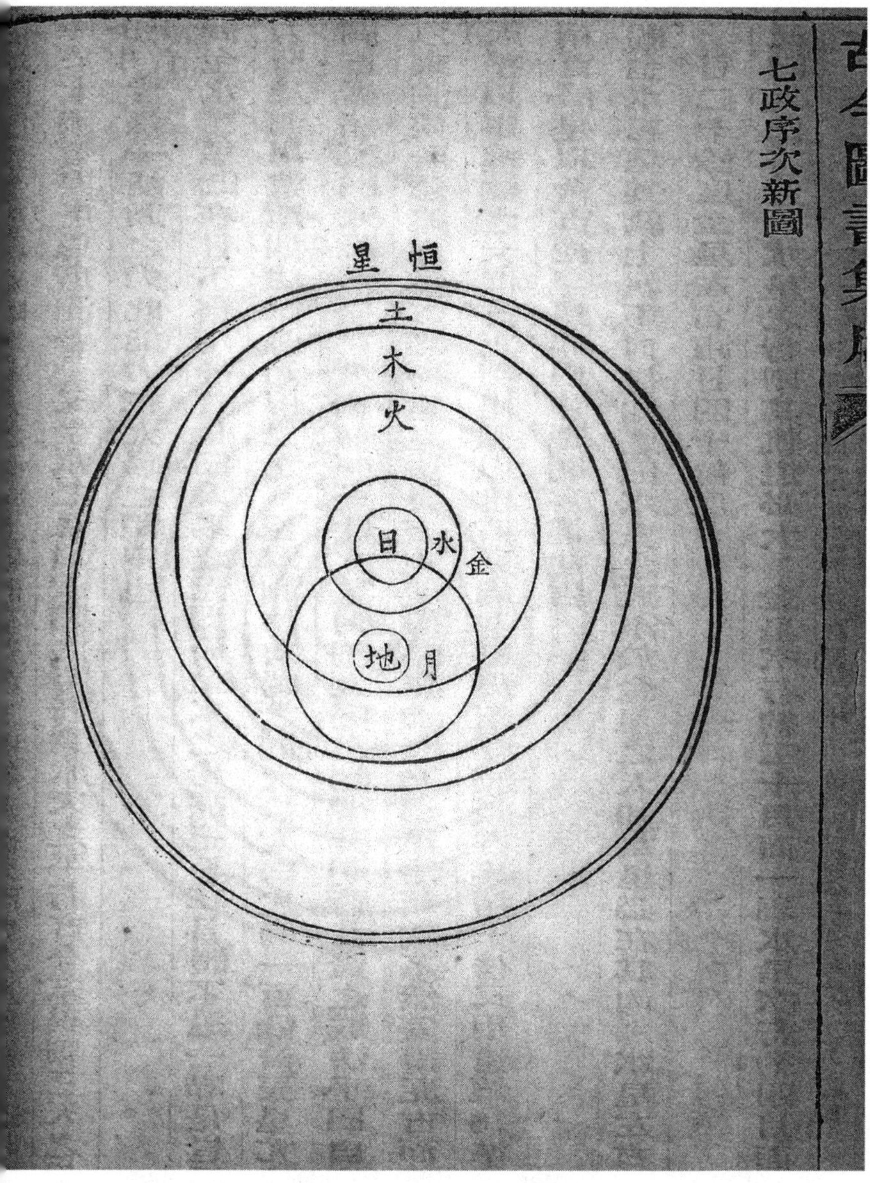

FIGURE 2.2. Tychonic cosmography: "New illustration of the order of seven luminaries/polities." The geoheliocentric illustration shows that both the Sun and the Moon revolve around the Earth. The rest of the luminaries revolve around the Sun, which is at the center of the illustration. The concentric orbs centered on the Sun from inner to outer are Mercury, Venus, Mars, Jupiter, Saturn, and Stellar Sphere. From *Gujin tushu jicheng*, 65–4.

Chinese and Muslim astronomers and astrologers who had long populated the Astrocalendrical Bureau.[39]

Yet the court never considered this new science to be an essential insight to the imperial cosmology or orthodoxy, viewing it instead as an ad hoc conceptual device designed for a specific task, useless for other purposes. It was nothing, thought the Manchu and Chinese officials, but a set of documents, assumptions, and approximations—as useful for state rituals as for horoscopes and almanacs. The literati maintained a firm distinction between the ontological and cosmological reality of the indigenous tradition and their heuristic deployment of the Tychonic model. The Qing court had adopted Jesuit mathematical astronomy exclusively on pragmatic grounds.[40]

ASTRONOMICAL ACCURACY

How was it that the Jesuit astronomers enjoyed technical superiority over their Chinese counterparts?[41] This certainly requires explanation, given the claim of historians of science that until the arrival of the Jesuits, China stood well ahead.

What advantages modern science enjoyed in the Protestant West were limited, said Joseph Needham, to its *methodology*: "When we say that modern science developed only in Western Europe at the time of Galileo in the late Renaissance, we mean surely that there and then alone there developed the fundamental bases of the structure of the natural sciences as we have them today, namely the application of mathematical hypotheses to Nature, the full understanding and use of experimental method, the distinction between primary and secondary qualities, the geometrization of space, and the acceptance of a mechanical model of reality."[42] The "fundamental bases" were, in fact, only partially shared by the missionaries in China. The spirit of modern science—including many of the items in Needham's list—was not systematically introduced to seventeenth-century China.[43] What then made Jesuit sciences superior?

There are at least two contexts from which to assess astronomy. The first is the long-term tinkering with calendars, almanac, and predictive models that served the state rituals for at least two millennia.[44] Two waves of foreign experts provided invaluable assistance to the imperial state by significantly improving their ability to forecast celestial events. After the

Muslim astronomers who served the Mongols and the Ming came the Jesuits.

For the second perspective, we must see through the Jesuits' eyes to understand why they were ready to mend the Chinese astronomical system. The Jesuits, who came from the Latin West, did not benefit from the scientific revolution taking place in Protestant Europe. Their thinking centered on the calendar reform initiated by Pope Gregory XIII in 1582. As it happened, concerns with administrative reforms drove both the Vatican and the Qing court to improve their grasp of heavenly cycles. The Jesuits did not bring the scientific revolution to China; they brought instead the gospel of the Counter-Reformation and were happy to share the calendar reforms then so important to the Catholic Church.

Accurate predictions had long been indispensable to imperial state rituals in China. Between 724 and 762, according to the calculation provided by NASA's Goddard Space Flight Center, twelve solar eclipses occurred in Chang'an, the capital of the Tang dynasty. The astronomical system then in practice, called the "great extension" (*dayan li*), predicted those but added a dozen others, amounting to a 50 percent success rate. After the fact, the official historical record—*The New History of the Tang* (Xin Tang shu) and *The Old History of the Tang* (Jiu Tang shu)—recorded sixteen eclipses, 75 percent of the true total.[45] The Tang imperial state revised its models fifteen times due to the low success rate in predicting eclipses. The Song state modified its models seventeen times, and the next dynasty, the relatively brief Yuan, applied five revisions. In each instance, the whole system was called into question. This was natural, as the success rate for predicting eclipses dropped below 50 percent, but it is also true that feuding cliques often made the Astrocalendrical Bureau their scapegoat.[46]

Imagine that an ephemeris issued by the bureau declared on such-and-such a date, "Mars remains in the vicinity of the Xin constellation." Such a forecast could have had tremendous consequences, as judicial astrologists associated Mars with chaos, sickness, death, and war. The Xin constellation (near Scorpius) was equated with the Palace of Light (Ming Tang), the symbolic home of the emperor, where he both resided and performed state rituals. A Martian disturbance in Xin suggested the violation of the imperial dwelling; no emperor wished for such an omen. According to modern calculations, of the twenty-three occasions on which such an infelicitous astrological reading was recorded in the standard dynastic histories, seventeen never occurred. Moreover, between 300 BCE and 1600 CE, Mars actually visited the vicinity of Xin thirty-eight times, most of which were not recorded in the standard histories.[47] It would appear that the appearance of felicitous and infelicitous omens in imperial ephemerides (and standard histories) had as much to do with political circumstances as astronomy. Such precedents weighted heavily on Qing mathematical astronomers, who had to walk a fine line between astronomical accuracy and political expediency.

When the Mongols overthrew the Song and seized control of China, they established a dynasty called the Yuan—meaning "great beginning"—a crucial character in *The Book of Changes* (Yi jing). Their Muslim and Chinese mathematical astronomers made a breakthrough in predicting eclipses: they called it the "season-granting calendrical system" (*shoushi li*).[48] The Ming dynasty adopted that system with trivial modifications, renaming it the "great concordance calendrical system" (*datong li*). For the next two centuries, the Ming imperial house constantly tinkered with that system in order to maintain an acceptable success rate for the prediction of eclipses. In 1450 the court astronomers missed a lunar eclipse, a grievous mistake embarrassing to the Jingtai emperor (r. 1449–57), who was not able to perform the salvage and protection ritual (*jiuhu*) in time. The same thing happened again in 1479; the Chenghua emperor (r. 1464–87) excused himself and his astronomers, noting, "Heaven is so subtle that it goes beyond human comprehension."[49] Four years later the court astronomers proposed a revision to the great concordance system; it was not approved. From that time forward, eclipses regularly embarrassed the so-called experts. No eclipse, whether lunar or solar, was accurately predicted in 1517, 1518, or 1520.[50]

The arrival of the Jesuits meant accurate predictions were available from 1644 forward. By 1668 the Astrocalendrical Bureau, by then in Jesuit hands, was routinely providing predictions for all state rituals. All astronomical instruments, including Jesuit telescopes, quadrants, sextants, and armillaries, were classified as "etiquette instruments" (*yiqi*) and were under the control of the Ministry of Rites, responsible for performing the rituals of salvage and protection when either a solar or lunar eclipse occurred.[51] The Astrocalendrical Bureau reported the impending eclipses to the minister of rites, who in turn sent both Manchu and Chinese officials to the imperial observatory at the predicted time to measure the eclipse's length. Runners and clerks were also commanded to perform the appropriate salvage and protection rituals—ceremonies performed in the location where the eclipse was visible to demonstrate the imperial possession of heavenly mandate and calm the local population (fig. 2.3). Afterward, the clerks would compare the measurements of the eclipse they had made with the prediction.[52] Any eclipse that lasted less than one minute did not require a ritual performance. The imperial calendar was considered accurate when this careful batch of tests supported its predictions.

The Jesuit success in China was not accidental, as the challenges the Catholic Church and Qing China faced in making accurate calendars were very similar. Most historical discussions of the rise of modern astronomy focus exclusively on the movement of the five visible planets, lingering on Mars retrograde and the various phases of Saturn. This observational astronomy, which drew its questions from the heliocentric cosmography that swept through early modern Europe, held little interest for either the Catholic Church or the Qing, both of which hoped above all to make reliable calendars and had no interest in debating cosmology or cosmography.[53]

For the Qing state, perennially concerned with when to insert an intercalary month, a precise account of the tropical year and the clear demarcation of the tropical and sidereal years were crucial. Let me begin with the former.

Mathematical astronomers had traditionally measured the length of the tropical year (also known as the solar year) from one winter solstice to the next.[54] They referred to that period as the "year numerator" (*suishi*). The length of the tropical year served as the baseline for the entire system. As the length of the tropical year, when measured carefully, is 365 days and a fraction, it cannot be divided neatly into lunations, which average about 29.53 days. The Chinese solution, similar in practice to the leap years of the West, involved adding an extra month to the calendar from time to time.

The tropical year was divided into twenty-four parts, each of which corresponded to a distinctive manifestation of *qi,* the basic unit of energy that informs all of matter and its transformations. The rituals carried out at the Qing court—like many other rituals carried out in local areas—were meant to bring human life into accord with the cosmic operation of *qi*.[55]

Those entrusted with dividing the tropical year into smaller units faced a peculiar challenge: when observed in North China, the Sun's movement

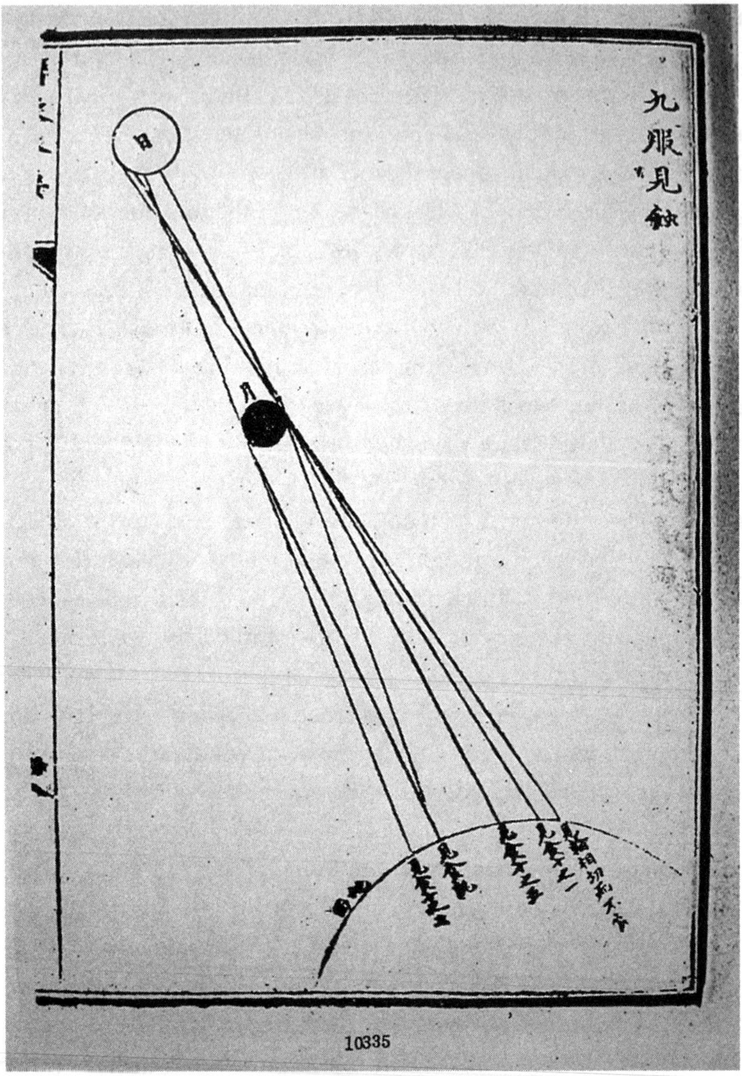

FIGURE 2.3. Solar eclipse as viewed from different locations in China: "Viewing solar eclipse from various locations outside the imperial capital." The phrase *jiufu* was a classical allusion from the *Rituals of Zhou* (Zhouli) and refers to nine locations of gradation of distance from the imperial capital. The illustration shows how the relative movement of the Moon (in the middle and black) gradually blocks the Sun for the viewers on the terrestrial sphere. In the bottom-right corner, it marks the successive stages of solar eclipse (from right to left) as: the solar and lunar rings tangentially meet but no eclipse, one-tenth of a solar eclipse, half solar eclipse, full solar eclipse, and half solar eclipse (again). From Li Guangdi, *Lixiang benyao.*

appears to speed up or slow down in different seasons. As early as the thirteenth century, Chinese and Islamic astronomers calculated this variable speed and made allowances for it in the season-granting calendrical system.[56] Because the Jesuits had studied the problem of identifying the same day on the Julian solar calendar and the Hebrew lunar calendar, they were prepared to cope with the Chinese hybrid of lunar months and solar years.[57] They never mentioned that the unequal motion of the Sun had been noticed long before by Chinese astronomers—likely in the thirteenth century.

The second technical issue—the difference between the tropical and the sidereal year—was thornier because the length of the latter was not a constant. Both ancient Greek and Chinese astronomers observed that the location of the Sun during the equinox (where its path crosses the celestial equator) had changed over history. Only those who could observe the equinoxes with relative accuracy, devise a reliable method for charting them, and keep records for a long time could make this discovery.[58] The problem was how to account for the variations and make the appropriate calibrations that would permit the calendar system to work properly.[59]

European Renaissance astronomers had devised an explanation called "the precession of equinoxes." We now know that the precession is due to the gyroscopic wobble of the Earth's axis, which takes 26,000 years to complete a full circle. The tilt of the Earth relative to the plane described by its orbit stays roughly constant at 23.4 degrees, but the planet's orientation is always changing.

One consequence of precession is that the North Celestial Pole (the North Star) appears to move around the sky. Further, precession introduces a difference between the sidereal and tropical years. Chinese astronomers called this variable "annual difference" (*suicha*). Barely detectable within a given year, its cumulative effect turned out to be considerable.

For the Qing state, the need to measure the length of the tropical year, or "year numerator," and the cumulative error caused by the precession of equinoxes over long spans of time, or "annual difference," could not be ignored. Literati weighing the merits, limitations, and potential pitfalls of Jesuit mathematical astronomy frequently revisited these two technical issues. Since they were thrown into sharp relief by the marriage of Jesuit mathematical astronomy and Qing state rituals, scholars deeply interested in classical and historical knowledge scrutinized them time and again.

CRIMINAL PERSECUTION

Early on, the Jesuits won the trust of the court by displaying remarkable skill at predicting solar movements and measuring the gap between the tropical and sidereal years. Then suddenly, in 1664, a commoner with no lofty connections accused the emperor's new darlings of violating his trust. Blindsided, the foreign missionaries barely managed to muster a defense.

Ultimately, it wasn't a scientific shoot-out but a criminal prosecution that brought down the Jesuits. Their accuser was a litigation master (*song-shi*) who specialized in drawing up complaints (*zhuang*) against officials. As a group, litigation masters had failed to rise above the lowest rungs of the examination system and operated outside the state apparatus. They were skilled at exaggerating their cases and manipulating the system for the advantage of their clientele—sometimes bringing down on themselves the wrath of the state. Yang Guangxian, descended from a hereditary military family long based in Huizhou, was one such litigator. He left his ancestral home for Beijing in 1637, determined to use his litigation skills in the factional struggle that broke out in the late Ming. In no time, however, he got into trouble and was exiled to the northeastern border near Manchuria. After the dynasty fell in 1644, Yang returned to the capital and began to curry favor with Manchu noblemen. Then, in 1659, when the political atmosphere seemed right, he launched his attack on the Jesuits in the Astrocalendrical Bureau.[60]

This assault succeeded only because it occurred during a crucial political window, when no emperor ruled the realm. In 1659 and 1664 Yang submitted memorials to the Ministry of Rites, accusing the Jesuits of sedition, of plotting against the universality and legitimacy of the great Qing dynasty. Such charges against a powerful Jesuit faction would have backfired when the young Shunzhi emperor continued to follow the cosmopolitan policy of his uncle Dorgon (1612-50). But he died of smallpox in 1661, and four Manchu noblemen—Soni, Suksah, Ebilun, and Oboi (ca. 1611–69)—were appointed regents to Xuanye, the Kangxi emperor (r. 1661–1722). The Oboi regency, named for the most powerful of the regents, consolidated power swiftly and initiated a full-scale review of state institutions.[61] Breaking with the cosmopolitan rulership embraced by Dorgon (1612–50), who had served as regent to the young Shunzhi emperor, Oboi signaled a return to a stern interpretation of the "Manchu way."[62] Yang managed to win the patronage of Suksah.[63] It was during the stern Manchu regency that Yang initiated his persecution of the Jesuits.

In the 1664 complaint, Yang characterized the foreigners as sorcerers who practiced their evil arts to hoodwink people and injure the Qing. Specifically, he accused the Jesuits of violating two important statutes in *The Great Qing Legal Code* (Da Qing lü li), which differed from *The Great Ming Legal Code* (Da Ming lü) only in details.[64] Submitted to the Ministry of Rites, the complaint was immediately accepted and the prosecution of Schall and his associates ensued.

We should not underestimate the uncertainties the Qing state was facing at this time; they weighed heavily on the minds of the Manchu conquerors. Banner troops continued to wage war against the rump Ming dynasty in the south and a maritime regime based in Taiwan absorbed the navy. The alliance with the Zhungar Mongols had not prospered, and Russian forces had made their way into Chinese Turkestan and Manchuria. Moreover, while an accurate calendar was of great utility to the newly established dynasty, the threat of political sedition could always be counted on to trigger an overwhelming response.

Yang began his complaint by affirming that no hierarchy could function properly without a legitimate leader: a family needs a patriarch and a dynasty needs an emperor. Those who betrayed their patriarch or emperor were traitors who, according to *The Qing Code*, should receive the death penalty. Consider the case of Jesus Christ. Was he not a rebel and a traitor to Judaea? Had he not been found guilty by a Roman court and crucified for sedition? In a parallel argument, Yang contended that Xu Guangqi had been induced to present the seditious Schall to the Ming court by the thought that in doing so he might lay hands on the marvelous machines the foreigners had brought from overseas. The "Jesuit problem," according to Yang, was inherited, initiated at the end of the Ming dynasty. Since then, Schall had performed as a mathematical astronomer in public but practiced his malign religion—founded on treason—in private. Now, proclaimed the litigation master, the true intent of the Jesuits had come to light.[65]

To prove seditious intent, Yang presented a remarkable array of evidence. He pointed to *An Outline of the Dissemination of Heavenly Learning* (Tianxue chuangai), published by Schall's follower Li Zubai (d. 1665). In this piece of propaganda, Li argued that all humans were descendants of those Judeans who survived the universal flood. The Chinese, this Christian convert had contended, were no exception.[66] As to the Jesuit organization, Yang itemized thirty churches that had been built in some of the Qing empire's most strategic locations; he proposed that the churches secretly harbored foreigners who had come illegally to China from overseas. Drawing parallels to the secret societies whose rebellions had toppled previous dynasties, Yang spoke of a Christian organization that extended throughout China proper and kept itself busy printing immoral books that had deluded the naïve and the gullible. The Jesuit conspiracy was revealed when the foreign astronomers boasted that the imperial ephemerides were made according to "new methods from the Western Ocean." Was it not outrageous that the Jesuits could accuse the Qing rulers of relying on the achievements of a foreign people in order to achieve their political legitimacy? This line of argumentation exploited the deep feelings of the Manchu nobles' and bannermen's ambivalence after the Manchu conquest, including the adoption of the Confucian canon as the imperial orthodoxy.[67] Thus the experienced litigant brilliantly exploited his Manchu patrons' suspicion toward Europeans.

To gather evidence, Yang had sent a spy, posing as a Catholic convert, to infiltrate one of the Jesuit churches. The faux Christian collected religious paraphernalia (including a piece of golden plate and an embroidered sachet used to identify members of the flock), a treasonous book, and a schedule of meetings. All these items were submitted as evidence to the Ministry of Rites along with Yang's complaint that a foreign organization, whose intent was manifestly malign, had been expanding across China for the past twenty years. Although the forces of the great Qing could easily outgun this quite pathetic group, Yang insisted that the only possible course of action was the elimination of the ringleaders before they could do any damage. Yang then cited statutes 255, "Plotting Treason," and 256, "Writing Books on Sorcery or Speaking about Sorcery"—surely the Jesuits would be found guilty of those.[68]

Calling sedition a serious crime is an understatement. Among the Ten Abominations (Shi E) listed in the preamble to *The Qing Code* was "plotting treason." Referring to the Ten Abominations alone would have afforded a credible litigant an immediate hearing, as they connected legal thought to the innermost core of Chinese culture and are in fact drawn almost wholly from *The Tang Legal Code* (Tang lü) of 653. What counted as sedition, however, was quite another matter. Efforts to find treason in appeals to the spirit realm had often come to naught.

Statute 255 reads as follows: "In the case of anyone who plots treason (that is to say, one who plots to betray his own country and secretly serves another country)—when there is joint plotting, no distinction is made between principals and accessories—all are beheaded."[69] Statute 256 (the commentary has been set in parentheses) reads:

> Those who make auguries or write books on sorcery or speak about sorcery, as well as those who circulate and use them and disturb the multitudes, will be beheaded (after a delay). (Those persons who received the material will not be punished. If [the materials] reach fewer than three people, the penalty [imposed on those who distributed them] will be exile to three thousand *li*. The penalty will be based upon an examination of the circumstances.) If one who has received such magical books keeps them in his possession, hiding them instead of delivering them to the authorities, the penalty will be one hundred strokes of the heavy bamboo and three years of penal servitude.[70]

Ever since China was first united under the Qin dynasty in 220 BCE, ruling houses had periodically been threatened by peasant rebellions organized by cult leaders who claimed to have acquired supernatural arts. Over time, many laws had been designed to eradicate such threats. Some of them addressed the village practices around which cults had historically emerged: so long as a ritual remained confined to one village it could be overlooked, but once it became a means for mobilizing large numbers of peasants, the state would feel compelled to step in. This goes some way to explaining why many villages in North China developed unique and local *sai* (lit., competition) rituals and operatic forms. The imperial states left them alone because, despite the cultural scope, complexity, and historical depth of their operatic performance, these rituals remained fairly isolated and "autarkic" in their region.[71] In other words, these rituals would not turn into organizing forces of peasant rebellion or potentially threaten the stability of the imperial state.

When Yang Guangxian cited both statutes in his complaint, he framed the Jesuits, placing them in the same category as the White Lotus rebels who had overthrown a foreign conquest dynasty, the Mongol Yuan. Had the Jesuits not adopted a military command structure for their organization? Offered a glimpse of a terrifying future, Oboi and his fellow regents acted.[72]

By transmitting Yang's complaint to both the Ministry of Rites and the Ministry of Punishment, the regents signaled that the Jesuits were to be charged with high treason and sorcery. It is important to note the legal distinction between two seemingly similar criminal categories: "Writing Books on Sorcery or Speaking about Sorcery," which appears in the Statutes on Punishments, and the crimes associated with sacrifices in the Statutes on Ritual (*lilü*, offenses within the purview of the Ministry of Rites).[73] The most important of the latter is statue 162, which is aimed at individuals who communicated with spirits outside the pantheon of imperially sanctioned deities. There is no suggestion that such heretics might set out to overthrow the state. Evidently the Qing rulers believed that some sorcerers were more dangerous than others. Still, as the panic surrounding tales of "soulstealers" showed, the legal line between those who would launch a rebellion and those who might simply harm people was vague enough to permit the imperial power to cross it at will.[74]

In 1715 the scholar charged with appending comments to a new edition of *The Qing Code*, Shen Zhiqi, duly noted the distinction between the two statutes: those sorcerers prosecuted under the statute pertaining to ritual intended to hoodwink, whereas those prosecuted under the statute pertaining to punishment intended to harm vast numbers of people. The former carried the more lenient penalty, as one would expect. Yet both statutes used identical language to describe a unique form of communication with the spirit world—mathematical astronomy and judicial astrology.

A law I have not yet mentioned is Statute 165, "Keeping Forbidden Books and Studying Mathematical Astronomy without Permission." While it could not have been applied to Schall and his colleagues—for reasons that become instantly clear—it is of considerable interest for the light it sheds on Qing thinking about astronomy:

> It is forbidden to conceal and keep for oneself objects and devices used to observe the heavens . . . diagrams of prophecies (books with symbolic charts that can be used to predict the dynasty's stability), other banned

> books, (paintings of) the rulers of former dynasties, and gold or jade imperial tallies and seals. (He who does not declare them to the government) will receive one hundred strokes of the heavy bamboo. The same penalty applies to all who study mathematical astronomy in private—an exception is made for students of mathematical astronomy at the Astrocalendrical Bureau. Moreover, the offender will be fined ten taels of silver, passed along as a reward to the denouncer. (All devices, utensils, and other objects will be forfeited to the government. Those who manage to complete a course of astronomical study without receiving permission will be sent to pursue further studies at the Astrocalendrical Bureau.)[75]

The primary responsibility of the Astrocalendrical Bureau had long been the production of the ephemerides and almanacs used for scheduling major imperial occasions. The part the Jesuits played in producing reliable calendars, then, made them essential to the Qing state apparatus. Statute 165, in contrast, was aimed at those who were not a part of the Qing state apparatus, those who might, in theory, use mathematical astronomy to claim the mandate of Heaven. The state was quite determined to retain a monopoly on mathematical astronomy.

But as valuable as the learned Jesuits had become to the smooth operation of imperial ritual, Yang happened to lodge a combination of complaints at just the right moment to bring about their downfall. At a time when the future of the Qing state was not certain, a heterodox group that had a national network must have seemed like an intolerable threat. Yang's attack had little to do with the technical details of mathematical astronomy. Instead, he seized the moral and intellectual high ground, charging the Jesuits with high treason and accusing them of using their scientific gifts in order to advance their evil religion. He mentioned European science only to link it to the larger crime: the subversion of the political and ritual order of the Qing dynasty.[76] In the brief interim while Yang was in charge of the Astrocalendrical Bureau, he recruited those Muslim astronomers and astrologers previously dismissed by the Jesuits. Their incompetence in maintaining minimal accuracy to predict solar and lunar eclipse had been repeatedly revealed over the course of the Ming. Therefore Yang's persecution of the Jesuits was primarily a legal and political move, coupled with a very risky bet on technical matters. Adam Schall died in 1666 in the imperial jailhouse, aged seventy-five. The technical weakness of Yang's persecution soon became obvious and the Jesuits returned to the Qing court with a vengeance.

AFTERMATH: ONSET OF DISCURSIVE BATTLES

At the time of Schall's death, the Qing ruling order was changing. The young Kangxi emperor purged the Oboi faction and seized power in 1667. He dismissed many associates of the Oboi faction. Yang Guangxian, who persecuted the Jesuits under Suksah's patronage, was an intriguing case and could not be dismissed right away. The emperor therefore called for an astronomical contest that would draw back to the Forbidden City Schall's younger colleague, the Belgian Jesuit Ferdinand Verbiest (1623–88). Then, before the emperor had learned how to handle the Oboi regents and Yang's faction, a crisis arose in South China: Wu Sangui (1612–78), a Ming general who had been rewarded by the Qing with a princedom for turning against his former masters, turned again and took up arms against the Qing. Wu allied with other Ming traitors to launch a military offensive; the conflict became a quagmire, dragging on from 1673 to 1681.[77] Memories of Yang's accusations against the Jesuits started to seem, by comparison, slightly ridiculous.

The Kangxi emperor could not help but distrust anyone associated with the Oboi faction, and he concluded that the Jesuits posed neither a political nor a military threat to him. One of his first actions after seizing the reins of power had been to issue a formal posthumous pardon of Schall, and he composed a laudatory epitaph of the astronomer who had done much to legitimate Manchu rule. To restore the reputation of the Jesuits, in 1668 the emperor mounted a bit of political theater: he arranged a competition between Verbiest and the Muslim astronomers appointed by Yang to see who would more accurately predict the length of the shadow of the imperial gnomon—a gigantic sundial in the Forbidden City. This was clearly a setup staged to discredit Yang. It was also designed to erase the image of Jesuits as seditious foreigners, recasting them as the more gifted peers of Chinese astronomers. Surprising no one, Verbiest won the contest.[78] He then assumed control of the Astrocalendrical Bureau, while his opponent was cashiered. A humiliated Yang set out for his home in Huizhou, but he died en route. Like his victim, Schall, the old litigation master had lived seventy-five years.

It may have been political theater, but the Jesuits certainly possessed astronomical tools that surpassed those of any Chinese rivals. At the Astrocalendrical Bureau, Chinese converts assisted Verbiest and Magalhães. And from the day of the Jesuit triumph over Yang until the 1840s, the practice widely adopted in the Qing bureaucracy of assigning a Manchu and a non-Manchu to share a post was applied to the top position at the Astrocalendrical Bureau, where a Jesuit and a Manchu shared responsibilities. Once this concession to appearances had been made, the supremacy of Jesuit mathematical astronomy was never questioned again.

For a time, Verbiest and Magalhães worked together at the court. And—the last step in the complete rehabilitation of the Jesuits' reputation—each published a lengthy and strenuous rebuttal to *I Cannot Do Otherwise*, Yang's 1665 account. In his essay, Magalhães offered countless bitter criticisms of Verbiest, which was part of the so-called Rites Controversy among various Catholic orders in Qing China (see chapter 3 for more details). Here is one example: "He wishes only to demonstrate that using astronomy to make calendars is superstitious and execrable, even the work based on true and natural mathematics."[79] In this sense, Yang's persecution should also be seen as an effective intervention to the perpetual bickering among the Catholic orders in China.

In Verbiest's critique of *I Cannot Do Otherwise*, he paid no attention to the charges of sedition. The Jesuits promoted their mathematical astronomy as a system that revealed God's glory, and they shared it only to convince the emperor and his officials that they would ultimately need salvation. Yang had keenly perceived the missionaries' agenda. Verbiest bypassed this criticism and focused exclusively on the impartial and empirical measurement of heavenly movements.[80]

"The study of celestial patterns," Verbiest proclaimed, "is the practical politics [*shizheng*] of our dynasty." And he urged the emperor to recognize that scholars had long studied such patters for practical reasons. Verbiest repeatedly emphasized that mathematical astronomy made possible imperial ephemerides and almanacs that were highly reliable. Celestial patterns—such as solar or lunar eclipses, the Sun's movement along the ecliptic, and the orbits of the five planets—transcended factional politics. It made no difference if the astronomer was Jesuit, Muslim, or Chinese—so long as he made impartial observations and applied Tychonic cosmology to verify them, the "method would be valid." The sky was up above:

if a particular personal agenda was not in the way, then anyone should be able to observe that sky and see the same patterns. But rather than employ reliable methods, Yang had allowed ambition and selfishness to obscure essential truths. He had gone so far, Verbiest insinuated, as to use his arguments to manipulate the imperial household.[81]

For the Kangxi emperor, observational empiricism took priority over textual empiricism. In the Jesuit's diatribe against Yang he wrote, "Measurement and verification are the ground on which my entire defense rests."[82] While he knew that most literati had no practical understanding of astronomy and knew only what was written about the heavens in historical records and the Confucian classics, his priorities directly challenged the sanctity of classical scriptures as well as the authority of the standard dynastic histories. He likened measurement, which he saw as a potential bridge between Tychonic cosmology and Chinese notions of celestial patterns (*hetian*), to the detachment innate to the highest imperial authority. Yang's scholarship, by contrast, he dismissed as "the satisfaction of human desires": selfish, divisive, and suited only to the pursuit of personal advantage. Was Verbiest picking up on a tension that ran through early Qing scholarship, the growing difficulty of reconciling an increasingly quantified view of the sky with the need to trace all truth back to the ancients?[83]

To convince the emperor of the superiority of Jesuit mathematical astronomy, Verbiest stressed the importance of accurate data collected over a long period. But he distrusted and dismissed the data recorded in the dynastic histories.[84] Tychonic cosmology had been built up by minutely comparing the positions of hundreds of stars observed from Denmark over significant spans, and it was quite applicable to the movement of celestial bodies in China.[85] As Verbiest pointed out in a number of treatises—among them "The Solid Proof That the Earth Is Spherical" (Di wei yuanxing shizheng), "Our Defense of Measurement and Verification Is Based On Equatorial Coordinates" (Bian yi chidao ceyan), and "The Accuracy of the Old and New Calendars" (Xingjiu erli shumi)—mathematical astronomers had observed the sky and amended their calendars over several millennia, and most of that work had occurred in the West.[86] Such procedures had grown reliable out of "long experience," Verbiest contended, and the details of celestial patterns had grown apparent over time.[87] As to Yang's inaccurate division of the tropical year, his false claims about a fictitious element he called "purple ether" (*ziqi*), his reversal of the sequence of the Shen and Zui constellations, his absurd techniques for observing the equally absurd ether, and his acceptance of Rahu (Luohou) and Ketu (Jidu), two imaginary planets drawn from Buddhism that were used to account for lunar eclipses—Verbiest picked them off one by one.[88] Few classical scholars could have read his demolition of the twenty-four divisions of the tropical year, a standard element in classical scriptures, without wincing.

It is no surprise, then, that the two grounds to which Verbiest turned in his defense of Schall's methods—empiricism and the Tychonic method—did not carry equal weight for the emperor and his officials. All astronomers believed themselves exceedingly careful in their observations, so the suggestion that Chinese and Muslim astronomers had simply failed to keep good records won few over. But of the many Western techniques sampled by the Qing court, spherical trigonometry, in particular, left the deepest impression. More importantly, spherical trigonometry—a branch of mathematics and geometry that deals with the relationships between trigonometric functions of the sides and angles of triangles and polygons defined by a number of intersecting great circles on the sphere—had developed from Damascus in the thirteenth century to Rome in the sixteenth century.[89] From the vantage point of classical scholars in China, the vicissitudes and nuanced difference in astronomical observations did not at all vindicate the Tychonic astronomy. Spherical trigonometry, on the contrary, could be traced back to the previous global conjunction when the Mongol conquerors brought Islamic astronomy to China. It was the development of spherical trigonometry that mathematical astronomer Mei Wending (1633-1721) attempted his grand synthesis and our protagonist Dai Zhen excelled in the 1750s. As the discursive battles between the Jesuits and literati raged on in the eighteenth century, spherical trigonometry became a new link between cosmology and imperial politics. The most important source of imperial power, namely the cosmological claims, remained outside missionaries' purview.

CHAPTER 3

SEARCHING FOR TRUTH IN THE ORIGINS OF CIVILIZATIONS

Orthodoxy cosmology in the Qing centered on the hermeneutics of *The Book of Changes*. According to the imperially commissioned and sanctioned *Book of Harmonizing the Times and Distinguishing the Directions* (Xieji bianfang shu, completed in 1739), orthodox cosmology should include the fundamental cosmic elements (*yin* and *yang*) and five cosmic phases in the natural world coexisting with ghosts and spirits in the supernatural world. The cosmological techniques included the system of calendars, almanacs, personal horoscopes, astrology, numerology, and geomancy to select auspicious from inauspicious names, events or sites.[1] All cosmic signs and techniques can be traced to the hexagrams in *The Book of Changes*, which are the permutation of cosmic elements to abstractly represent ten thousand things in both the natural and supernatural worlds.[2]

The triumph of Jesuit science in the Qing court at the beginning of the eighteenth century must be considered in the context in which the Catholic evangelical practices were kept out of the orthodox cosmology sanctioned by the Confucian canon from central to local governments: the accurate prediction of celestial events, a knack for survival in the face of political and legal persecution, and the separation of its technical offerings from its proselytizing agenda. For the Jesuits, after all, imparting a scientific education was, in itself, trivial; their goal was the religious conversion of imperial elites. Their proselytizing agenda—the subject of a vast literature—should be carefully scrutinized in terms of how they meticulously avoided conflicts with the orthodox cosmology.

The peculiar French Jesuit Joachim Bouvet (1656–1730), who ventured to study *The Book of Changes* and whose career demonstrates how cosmological claims were manipulated to serve political power, was an exception. Because he won the Kangxi emperor's trust, Bouvet was the only Jesuit who had a real shot at convincing him of the correctness of his cosmological claims—albeit cloaked in a Neo-Confucian vocabulary. The intellectual moment is significant because Bouvet's arguments embodied the essence of cosmological claims rejected by the Chinese scholars in the eighteenth century. Dai Zhen and his followers would simply consider it as lofty but unconvincing rhetoric advanced by a heretic sect. Yet, the way Bouvet subsumed mathematical astronomy in the essential dogma of Catholicism mirrors exactly how Chinese scholars linked technical know-how to classical knowledge. When cosmological claims were contested in the political arena, however, the Jesuits rarely stood a chance.

Placing Bouvet side by side with contemporary Chinese and Islamic scholars and mathematical astronomers reveals the scope and variety of intellectual calisthenics to claim the cosmological origins, as Dai Zhen and his followers moved to a position where all truths were universal and origins were not so important. Such a cosmopolitan (some might even say "universalist") position was not common in the early modern world. Instead, groups tended to make fierce claims to originality. This habit was widely shared among almost all intellectual elites across Eurasia. What is more interesting, however, is that European missionaries in Qing China shared the playbook with their Islamic and Chinese competitors. They competed for imperial favors and patronage using exactly the same argumentation to the cosmological originality.

JOACHIM BOUVET

One of the six missionaries sent to China by the French Académie des Sciences (founded in December 1666) and dubbed "Mathématiciens du Roy" (the King's Mathematicians), Bouvet strayed far from the standard creed while bending Qing imperial orthodoxy. He concocted a grand myth regarding the postdiluvian arrival of the word of God in China, which were messages encoded in the Confucian classics. This connection, Bouvet felt, elevated China in the civilizational hierarchy. At the same time, Bouvet inadvertently endorsed a cosmopolitan consciousness that had just appeared on the intellectual horizon. Bouvet consistently and simultaneously exhibited an aversion to early modernity, his thinking heavily influenced by the Counter-Reformation and early Christian theology.[3] Many historians would stress the fact that Bouvet's proposal was merely an outlier and extreme case in the spectrum of early modern European thought. This may be a valid point but it is quite irrelevant to our story here: Bouvet had the ears of the Kangxi emperor! His political proximity to the center of imperial power made him one of the most significant players regarding Catholicism and mathematical astronomy in the Qing court.

Launched in 1600, the China mission of the Society of Jesus struggled throughout its history to win the trust of members of the elite and the imperial household. As religious newcomers, Jesuits had to contend with Buddhists, Daoists, Muslims, and members of millenarian cults. Routinely called on to prove their loyalty to the Qing, the Europeans would eventually run pell-mell into a mammoth intellectual and political battle.

Time and again, the work of the missionaries had terrible consequences. In 1664, inspired by the Jesuit Adam Schall, Chinese astronomers at the Astrocalendrical Bureau declared that their distant ancestors were descendants of Noah, a claim that resulted in their execution. One of them, Li Zubai, had suggested that the first inhabitant of China, the legendary Fuxi, belonged to the lineage of Adam and Eve.[4] Eight years later the Jesuit Louis Buglio, along with his colleague Verbiest, composed a syllogism to prove that Chinese people were descended from Europeans. If the Chinese could not be traced back to people living elsewhere, all others must have descended from Chinese stock. But other people did not descend from the Chinese, thus the Chinese must have descended from others.[5]

After examining ancient records in both China and the West, Buglio concluded that there was no evidence of a Chinese traveler reaching the West; there were records only of early Christians traveling to the East. These were descendants of the survivors of the biblical flood, and Buglio reasoned that they must have arrived in China before "true religion and culture" (code for "Christian civilization") had been established. Thus, China achieved its high level of civilization around 1040 BCE because at that time Judeo-Christian thinking had worked its way East, influencing rituals and music, literature, art, and morality. The Jesuits, Islamic scholars, and Chinese literati all asserted the existence of a universal truth; each claimed to have discovered it first, only to struggle to put it into language acceptable to others. Often cited was the Song dynasty thinker Lu Jiuyuan (1139–92): "If a wise man arose from the Eastern Sea, at one end of the world, his profound consciousness would permit him to comprehend the world and deduce a set of abstract principles for understanding the universe. If another wise man arose from the Western Sea, at the opposite end of the world, he would comprehend the world with the same profound awareness and arrive at the same set of universal and abstract principles."[6] Authors who quoted this passage often spoke of a founding moment for universal truth, only to provide a chronicle of the process by which the true civilization or religion became corrupt. But in the end there was always redemption or restoration.[7]

Buglio offered a specific scenario and attempted to persuade the literati that the civilizing mission of the ancient sages was a prelude to the missionaries' efforts in modern times—and naturally he mentioned Lu Jiuyuan. Literati converts to Catholicism provided a more nuanced theory of the Western origins of Chinese civilization. In Hangzhou in the early 1660s, for example, Liu Ning, who lived until 1715, circulated a manuscript entitled *Records of True Awareness* (Jue si lu), in which he defended Christianity against Buddhism. The first two sections, "On Fundamentals" and "The Name of Our Lord Did Not Come from the West," are pertinent here.[8] Liu, unlike Buglio, argued that the transmission of true value—the way (*dao*)—originated in ancient China; only after a period of chaos had the way become obscured.

According to Liu, the people of China would never have grasped the message of the ancients without the Jesuits, who shared literati values and sympathized with the imperial orthodoxy, unlike previous visitors from the "western regions" (*xi yu*).[9] References to the "western regions" appear in Chinese records dating back to the Han dynasty (206 BCE–CE 220), when they referred to Central Asia; the term was applied, after the consolidation of the Ming regime, to the lands of outsiders such as Mongols, Tanguts, Muslims, Tibetans, Persians, Turks, Nestorians, Jews, and Armenians. Insisting that the Jesuits were not outsiders, Liu treated the order as a link in the transmission of orthodox literati values.

Zhang Xingyao (1633–1715), a friend of Liu's who had also converted to Catholicism, wrote a preface for *Records of True Awareness* in 1702. He felt that the literati of the day had failed to comprehend the true wisdom of the ancients; furthermore, they had failed to grasp that Buddhism was evil whereas Confucianism and Catholicism complemented each other.[10] Zhang cited Lu Jiuyuan's universalist adage to stress the importance of understanding the wise men from afar.[11]

Each scholar offered a narrative that joined Catholics and Confucians, implying an ecumenical toleration. Yet this compromise did not go unchallenged for long. Bouvet, whose ideas extended those of Liu and Zhang, began to circulate a dangerously radical narrative at the Qing court—the fuse had been lit.

A THEOLOGIAN GOES TO CHINA

Three years after their arrival from France in 1688, Joachim Bouvet and Jean-François Gerbillon (1654–1707) began serving as the Kangxi emperor's mathematical tutors, using Manchu to teach him basic trigonometry and acting more generally as loyal courtiers.[12] All three were contemporaries and soon became intimates.[13] Bouvet, in particular, developed a powerful affection and admiration for the emperor.[14]

It was not mathematics but what has been called "ancient theology" that truly fascinated Bouvet. This was a blend of Neo-Platonism, Christianity, and Gnosticism that dated back to the first century of the common era.[15] Some early Christian texts, such as Clement of Alexandria's *Letter to Eusebius of Nicomedia*, purportedly made use of pagan texts such as the *Hermetica*, the *Orphica*, the Sibylline prophecies, and the Pythagorean *Carmina aurea*. While the authors of these works—Hermes Trismegistus, Orpheus, and Pythagoras—were strangers to Christianity, the texts were believed to offer profound insights into monotheism, the Trinity, and the Creation, among other core Christian tenets. Various theologians in early modern Europe claimed that this ancient pagan theology derived from Moses; others believed its origins lay further back, and they ascribed central ideas to Noah and his "good" sons, Shem, Ham, and Japheth, and to antediluvian patriarchs such as Enoch and even Adam himself.[16]

By 1700 theologians at the Sorbonne were decrying many of the ideas of one of ancient theology's principal theorists, Louis Le Comte (1655–1728), whose *Nouveaux mémoires sur l'état présent de la Chine* (New memoirs on the present state of China) was published in 1696. Le Comte, like Bouvet, had served in the Kangxi emperor's court and returned to France convinced that the ancient Chinese had received a true understanding of the Christian God from Noah's descendants, only to have that dogma polluted by Buddhism. The most striking feature of his defense of the Jesuits' mission was a series of comparative readings of religions around the globe. But Le Comte's critics were unconvinced by the pictures he drew of Christian worship in ancient China: they saw no evidence of grace or morality, and the notion of a Confucian Heaven struck them as abhorrent.[17] Prominent theologians soon condemned Le Comte's book. This had profound repercussions for Jesuit activities in China. Pope Clement XI's condemnation of the Chinese practices, such as the equivalence of *Deus* with Chinese Heaven and the worship of Confucius and ancestral temples, proved fatal to the Jesuit interpretation of Confucianism, and this "Confucian-Christian symbiosis" came to a sudden end.[18]

BRINGING CHINA INTO CHURCH HISTORY

At the core of Bouvet's ancient theology was an account of the unfolding of sacred history, a process that revealed God's plan for all mankind. That history was divided into three stages. During the first stage, from Creation until the time of Moses, a "law of nature" was in effect. During the second stage, which lasted from the death of Moses to the coming of Christ, humans were guided by the "written law." During the third and final stage, extending from Christ's day to Bouvet's day, they were subject to the "law of the Gospel." Since humans had disappointed God at a number of moments during the second stage, redemption was needed. Naturally, the Church was the vehicle of redemption, and the Jesuits in China saw themselves as the human representatives of that vehicle.[19]

The Kangxi emperor, whom Bouvet educated and served, sent him back to Europe in 1697 as his envoy, part of an effort to persuade Jesuit astronomers to join the mission in Beijing. Upon his return, however, Bouvet was caught in the crossfire of the Chinese Rites Controversy.[20] A long series of compromises, adjustments missionaries made to the beliefs of the people in whose land they lived, prompted the China mission's vicar apostolic, Charles Maigrot (1652–1730), to write a "Mandatum seu edictum." Sent to the Holy

See, this attack on his fellow missionaries ignited a theological debate at the Sorbonne, which in turn inspired Bouvet to write a defense of *The Book of Changes*, the foundational work of Chinese cosmology. After admitting that his views "do not represent the overall sentiment of the missionaries [in China]," he went on to say: "Until now, most have seen the book entitled *Yi jing* [i.e., *The Book of Changes*] as an assortment of superstitious doctrines without any real substance. But I believe it has been my good fortune to discover an irrefutable road for explicating the legitimate principles of Chinese philosophy, which, upon being well heard, will prove to be as sound at least as that of Plato or Aristotle, and I dare state that the proposition condemned by M. Maigrot is one that appears very true to me."[21]

Bouvet was in good company, as some of Europe's most remarkable thinkers were discovering the appeal of Chinese thought at just this time. For Gottfried Wilhelm Leibniz (1646–1716), who published *Novissima sinica* (Latest news of China) in 1697, the land of Confucius and Mencius remained a consuming interest until his death. In 1700 Leibniz (one of Le Comte's champions) wrote to Bouvet: "I find it strange that there should be such an outcry against your colleagues, who have written that the ancient Chinese had the true religion. What harm is there in that? Even if it were false, is it a mistake that would have dangerous results? Not at all."[22] Leibniz's interest in ecumenism, his belief in mathematical and linguistic emblems, and his support of the Jesuits in China greatly encouraged Bouvet's own intellectual pursuits.[23] Still, most of the Enlightenment savants whom Bouvet encountered during a visit to Europe that lasted from 1697 to 1698 found his philosophical proclivity either suspicious or exaggerated. Only when he returned to China did he find a large supportive audience.

During his sojourn in Europe, Bouvet confided to an unknown correspondent (likely his superior) that he had discovered the key to Chinese philosophy in its seminal text. He hoped to show the truth by analyzing the enigmatic figures of the book *Yi King* [i.e., *The Book of Changes*] under which Fou-hi [i.e., Fuxi, China's mythical founder of monarchy and first philosopher] encoded his principles, stating: "As a matter of fact, as I do believe that nothing could be more appropriate to dispose the spirit and heart of the Chinese to embrace our Holy Religion than to show how much it is in conformity with the principles of their old and legitimate philosophy (as for the modern one, I agree that it is not healthy), I will work to complete some memoirs that I have prepared on this matter."[24]

Jesuit missionaries were determined to bridge the gap between "our Holy Religion" and "their old and legitimate philosophy"; that led to the Rites Controversy. Matteo Ricci set the China mission paradigm to "accommodate" certain characteristics of Chinese Heaven as divine. By focusing exclusively on one work, *The Book of Changes*, Bouvet went beyond Matteo Ricci by claiming *The Book of Changes* itself was a divine revelation. He did indeed complete his *Mémoires*, which included annotated translations of selected passages from China's classical scriptures.

In 1700, soon after returning to the Kangxi emperor's service, Bouvet wrote to Charles Le Gobien (1671–1708), whose *Lettres édifiantes et curieuses* (Edifying and curious letters) is one of our most important sources of information about the history of Catholic missions: "I believe that one must suppose, as a certainty, that the canonical books and the Chinese characters are much older than the Chinese themselves, and that they are faithful monuments of the oldest traditions, such as those the common Fathers of all nations left with their descendants, and that the Chinese preserved these monuments more carefully than others did."[25] Bouvet proposed that God's message was to be found in the ancient texts of all civilizations—Egyptian hieroglyphics, the mystical doctrines of the Kabbalah, and the hexagrams of *The Book of Changes*.[26] The Sorbonne's condemnation of Le Comte, however, meant that such an allegorical and comparative reading of ancient sources would find almost no audience among Jesuits in Europe. Nevertheless, Bouvet expected his work to make an important contribution to the recovery of antediluvian truth, and the Kangxi emperor seems to have championed his philological approach, different though it was from Verbiest's emphasis on measurement and verification. When the Rites Controversy flared up in 1702, Bouvet devoted himself to elucidating the classical scriptures. He believed his sole duty was to sway the emperor; the pope's views were of no concern.

On December 5, 1701, Pope Clement XI (pontificate 1700–1721) appointed Charles Thomas Maillard de Tournon (1668–1710) legate to the East Indies and imperial China. This legation was sent to resolve conflicts among the missionaries and to enforce the decision of the Holy Office against the further toleration of the Chinese rites among the Chinese converts. Tournon had been in China for a year when he issued a decree at Nanjing on January 25, 1707, obliging all missionaries to stop worshiping Confucius, to oppose the nearly universal ancestor worship that was essential to Chinese culture, or face excommunication. In response, the Kangxi emperor forced the Jesuit missionaries to pledge loyalty to the Qing court and began monitoring their communications. Both sides felt that all room for compromise had vanished.

In anticipation of new communications from Rome, the Jesuits at the court anxiously prepared for the worst. In October 1705 Claudio Filippo Grimaldi (1688–1707) explained to the general of the Society of Jesus, Tirso González de Santalla (1624–1705), that because Bouvet had located Christian messages in Chinese scriptures more effectively than Ricci had, the only reasonable explanation for Tournon's actions was a conspiracy of those provincial Jesuits who had his ear.[27] The Jesuits in the Qing court attempted to present Tournon as a figurehead who could not comprehend Bouvet's accomplishments and indirectly hinted that Tournon had been out of touch with Chinese reality. Led by Bernard Kilian Stumpf (1655–1720), who would take over the Astrocalendrical Bureau in 1711, the Jesuits compiled between 1701 and 1703 a long list of passages from the classical scriptures designed to link the Chinese word for Heaven with the idea of a Christian God. Reworking, annotating, and extending this compilation, Bouvet produced a book called *A Mirror of the Ancient and Modern Worship of Heaven: Essentials of the Heavenly Learning* (Gujin jing tian jian tianxue ben yi).[28] Tournon ordered the Jesuits in the Qing court to stop circulation of this book. The Chinese version of this work definitively helped Bouvet win over Kangxi's support. But the Latin translation, ordered by the Kangxi emperor and completed in 1706, was presented to Tournon as the justification of the routine Jesuit practice in China. This gesture had no doubt angered Tournon, who understood it as the defiance from the Jesuits in the Qing court.[29]

The form of Bouvet's *A Mirror of the Ancient and Modern Worship of Heaven* was as revealing as its content. He described the manuscript to Leibniz as "a complete collection of all the most beautiful expressions which are found in the classical books of China relating to the divinity, *arranged in the form of a catechism*."[30] When presenting it to the emperor, Bouvet explained that the true and omnipotent Lord was not entirely revealed by any single Chinese text. Rather, each disclosed a fragment of holy truth. By collecting these messages and using a "set of mirrors"—the implicit comparison was to a telescope—to decode them, a complete image might be assembled. The first part of *Mirror* contains forty-two articles of Catholic dogma, followed by a collection of quotations from and interpretations of the Five Classics and the Four Books. The second part has forty-one articles, more or less corresponding to the first part, each followed by three quotations drawn respectively from peasant customs (*minsu*), literati conventions (*shisu*), and the classical scriptures (*jingwen*). This variety, said Bouvet, conveyed a complete picture of the Lord, governor of natural and human realms.[31]

Moreover, Bouvet exhibited none of the racism displayed in Buglio's account of the biblical hierarchy of the races; instead he focused on what all ancient cultures had in common. He argued that a reverence for Heaven (*jing Tian*) had always been the most significant component of Confucianism and the basis of the Chinese social order.

Bouvet had been granted access to interpretations of the classical scriptures in the imperial collection, works given to the Kangxi emperor by outstanding scholars, as well as to a compilation of ancient texts entitled *A Reflecting Lake of Ancient Prose* (Guwen yuan jian). From the latter text, which had been written at the court in the 1680s, Bouvet retrieved interpretations of the philosophy of Mencius and Zhuangzi.[32] He mined these sources for classical references to a transcendental and omnipotent power, while distancing himself from orthodox Confucian scholarship.[33]

Despite the best efforts of Bouvet and other Jesuits in the Qing court, Tournon's confrontation with the Kangxi emperor resulted, in 1707, in a dramatic setback for the China mission. Mocked by the Manchu ruler, Tournon, who spoke no Chinese, called for support from Maigrot, at the time vicar apostolic of Fujian. A bitter foe of Bouvet and other Jesuits who adapted to Chinese rites and terms, Maigrot was prepared to reply to the emperor. A confrontation took place in the presence of court Jesuits and officials, and Kangxi told Maigrot he had been deceived by charlatans.[34] Viewing Chinese customs and rites from a Christian perspective, he continued, would be as absurd as judging European customs from a Chinese perspective.[35] Maigrot and Tournon refused to accept this and the following year they were deported to Macao.

THE BOOK OF CHANGES AS DIVINE MESSAGE

When Tournon died in Macao three years later, Bouvet became even more defiant of Rome. He believed *The Book of Changes* could settle all of the troubles that had arisen over rituals. Defying his superior in the China mission, he told the Kangxi emperor in 1711 that he planned to search for the hidden and encrypted divinity of *The Book of Changes*.[36] For more than a year he presented the emperor with a series of exegetical readings, now preserved among his Chinese manuscripts held in the Vatican Library.[37] Essentially, their content may be divided into three major components: sacred history based upon the Septuagint, tripartite messianic cycles, and the cosmological iconography of the hexagrams in *The Book of Changes*.[38] Each lecture was meticulously presented in the philosophical vocabulary of Neo-Confucianism. Bouvet's wholesale reduction of the natural universe and all human history to messianic prophesies went back to patristic and medieval traditions of sacred history and cosmological iconography.

By 1712 Bouvet was writing with bitter resignation that if only his superiors had not forbidden him to publish his messianic interpretations of China's classical scriptures, he would "now dare, in the Far East, to undertake something similar to what, about thirty years ago, the most illustrious Huet, master of the deepest erudition and most shrewd investigator of both sacred and profane antiquity, achieved in the West with a series of comparisons between these two kinds of antiquity, namely an evangelical demonstration, which by arguments drawn from the pagans themselves, should establish the credibility of all the mysteries of the Gospels."[39] Pierre Daniel Huet (1630–1721), a French Jesuit, had in 1679 published *Demonstratio evangelica* (Proof of the Gospel). The book "required a sophisticated understanding of epistemology, detailed knowledge of the history of geometry, a mastery of ancient languages, knowledge of ancient civilizations, and skill in biblical exegesis."[40] The divine qualities of the pagan gods, Bouvet asserted, "had been veiled from the beginning by the inventor of the Symbolic Theology, using a device as holy as it was ingenious, under these different images, as under so many emblems or sacred enigmas," objects that Satan and human perversity later changed into idols. Huet's mistake arose from his failure to see that Moses was "one of the most perfect figures" of the Messiah. Bouvet's enterprise outshone Huet's, he claimed, because "the traditions of the Chinese are without doubt far purer and nearer to our sacred mysteries, than those of the Western pagans."[41]

In *Speculum christianae religionis in triplici lege naturali, mosaica et evangelica* (The Reflection of the Christian religion in the three-fold law of nature, Moses and the Gospels, published in 1663), Paul Beurrier (1608–96) had argued that Christianity is confirmed by law of nature, law of Old Testament (mosaic) and law of New Testament (evangelica). He had also presented a theory of the three ages of man: the Age of Innocence, the Fall, and the Redemption. Bouvet applied these in *A Key to "The Book of Changes"* (Yi yao), where he described a first stage "prior to Heaven" (*xian Tian*) during which an omnipotent God created the world. In the second stage, "posterior to Heaven" (*hou Tian*), humans were punished by and isolated from the omnipotent God. In the third stage, which completed the circle, returning to the first stage, Bouvet described the return to Heaven and human salvation. These three stages not only made sacred history an apocalyptic cycle, they shed light on Bouvet's position in the imperial court, for the emperor was about to bring the third stage into being.[42]

Although his tripartite model was foreign to the literati, Bouvet discovered that he could lend his model credibility by linking it to the ancient Chinese idea of cosmic and dynastic cycles. So he identified the reign of the legendary emperor Yao, considered the first sage king, as precipitating the Fall. He equated the condition "prior to heaven" with the biblical Garden of Eden, and the condition "posterior to Heaven" with the long and dark decay of the human world following the Fall. He offered a corresponding reading of two consecutive *Book of Changes* hexagrams: *pi* (obstruction) and *tai* (piece). In this way, Bouvet distinguished the dynastic, ontological, epistemological, historical, and figurative stages before and after the Fall (table 3.1).

The intellectual core of Bouvet's sacred history differed significantly from his previous biblical genealogies. Instead of the Latin translation of the Bible (the Vulgate), he chose the Greek translation of the first seven books of the Old Testament (the Septuagint), because the latter dated the Creation and the Flood earlier, making it a closer match with the flood mentioned in *The Book of Documents*.

By 1713 Bouvet had stopped corresponding with Jesuits in Europe: his only concern was convincing the emperor. Because the shaky ground of

TABLE 3.1. Bouvet's binary distinction, drawn from the Neo-Confucian vocabulary and *The Book of Changes* to code his messianic message

Classification	***A priori* (before the Fall)**	***A posteriori* (after the Fall)**
Dynastic	Before the legendary emperor Yao	After the legendary emperor Yao
Ontological	Prior to Heaven (*xian tian*)	Posterior to Heaven (*hou tian*)
Epistemological	Learning of Way or Mind (*daoxue* or *xinxue*)	Learning of the cosmos (*tiandi wanwu*)
Historical	The Age of Innocence	The Fall
Figurative (represented by two consecutive hexagrams): when *Pi* goes away, *Tai* will come, and vice versa	*Pi* ("obstruction" or "standstill"): Its inner trigram is earth (*kun*), and its outer trigram is heaven (*qian*)—the opposite of *Tai*	*Tai* ("peace" or "greatness"): Its inner trigram is heaven (*qian*), and its outer trigram is earth (*kun*)—the opposite of *Pi*

European biblical chronology no longer mattered to him, Bouvet could offer the emperor an interpretation that differed from those he shared with his correspondents in Europe, particularly Leibniz.[43] He abandoned his mathematical and numerological analysis of the binary system in *The Book of Changes* and turned exclusively to a philological analysis of the book's key words. Nonetheless, he was wise enough not to repeat the mistakes of his predecessors, such as Schall and Verbiest, and he carefully avoided drawing a connection between Noah's sons and the earliest Chinese people.

This project was bluntly rejected in 1710, a great blow to Bouvet's intellectual integrity. The emperor challenged Bouvet's use of biblical sources, his Chinese linguistic competence, the comprehensiveness of his presentation, and his ability to decipher *The Book of Changes*.[44] On this last point, the emperor then asked his capable Manchu interpreter, Hesu, who translated popular Chinese novels into Manchu and facilitated every conversation between Kangxi and Bouvet, "Can *you* understand *The Book of Changes*?" Hesu replied, "Although I cannot fully grasp *The Book of Changes*, I can still vaguely interpret passages by following the commentaries on each hexagram. When I read Bouvet's work, it eluded me. Bouvet's illustrations resemble ghosts and spirits, and some of them look like flowers. Although I could not make sense of it, I found it quite funny. But it was unfortunate that Bouvet claimed that so many Chinese books were Western doctrines [*Xiyang jiao*]." The emperor endorsed Hesu's criticism.[45]

Until the emperor rejected it, Bouvet's grand project had been seen by many as a supplement to the major imperial project, entitled *A Synthetic Interpretation of "The Book of Changes"* (Zhouyi zhezhong), that the Kangxi emperor had commanded Li Guangdi (1642–1718) to compile. That project had in turn grown out of a larger imperial mandate to produce a standard set of interpretations of all classical scriptures, to be added to the curriculum of the civil service examinations. Relying heavily on Zhu Xi's *Original Meanings of "The Book of Changes"* (Zhouyi ben yi), Li published *Synthetic Interpretation* in 1715 with full imperial endorsement.[46] Bouvet's analysis of *The Book of Changes*, on the other hand, was rejected as incomprehensible.

Rather than retreat, Bouvet summoned new allies. The first, Jean-François Foucquet (1665–1741), reached Beijing in 1711; a second, Joseph de Prémare (1666–1736), arrived in 1714.[47] Under the emperor's supervision, the trio of French Jesuits conducted intensive research on *The Book of Changes*. Soon, however, things fell apart. Foucquet and Prémare could no longer stand Bouvet's single-minded pursuit. After a series of bitter disputes, in 1716 Prémare passed along some of Bouvet's manuscripts to Paris, noting, "This kind of writing shows what is going on in the head of this good old man, better than anything one might say to you about it. . . . I hope that a look at these notebooks will prove that the emperor was right when he said that Father Bouvet had gone mad. This can happen to anyone, and it often happens to people who have spent thirty or forty years among chimeras."[48] More broadly, Prémare went on to blame Bouvet for the French mission's failure to make progress at the court: "Every one of these fifty-plus notebooks is filled with extravagant ideas about Chinese books and errors regarding our holy religion. To try to prove it would be a vast and awkward enterprise. . . . Obsessed by his own ideas, he listens to no one who does not agree that his *Book of Changes* is the most beautiful, the greatest, and the most learned, saintly, and salutary work ever written by any human being. The more he is attacked, the more unyielding he becomes, because he is convinced that such attacks are the work of the Devil."[49] Prémare made it clear that Bouvet had not only failed to secure either imperial patronage or the support of the literati, but had also driven a wedge between himself and his fellow Jesuits. Nevertheless, Bouvet never once flinched in his devotion to his mission, even after the emperor dismissed his research as ridiculous and Prémare called him a madman. Committed to his project right to the end, Joachim Bouvet died in Beijing in 1730, aged seventy-four.

MEI WENDING

While Bouvet pursued his quixotic goal, Chinese scholars sought to incorporate the origins of Jesuit mathematical astronomy into a universal Chinese civilization. For example, Wang Xichan (1628–82) claimed that Europeans had merely tacked a few additional techniques onto a system they had received from China. Mathematical astronomers at court and some literati, he explained, were attracted by the seeming novelty of the Jesuit sciences; only because they did not understand the true implications of ancient calendars did they view the Western ways as unprecedented. Wang then listed the correspondences between the terms used by Jesuits and the terms in ancient calendars.[50]

Mei Wending (1633–1721), the most accomplished mathematical astronomer in eighteenth-century China, played an important role in rewriting the origins of Jesuit mathematical astronomy. A scholarly eccentric from Xuancheng, in Anhui, he lived in Nanjing from 1675 to 1680. During that period he met Muslim astronomers and acquainted himself with both Persian and Chinese astronomy.[51] Despite the political uncertainty of the time, Mei took the provincial examination four times, failing on each occasion; he was never quite able to shake off this series of stinging blows.[52] Nevertheless, he was far from a bitter recluse and socialized with Ming loyalists in the lower Yangzi, many of whom were mathematical astronomers.

In 1673 Mei worked with Shi Runzhang (1619–83) to produce a gazetteer for Xuancheng.[53] Sixteen years later, Shi invited Mei to Beijing, where he helped him find work on the *Ming History* (Ming shi) project, working for Wan Sitong (1638–1702). There Mei met Zhu Yizun (1629–1709), a pioneer of evidential scholarship, and the two became good friends. Although Shi and Zhu broadened his understanding of the Confucian past, technically Mei remained unqualified to work on the *Ming History* project. Among his collected works is a rhapsody (*fu*) on the same theme set for the special civil service examination held that year—cosmic models and astronomical instruments (*xuenji yuhen*)—as if to prove his eligibility to work on the astronomy section of the *Ming History*.[54] Lu Longqi (1630–92), a scholar selected by the Kangxi emperor to offer him a daily lecture, praised Mei for contributing to mathematical astronomy and using it to enhance the Confucian canon.[55] His friends in the lower Yangzi were equally approving.[56] Far from an obscure erudite, Mei mixed in the polite society of his day.

By the time he traveled to Beijing in 1689, Mei had adopted an eclectic approach toward Jesuit mathematical astronomy, as he revealed in *A Synthesis of Chinese and Western Mathematics* (Zhong-Xi suanxue tong), written in 1680. (In spite of the title, he advocated the combination of *three* traditions: Chinese, Western, and Muslim.)[57] It was on the strength of this and other works that Mei had managed to land temporary work under Wan, but he did not last long.[58]

A meeting with Li Guangdi changed both men's lives. Li was serving as grand secretary to the Kangxi emperor, who routinely consulted him on a wide range of questions. A conversation on the topic of mathematics and astronomy that took place in 1689 had proved humbling for the learned official. The emperor then instructed Li to read the historical records with an eye to "correct principles," meaning Zhu Xi's principles of nature, the natural philosophy of orthodox Confucianism. As it happened, Mei was able to fill in the gaps in Li's education, and he ended up spending several years composing fifty essays in response to a series of questions the emperor had asked his advisor.[59]

Charged with overseeing a number of major ideological compilations, Li published Mei's essays under the title *Questions in Mathematical Astronomy* (Lixue yiwen) and submitted the work to Kangxi in 1702, one year after the emperor had formally demanded such a text. In his preface, Li wondered, "Is producing a work like this really so urgent?" He replied, "If this was a topic relevant only to mathematical astronomers, then this kind of technical undertaking would not be so urgent. But Mei Wending means to clarify [essential] principles. Could it ever be said that the explication of Confucian principles was not an urgent task? Is ignorance of principle permissible?"[60] Although Li thus elevated Mei's work to the level of the study of principles—the core of imperial orthodoxy—in his memoirs he dwelt on its technical value.[61]

From Mei's perspective, the book had been written to suit Li, who had suggested that he build on *New Cosmographic Studies* (Gexiang xinshu) by Zhao Youqin (1271–1335?), providing a comprehensive explanation of both cosmography and the allied science of mathematical astronomy. Zhao had used the natural philosophy of Zhu Xi, and particularly his orthodox Confucian language, to redirect cosmology and cosmography into the system of "nature and principle" (*xingli*).[62] To the extent that Zhu had studied cosmology and mathematical astronomy, he had kept them in the domain of *qi* (the material world) as opposed to that of *li* (the world of principles). Such an ontological distinction was crucial to Zhu's philosophy.[63] By recommending *New Cosmographic Studies*, Li revealed to Mei that he had failed to see that Zhao's cosmological theory remained qualitative rather than quantitative. More importantly, Li seemed to be asking Mei to adopt a Tychonic model of the cosmos.

In some respects, *Questions in Mathematical Astronomy* followed the Ptolemaic model, which placed the Earth at the center of the cosmos, surrounded by the Sun, the Moon, and the five planets with their epicycles. However, Mei also adopted features of the Tychonic model—which posited that the Moon revolved around the Sun—and even included Keplerian ellipses. He argued that an accurate calendar could only be crafted once the system generating it had passed repeated tests.[64] As the mathematical working of the cosmos was harmonious, orderly, and beyond human intervention, astronomical knowledge was pursued in order to comprehend the mandate of Heaven, which Mei viewed as a self-regulating celestial pattern rather than a moral precept. Only the legitimate ruler was granted the privilege of full comprehension, omniscience that enabled him to maintain order. This turned mathematical astronomy into an indispensable part of Confucian learning.[65]

Mei's argument has historical, geographical, and methodological dimensions. In an essay entitled "On How the Moderns Became More Accurate Than the Ancients in Mathematical Astronomy" (Lun lixue gushu jinmi), he explained, "Contemporary scholars state that a permanent and consistent method should have grown out of ancient mathematical astronomy [long ago]. That method was unfortunately destroyed by fire during the Qin, and it could not be reconstructed. Therefore, mathematical astronomers have frequently changed their methods in an effort to reconcile cosmic models and mathematical astronomy. They hope to recover what was lost in ancient China."[66] A flexible method, Mei argued, was needed to cope with an increasingly dense accumulation of empirical data.[67]

Mei offered two arguments in support of his claim that "constant renovation" in mathematical astronomy ruled out the creation of a singular, unchanging method. First, the ancient sages had stated that the ontological foundation of mathematical astronomy was the radical change (*ge*) hexagram in *The Book of Changes*.[68] As the authoritative "Tuan" commentary stated, "Radical change functions like Water and Fire trying to extinguish each other, like two women living together who find themselves at odds. This is called *ge* (radical change)." The commentary continued, "Only on the day when it [radical change] comes to an end does one begin to enjoy trust. This means, when the change occurs, people trust him (the one who brought about radical change). Such a person brings joy through civility and enlightenment, and he shall prevail thanks to his practice of righteousness. If the outcome of the change was righteous, any regret would disappear."[69] In Mei's reading, the gloss indicated that mathematical astronomers would be responsible for predicting the pattern of celestial and terrestrial changes, thus contributing to the sovereign's civility and enlightenment by elucidating the mandate of Heaven.

Mei's second line of argumentation was drawn from a controversial passage by the ancient mathematical astronomer, geographer, and classicist Du Yu (222–284): "Creating an astronomical system is a matter of interpreting celestial patterns to discover their inherent (technical features), rather than forcing conformity so that predictions will be validated (by these patterns)."[70] In Mei's view, if the model of the cosmos did not accord with the celestial patterns, it had to be altered to approximate the mandate of Heaven. This line of reasoning underpinned late Ming policy questions of mathematical astronomy in the civil service examinations.[71] As Mei wrote for an audience of lower Yangzi literati familiar with such questions, the argument for validating the primacy of celestial patterns appealed to both literati, who advocated this line of inquiry in "concrete studies," and the emperor, who advocated the appropriation of Jesuit mathematical astronomy.[72]

The model of the cosmos was not only subject to technical adjustments, Mei claimed, but also depended upon the accumulation of empirical knowledge. Since celestial patterns had gradually altered over long periods of time, he argued, no one could have devised a permanent model. The cycles of days, months, and years might have been detectable, but the gradual movement of Mars, Jupiter, and Saturn around the ecliptic would not have been. These planets usually appear to move eastward through the constellations: both Mercury's and Venus's passage through the zodiac take, on average, one year; Mars's passage averages 687 days, Jupiter's twelve years, and Saturn's twenty-nine years. The crucial fact for Mei was that the star that stood at the center of the celestial sphere in *The Book of Documents* moved one degree every sixty to seventy years, and completed a full circle in about twenty-six thousand years—a phenomenon known as annual difference (*suicha*, referring to the precession of the equinoxes). How, Mei asked, could ancient sages, however brilliant, have detected these movements without access to empirical data collected over time?[73] In summary, Mei argued that methodological adjustments, the accumulation of knowledge about the heavens, and the incremental improvement of the cosmic model were all indispensable to measuring and predicting celestial patterns.

Although mathematical astronomy is universal in the sense that everyone studies more or less the same celestial patterns, progress had been influenced by cultural demands. Mei believed that China was best suited to effect a compromise among the various systems because it embodied the summation of accomplishments from all regions. For example, the Ming dynasty's Yongle emperor (r. 1402–24) dispatched a fleet led by a Muslim eunuch, Zheng He (1371–1433), to the Indian Ocean in the early fifteenth century, only to learn of many previously unknown countries. Zheng bestowed the name "Western Ocean" (Xi Yang) on the area.

After Europeans entered China, they claimed they had come from the "Great Western Ocean" (Da Xi Yang), by which they meant the Atlantic. When they called their mathematical astronomy "the new method from the Western Ocean" (*Xi Yang xin fa*), implicitly suggesting that it descended from the Muslims' "old method from the Western Ocean" (*Xi Yang jiu fa*), they were collapsing two regions into a single term. However,

Mei insisted on distinguishing the Muslim science from the Jesuit, placing the latter along a linear progression within the Chinese tradition.[74] By and large, though, the Chinese tendency was to lump the Jesuits together with all of the other countries and cultures west of China, including India, Persia, and Central Asia.

For Mei, the key to progress in both China and the West was the accurate measurements that permitted the aligning of cosmic models with celestial patterns. His essay "On the Similarity between the Chinese and Western Methods" (Lun Zhong-Xi erfa zhitong) opens with a declaration and a bold rhetorical question: "Clarifying the celestial patterns takes time, and making the calendar more precise also takes time. Why did we adopt new techniques from the Jesuit calendar and abandon all we had learned up to that time?"[75] Mei asserted a correspondence between sets of variables in ancient Chinese and Jesuit astronomy; this permitted him to hypothesize a single tradition—as "the moderns become more accurate than the ancients"—while simultaneously allowing for differences between the two sets of variables. Examples of such differences would be the trajectory of Mars, the location of the "Old Man Star" (Laoren Xing), the divisions of the year into fortnights or months, the movement of the five visible planets, and the path of the Sun.

This innovative line of reasoning allowed Mei to dismiss the Jesuit notion of God. In "On How the Moderns Became More Accurate Than the Ancients in Mathematical Astronomy," he wrote, "Our mathematical astronomy has become more accurate because we have accumulated data and improved techniques. But the Jesuits claimed that their mathematical astronomy remained consistent from antiquity up to the present. Is their mathematical astronomy bestowed by the 'Lord on high' [*shen*] as they claimed?"[76] His answer was a simple "No!" Mei then compared several major astronomical systems from "the West" in chronological order, including the Buddhist system articulated by the Indian monk Yixing, the Persian and Arabic system integrated into the calendar of the Yuan dynasty, and a modified system from the Ming dynasty. More accurate than any was the Jesuits' because it followed a series of systems derived from Ptolemy and revised first by Copernicus and then by Tycho Brahe. After the Europeans invented the telescope, they were also able to observe celestial patterns more clearly and to approximate their cosmic models more closely. Thus, the Jesuit system had been improved incrementally over a long period of time, rather than being bestowed outright by the "ruler on high."[77]

In 1705, after the Kangxi emperor read Mei's *Questions in Mathematical Astronomy* and Bouvet's *Essentials of the Heavenly Learning*, he issued two works of his own: "On the Mathematics of Triangles" (Sanjiaoxing tuisuanfa lun) and "On Sky-Measuring Rulers" (Liangtianchi lun).[78] In the second he declared Heaven's nature unfathomable and argued that, although the celestial spheres could not be measured directly, owing to the magnitude of Heaven, one could determine celestial patterns indirectly by use of a sundial. Thus, the ancients did not devise the calendar by accident but by measurements.[79]

The dispute that had broken out over astronomy in 1664 arose, said the emperor, from a lack of technical competence among those charged with preparing the court's calendars; that had led to a split between "contemporary methods" (*jin fa*)—he meant Jesuit astronomy—and "ancient methods" (*gu fa*). He argued that, although contemporary methods differed from ancient ones, both originated in ancient China, a land surrounded by barbarians on four sides. All of astronomy derived from ancient texts. The historian-geographers who described the peoples at the margins of the Chinese classical world and the early antiquarian accounts of astronomical instruments provided evidence of the diffusion of astronomy from the Chinese heartland to the periphery—including Europe and Persia.[80]

LIU ZHI

The Muslim experts, as outsiders and technicians, had a much longer history in the imperial court than the Jesuits. When the Mongols conquered China and established the Yuan dynasty (1267–1367), they brought with them Islamic astronomy. It is very likely that they included some of the mathematical astronomers who had worked at the Maragha observatory in what is now northwest Iran, built by the polymath Nasīr al-Dīn al-Tūsī (1201–74).[81] Under Mongol domination, both Islamic and Chinese mathematical astronomers enjoyed imperial support, but they worked separately at the Astrocalendrical Bureau. The sharing of Islamic methods began after the Ming dynasty expelled the Mongols in 1368. Only then did the Muslim astronomers employed at court have two important books translated: *Muslim Mathematical Astronomy* (Huihui lifa) and *The Astrological*

Book, Translated in the Ming Dynasty (Mingyi tianwenshu).[82] The first of these described the achievements of thirteenth-century astronomers, drawing on both Arabic and Persian sources. It provided the dates of eclipses, models of planetary movements, and a list of the days allotted to each of the twelve zodiacal constellations. *The Astrological Book* included detailed explanations of the position and movement of celestial bodies, all crucial to the judicial astrology practiced in the Ming dynasty. Still, this work was dismissed by Chinese literati as a mere product of "western regions" (*xi yu*).[83]

Although they were segregated, Chinese and Muslim astronomers both worked at the Ming court, making astronomical and astrological predictions for the Ministry of Rites. At this time, the dynamic scientific community at Maragha had abandoned the conception of a flat earth and improved on the Ptolemaic cosmos of concentric celestial spheres by developing a fully elaborated spherical trigonometry; its system of epicycles offered improved accuracy and predictive potential. These ideas were almost entirely absorbed by the Latin West but were never introduced to China.[84]

Moreover, Chinese and Muslim astronomers never collaborated at the Ming court: institutionally segregated, the staff of the Astrocalendrical Bureau was regarded as mere technicians, and all discussion of cosmography and cosmology, which could have significant political consequences, was forbidden. This peculiar relationship underwent a dramatic change after the Ming dynasty fell in 1644.

While some missionaries had arrived in China during the Ming, the Jesuits' rise to prominence coincided with the Manchu conquest. An early accomplishment was the elimination of the Muslims from the Astrocalendrical Bureau.[85] Still, a Muslim academic community flourished in the Lower Yangzi Delta from 1630 to 1730. Rendering a variety of Islamic texts in Chinese, the group produced a body of literature referred to as *Han kitab*, a combination of the Chinese word *Han*, referring to the Chinese language, and the Arabic word *kitab*, meaning books. Because of the use of terms associated with the Confucian tradition, some viewed these activities as a form of accommodation to local customs. As Jesuits attempted to reconcile Confucianism and Christianity, *Han kitab* scholars tried to show that Confucian and Islamic science and religion derived from the same set of principles.[86] They cited Lu Jiuyuan, arguing that Islamic doctrines were quite compatible with the Neo-Confucian philosophy of "nature and principle."[87]

The most prolific and profound contributor to the *Han kitab* movement was Liu Zhi (1660–1739). His trilogy, comprising *The Nature and Principles of Islam* (Tianfang xingli), *The Rules and Proprieties of Islam* (Tianfang dianli), and *The True Record of Islam's Ultimate Sage* (Tianfang zhisheng shilu, a biography of the prophet Muhammad), stood as the movement's eloquent pinnacle.[88] *The Nature and Principles of Islam* is based on four books by three well-known Sufi teachers: Najm al-Dīn Dāya, Azīz Nasafī, and Abd al-Rahmān Jāmī, who wrote in Persian and Arabic. For Liu, these four books were "foundational classics" (*ben jing*). Many of the literati of the Lower Yangzi Delta became familiar with Liu's digest.[89]

A great champion of the most famous of the *Han kitab* works was Xu Yuanzheng (*jinshi* 1685). In a preface to Liu's *Nature and Principles of Islam* dated 1708, Xu lamented that for more than two millennia scholars had vainly attempted to offer precise definitions of nature and principle, though Dong Zhongshu (179–104 BCE) and the Song dynasty thinker Zhou Dunyi (1017–73) had offered brilliant sketches: "Unfortunately, the Buddhists and Daoists blocked the way to a proper understanding of nature and principle and obscured sagely learning. This confused many scholars. The Cheng brothers and Zhu Xi attempted to salvage the meanings of nature and principle in the Song dynasty, but in the Ming dynasty Wang Yangming and his followers once again stirred up confusion and disorder."[90] According to Xu, when the Jesuits arrived in China, they followed the example set by Buddhists, Daoists, and Wang, proposing science and technology as an alternative path to grasping nature and principle. And the emphasis on success in the examination system distracted scholars from addressing the sorts of basic questions with which Zhu Xi had wrestled.[91]

Was it not then wonderful that scholars living in faraway Arabia, who wore unusual clothes, spoke a strange language, and operated in their own institutions, nevertheless had the same conceptions of nature and principle as the Chinese? Xu explained that Liu held the reader's hand throughout a poetic journey from being to reality. Inspiring indeed were Liu's discussions of the metaphysical transformation from "prior to Heaven" (*xian Tian*) to "posterior to Heaven" (*hou Tian*) and of the harmonic correlation between microcosm and macrocosm. More importantly, Xu continued, Liu used illustrations to demonstrate the operation of and logical relations between nature and principle.

Xu then posed a question: where did Liu's book come from? His answer: the ideas belonged to Arabian sages, and they had been preserved and enriched by the wise men who came after them. For Xu, Liu's compilation was a major contribution to the Confucian canon. It might have looked as if Liu wanted to disseminate Arabian teachings, but, in practice, the book's message was the true essence of Confucian thought. At the end of his preface, Xu exclaimed, "Oh, how broad was Liu Zhi's mind, and how great was his contribution!"[92]

PHANTOMS

The common enemy for the Catholics, Muslims, and, to some extent, Confucians, was Buddhism. By 1700 Buddhism was so entrenched in Chinese society, blended with Confucianism, indigenous Daoism, and local belief systems, that it had left a deep impression on both national and local elites. This helps us understand why Bouvet's single-minded pursuit appeared particularly heretical and aggressive to Chinese scholars. Islam, particularly in Liu Zhi's version, appeared less aggressive by comparison. Jesuits were welcome to flaunt the technical superiority of their mathematical astronomy, but their challenge to the transcendental claims of the Confucian canon could not be countenanced, and their descriptions of spiritual life contradicted the everyday experience of popular gods in the imperial capital.[93]

Ideas of various religious groups concerning the supernatural in Qing China are evident in stories of phantoms, ghosts, and spirits shared by classical scholars in the middle of the eighteenth century. If these scholars rejected Bouvet and Liu Zhi's claims, they were not moved by sheer bigotry—they distrusted indigenous ideas such as Neo-Confucianism too. They also distanced themselves from Mei Wending and his search for the origins of civilizations.

Unlike these scholars, Dai Zhen and his patron Ji Yun (1724–1805) whimsically used phantoms to satirize and ridicule the transcendental claims of both Neo-Confucian and Buddhist truths.[94] Examples from a remarkable collection called *Jottings from the Straw Hut for Reading Sub-*

tleties (Yuewei caotang biji), comprised of transcripts of several conversations between Dai Zhen and Ji Yun, have been selected to demonstrate the playful attitude of some toward Neo-Confucianism and Buddhism. The rather eclectic collections known as *biji* (lit. records of the brush) could include anecdotes, quotations, musings, speculations, criticisms, and whatever else the author cared to set down. The following stories that include phantoms appearing in scholarly studios highlight intellectual assumptions and convictions of several of the scholars discussed in later chapters.

The first story, told to Ji Yun by his friend Qiu Yuexiu (1712–73), suggests that Neo-Confucians and Catholics suffered from the same illusions. Qiu served as the head of three of the state's ministries before being appointed chief editor of the Four Treasuries imperial library—a post to which Ji was promoted upon Qiu's death. A prominent member of the national elite, Qiu may be considered representative of the national elites around 1770. His story concerned an imperial guard named Shi Dongcun, fond of scholarship and lofty discussions of principles and meanings.

> One summer night, Shi was strolling outside Beijing's Xuanwu Gate, just outside the city wall surrounding the Banner neighborhoods ringing the Forbidden City. He overheard two people chatting and suspected that they were fox phantoms. The first pointed to the north and said, "That building used to be a Confucian academy called Shoushan [Paramount Good]. Now it's a Catholic church. The astronomy, astrology, and fancy instruments of these Europeans are all beyond us. But their religious doctrines are basically a kind of Buddhism disguised as Confucian principles. I once eavesdropped at their church. When they came up against something they could not explain, they always attributed it to the Heavenly Lord. That's why Catholicism doesn't work in China. If we try to look closely at what they do, we become mesmerized."
>
> The second fox phantom said, "You think they are mesmerizing; I think they are foolish zealots. Their kings sent them to China by way of endless seas, and their marching orders are to convert us to Catholicism. If we consider their labors in terms of their goals, what they've been doing does not make much sense. And yet, since the days of Matteo Ricci, they've been arriving continuously, one by one. They don't seem likely to stop until they get what they want. They're sure to go mad!"

> The first fox phantom said, "You think these Europeans are foolish, but those who built the Shoushan Confucian Academy [i.e., Donglin partisans] are even more foolish! When the evil party [the eunuchs] was controlling the state, trying to get its hooks into righteous gentlemen, they gathered [at the Confucian Academy] for empty talk, an attempt to revile and edge out the evil party. In so doing, they actually gave the evil party an excuse to arrest them all. Why should they complain? Only Confucius could muster three thousand followers; Mencius only had a few. Those Neo-Confucians in the Song dynasty lacked Confucius's virtue, but they gathered their friends to form political parties. They fought and weakened the state, which perished as a result. The Donglin partisans repeated the same mistake: they pursued illusory fame and suffered the consequences. As we gaze at a building they left behind, how can we not raise the standard set by our predecessors?"
>
> The two fox phantoms exchanged glances and sighed. Turning, they caught sight of Shi Dongcun—and vanished. Shi commented, "While all of the scholars scrambled to become Donglin-style activists, fox phantoms, who live outside the human realm, furtively expressed their disapproval. Who was mistaken—the men or the fox phantoms?"[95]

In another passage from *Jottings from the Straw Hut for Reading Subtleties*, Ji offers his reflections on China's place in the world. The tone is unmistakably satirical:

> After one dies, one's soul should be registered in the afterlife. The Earth's circumference is 900,000 *li*, and its diameter is 300,000 *li*. There are many other countries on Earth. Their population should outnumber China's. Why is it that those who travel to the afterlife and return encounter only Chinese ghosts and no foreign ones? If Chisong and Guangcheng, who lived at the time of the Yellow Emperor, were truly immortals, why were they only mentioned in high antiquity? Why do we hear nothing about those ancient immortals these days, reading instead about recent arrivals? Are we all going to end up like that hypocrite Zhu Xi who opposed Daoism in public and enthusiastically studied the Daoist Wei Boyang in private?[96]

Ji went on to say that he had polled a number of "experts," including astrological officials and Daoist masters, on the question of the underworld's residents. No one could give him a satisfactory answer. His message was unambiguous: ethnocentrism does not make any sense. Hence the Sinocentric framework implied in Neo-Confucian moral philosophy had to be rejected.[97]

The following example addresses the potential trivialization of evidential scholarship. In response to a tale Ji had told him about scholars and phantoms, Dai recounted the following:

> During a candlelit debate, two scholars disputed the question of whether the *Spring and Autumn* (Chunqiu) had relied on the calendar of the Zhou dynasty or the Xia. When the conversation took a nasty turn, they heard someone standing outside the window sigh and say, "Zuo [Qiuming] lived in the Zhou dynasty, so he could hardly be unfamiliar with the Zhou calendar. Why are you two wasting your breath?" When the two scholars went outside, they saw only a servant boy sleeping under the window. Ji Yun reflected upon this episode and wrote the following: When scholars investigate antiquity by means of evidential analysis, they may use as many as 140,000 characters. How do they know that no phantom in the shadows is mocking them?[98]

Both Ji and Dai were acutely aware of the how certain philological discussions could be trapped in unproductive loops, and they warned their readers against such dead-ends. They were also the most important gadflies of the evidential scholarship movement in the second half of the eighteenth century. Compared with Neo-Confucians, Buddhists, Catholics, and Muslims, Dai and Ji possessed unusual intellectual convictions. It is precisely their different vision, not their methodology, that gave them the confidence they needed to cope with religious claims and distinguish themselves as Confucians marching to a different drummer.

The fourth example, another story from Dai, deals with moral vanity:

> A Ming dynasty scholar named Song traveled deep into the mountains of She County to look for a burial site. The sun was setting, and a storm was on its way. Spotting a cave, he headed into it, only to hear someone say, "A phantom inhabits this cave. Please leave."
>
> Song asked, "Why are you here?"
>
> The voice replied, "I am the phantom." Song asked him to step forward.

> The phantom said, "If we met, *yin* and *yang* would do battle in this cave. You would feel uncomfortably cold and hot at the same time. Why don't you build a fire between us to protect yourself? Then we can converse at a distance."
>
> Song asked the phantom, "Why do you live here? You must have a tomb."
>
> The phantom replied, "I was a magistrate under Emperor Shenzong of the Song. Disgusted by how officials stole from each other and fought for advantages, I resigned and went home. When I died, I implored Yan Luo, the ruler of Hell, to keep me from being born again as a human being. He allowed me to mortgage my salary in the next life in exchange for registering as an official in Hell. To my surprise, officials in Hell also stole from and exploited each other. So I resigned and returned to my tomb, only to find it crowded with countless phantoms whose constant coming and going was calculated to bother me. I fled to this cave. Although solitude, fierce winds, and chilly rain have proven very difficult to bear, the nuisances that go with an official career and the pitfalls that accompany everyday life would be much worse. I have lived here by myself for so long that I've completely forgotten how much time has gone by. I had not even met a phantom for years. I had become content, pleased that I had finally transcended all acres and could now live in absolute tranquility. When you showed up in my cave, it was truly a disturbing surprise! Tomorrow I shall move. Please do not come to find me."
>
> The phantom then stopped talking. Song asked his name. The phantom did not answer. Song had his brush and inkstone with him, so he wrote "Phantom Concealed" (*gui yin*) on the entrance of the cave and left.[99]

This final example also concerns a meeting of humans and phantoms. Once more the tale was told by Dai Zhen:

> A few generations back, a member of Dai's lineage lived in an empty house in a remote alley. It had long stood vacant, and many said it was haunted. But Dai's relative said, "I'm not scared!" At night, he made out a vague shape in the dark, an experience so eerie that he could feel it in his skin and bones.
>
> An enormous phantom appeared and angrily berated him: "Are you really not afraid?"

> "No!" he replied. The phantom proceeded to assume many different terrifying forms.
>
> After a while, the phantom asked him again, "Are you really not afraid?"
>
> He replied once again, "I'm not."
>
> The phantom seemed to grow calm, and it said, "I don't have to drive you away—I just don't want you to brag. If you'd said you were scared, I would have left."
>
> Dai's relative retorted angrily, "I really am not afraid of you. Why should I pretend to fear you? You can do whatever you want!"
>
> For a long time, the phantom tried to persuade him to give in. He was unmoved. The phantom sighed and said, "Over more than thirty years of living here I've never run into someone like you. How can I stay with such a moron?"
>
> And so the phantom vanished.
>
> Ji then presented two commentators who offered suggestions as to the lesson to be drawn from this episode. One said, "Fear of ghosts is normal. If this person had compromised and said he was afraid, he might have avoided all sorts of unnecessary trouble. Just imagine the terrible things that could occur if they continued to provoke each other like that." The other replied, "Those who possess powers as profound as the immortals can dispel phantoms and demons by means of their serenity and tranquility. I am not that powerful, so I would try to overcome a phantom by increasing my *qi*. So long as my *qi* remained vigorous, no phantom would be able to bother me. If my *qi* should seem to fade, then the phantom might be able to take advantage of my weakness. In the preceding story, the phantom vainly set a series of traps." Most people have agreed with the second commentator.[100]

All five stories show that phantoms, like humans, suffered from vanity and weakness just like humans. Some of them were evidently quite wise and could be counted on for acute political insights. Immortals for Dai and Ji were probably not so different from phantoms. When Matteo Ricci first arrived at China, he liked to debate with his fellow missionaries about whether the Neo-Confucians were atheists. What would the European missionaries have said about the empirical classical scholars like Dai and Ji of the eighteenth century?

The struggle between European missionaries and Chinese scholars pried open a new conceptual space for cosmological claims, one that stood apart from the existing orthodoxy cosmology formulated by the Neo-Confucians. At the same time, classical scholars could no longer ignore and dismissed the success and utility of the Jesuit sciences, which were now fully operational in the Qing court. The time had come to anticipate someone who could brilliantly incorporate the strength of the Jesuit sciences and devise a new cosmological and political philosophy based concretely on the sound interpretation of the classical scriptures.

CHAPTER 4

HOW TO BUILD A COALITION AROUND SCIENCE

After Yang Guangxian's persecution of the European missionaries, the ascendance of their astronomy was far from smooth. But by 1670 the Jesuits had consolidated control over the Astrocalendrical Bureau (Qin tian jian). At about the same time, the Rites Controversy waged among Catholic orders revealed to the Kangxi emperor that the Jesuits were quite capable of brawling with their fellow Christians—including Dominicans and Franciscans. To the emperor, the jostling among Catholic orders resembled the factional struggles endemic to the Qing bureaucracy. He managed the Jesuits as he would other court factions, albeit with a bit more caution as they remained loyal to a foreign ruler.[1]

NANJING OBSERVATORY

If we look closely at the prodding, testing questions that the Kangxi emperor tossed at one of his most trusted officials during a tour of the astronomical observatory in Nanjing (then Jiangning), we can learn something about how he viewed Jesuit science, and, taking a step further back, more about the influence of factional politics over imperial policy.[2] Explanations of the roles of identity and history in the constant refashioning of Manchu ruling ideology have not addressed the roles of cosmology and politics in the same process.[3] The imperial perception of the cosmos had something to do with careful measurements and increasingly accurate models, but it was just as conditioned by imperial politics. The ruling ideology in this regard was an intriguing outcome of the political maneuvering around not-completely-flexible images of the cosmos and nature.

Each conquest dynasty looked back at its alien predecessors and broadened the genealogical table in the Temple of Emperors, where the successive patriarchs of the imperial lineage were worshipped. It was here that ruling legitimacy and a set of guidelines for the orderly transmission of political power was derived. When the Kangxi emperor perused *The History of the Liao Dynasty* (Liao shi), the chronicle of a conquest dynasty like the Qing, he came across a passage in which the Khitan minister, Xiao Siwen, advised the Muzong emperor to declare a general amnesty because the Old Man Star (Laoren Xing) was visible.[4] The incident was of particular interest because the Manchus believed they were descended from the Jurchen—the tribes living in Manchuria who had conquered the Khitan of the Liao dynasty. Even so, the emperor disputed the authority of the history. After all, did not *Celestial Officials in the Historical Record* (Shiji tianguanshu) state, "There is a big star below the Wolf Star [Lang Xing]; we call it the Old Man of the Southern Pole [Nanji Laoren]. If the Old Man Star were visible, then the polity would be stable; if it were not visible, then wars would follow"?[5]

Although Xiao had been adhering to an ancient tradition in advising the Khitan emperor, Kangxi doubted that within the Liao territory, which included the North China Plain and the northern steppe, one could have seen the Old Man Star, which is normally visible only from more southerly latitudes. The Kangxi emperor was a pragmatic man and had no compunctions about correcting the historical record if it did not conform to his vision of imperial rule. Others certainly had more respect—often amounting to reverence—for historical records and the classical scriptures. As he was aware of this, the emperor arranged a theatrical performance for his high officials at the Nanjing observatory that constituted a public experiment.[6]

The performance came to involve many of the court factions, which drew on their social networks in the Lower Yangzi Basin. Such networks were crucial to receiving a lofty appointment following success in the civil service examinations.[7] The history of Qing factionalism began in the early 1660s, when two rival groups were established at court, both headed by members of the Manchu elite. Mingju (1635–1708), a noble Manchu Bannerman, recruited many Han literati to his cause; meanwhile, Songgotu (?–1703), a Manchu Bannerman who often advised the Kangxi emperor, recruited Manchus and Jesuits. In 1688 these factions underwent a significant mutation: Mingju lost imperial favor and new factions appeared, headed by Li Guangdi, Xu Qianxue (1631–94), and Gao Shiqi (1645–1703), among others.[8] In his memoir, *Conversations with Li Guangdi* (Rongcun yulu), Li portrayed himself as drawn into a vicious struggle between Xu and Gao, who had joined forces to oust Mingju only to become rivals after their success.[9]

The performance at the observatory took place in 1689, during a tour of the Lower Yangzi Delta. On the same day, the emperor gave an audience to two leading Jesuits—Jean de Fontaney (1643–1710), head of the French mission at Jiangning, and Jean Gabiani (1623–96).[10] Over the course of a conversation that touched on many astronomical puzzles, the emperor quizzed Fontaney and Gabiani about the Old Man Star. Was it visible in the Jiangning area? If so, how many degrees would it be above the horizon? And what about in Canton? Fontaney and Gabiani answered in general terms; later they provided detailed calculations. After putting the Jesuits on the spot, the emperor moved on to the observatory, where he mentioned the star to his trusted advisors Li Guangdi (at the time a chancellor in the Hanlin Academy) and Zhang Yushu (1642–1711), who confessed, "We know nothing about it!"[11] In his memoir, Li portrayed this incident as an attempt by his enemies to discredit him.[12]

The account of this episode in *The Veritable Records of the Qing* (Qing shilu) presents the conversation between the sovereign and his aides in some detail. To illustrate the importance of Verbiest's claim that astronomical measurement took precedence over historical records and even the classical scriptures, the emperor asked Li about the sequence of the constellations:

> Kangxi: "How many constellations can you recognize?"
>
> Li: "I cannot recognize all twenty-eight of them."
>
> The emperor ordered Li to point out the ones he knew.
>
> The emperor asked Li, "In the ancient system of mathematical astronomy, the list of twenty-eight constellations set the Zui constellation before the Shen, but our current system reverses the order. Why is that the case?"

> Li replied, "I do not understand this."
>
> Kangxi then instructed him, "If you use the instrument in this observatory to measure them, the Shen constellation does indeed come before the Zui. Hence you should have faith in our system."[13]

The Kangxi emperor essentially instructed Li to believe his own eyes. Was he thinking of the trials to which the Jesuits had been subjected in 1644 thanks to Yang Guangxian? Here, the Kangxi emperor seems to have been determined to show his aide that the Jesuits were correct not because of an abstract system introduced by suspicious foreigners but because anyone could witness the sequence. When Yang's faction had accused the Jesuits of altering the sequence of Shen and Zui, the Jesuits had pointed to their measurements and verifications.[14] Forty-five years later, the emperor made clear that measurements and verifications—embodied in Jesuit mathematical astronomy—would take priority over historical records.

The second part of the conversation reveals that the emperor concurred with Verbiest's claim that textual evidence alone was insufficient. To prove that, the emperor pointed out an inconsistency between ancient textual records and observable celestial patterns in the present—holding that the accumulation of small deviations became significant over a long period:

> Kangxi asked Li, "How far has the Centered Star [Zhong Xing] moved since the time of Yao?"
>
> Li replied, "According to previous scholars, it has moved about fifty degrees."
>
> Kangxi asked, "Do the stars [*hengxing*] move?"
>
> Li replied, "According to our new system of mathematical astronomy [i.e., the Tychonic model in which Sun, Moon, five planets, and stars were each located on one of the concentric spheres revolving around the Earth], the stellar sphere [*hengxingtian*, the outermost sphere on which all the stars were fixed] rotates eastward, but very minutely."
>
> Kangxi then instructed Li, "The reason that Guo Shoujing's instruments [from the thirteenth century] cannot be used today is that he did not know that the stellar sphere moves."[15]

What the emperor suggested here was that the Centered Star, which *The Book of Documents* treated as the linchpin of ancient Chinese astronomy, had moved so much that it was no longer a useful reference point. Even the Yuan dynasty's Guo Shoujing, whom most scholars revered as the greatest mathematical astronomer of all time, could not detect the accumulation of small errors because he was unaware of the precession of the equinoxes.[16]

As he continued speaking with Li and the other court officials who accompanied them, the Kangxi emperor attributed the inaccuracy of Ming-era astronomical instruments to a reliance on a theoretically immobile Centered Star—the very system that had been defended by Yang Guangxian. Yang had written his "Explanation of the Centered Star" (Zhongxing shuo) as a challenge to Jesuit mathematical astronomy, arguing that since the star had been selected as a reference point by the ancient sage king Yao it had to be reliable.[17] Certainly many had relied on it—to draft various dynastic systems of mathematical astronomy as well as the Ming dynasty's "Granting the Seasons" calendars. By ignoring this line of argumentation, the emperor implicitly assigned the Jesuit method priority over the Ming.

After stressing the significance of celestial patterns, the emperor unrolled a detailed star map, pointed to a big star in the south near the horizon, and told his officials, "This is the Old Man Star." Thereupon, Li immediately remarked, "According to our historical records, the Old Man Star indicates that we are living in humane and prosperous times." The emperor, who had anticipated the comment, replied, "If you calculate the position of the Old Man Star from the North Pole, then you should be able to see the Old Man Star from Jiangning. However, if you are in Beijing or farther north, then you would not be able to see the Old Man Star at all. The prosperity of the dynasty has nothing to do with the visibility of the Old Man Star."[18]

But the prosperity of the dynasty was connected to other heavenly events, for later that year the emperor said, "If our administration is at fault on Earth, Heaven will respond with calamities from above."[19] He was evidently willing to use traditional Chinese cosmology to keep local officials from sending in misleading reports on local conditions.

Using a theory of causation that has been called "moral meteorology," the Kangxi emperor in 1678 and 1679 linked heaven-sent disasters, such as drought, flood, earthquakes, and eclipses, to the virtue of the state. This notion of causation created problems for the Qing state. First, how were the moral causes of meteorological phenomena to be identified, especially as some, notably eclipses, were predicable? Second, what religious or philosophical meaning was to be assigned to Heaven's behavior? Why did Heaven respond to the people's woes in times of disaster by making them worse?[20] The conversation between the emperor and Li provides a good indication of the emperor's exceptional pragmatism applied to predictable events and detectable patterns within the purview of mathematical astronomy. Still, he did not hesitate to fall back on morality-based cosmology. This is what the emperor said in 1697, in response to a prediction from the Astrocalendrical Bureau:

> Although human beings can predict solar eclipses in advance, from ancient times, sovereigns have all nonetheless regarded these eclipses as warnings of which to be fearful. The general explanation is that reforming human affairs is how one shows reverence for portents from Heaven. Only commonplace rulers [whom Kangxi dismissed] ascribe eclipses to matter-energy-vitality and numerology [*qishu*]. Last year there were floods and earthquakes. Now there is going to be an eclipse of the sun. The significance of this is undoubtedly that it will be caused by a preponderance of *yin* [to tip the balance of *yang* and *yin*, which should ensure the proper working of the cosmos]. How can one maintain that such phenomena are not connected to the doings of humankind?[21]

Unlike his performance in 1689, the Kangxi emperor did not dismiss the political implications of the cosmic sign. This was not merely political opportunism, which does not fare well against cosmic changes and natural disasters. The reconciliatory posture, revealed in this quotation in 1697, endorsed a weaker version of moral meteorology but also left sufficient distance from the unpredictable conditions of the cosmos and nature. This ambivalent attitude toward moral meteorology continued to evolve and operate as part of the ruling ideology of the Qing emperors.[22]

At the same time, Li's limited grasp of mathematical astronomy should not be taken to mean that all scholars were equally ill informed. Take, for instance, a practical-minded scholar from the south, Qu Dajun (1630–96), who observed the Old Man Star from Canton, commenting that it was only visible in South China. Located near the horizon and visible to naked eye, the visibility of the Old Man Star serves as a reference point relative to an observer's location in China, similar to how an observer might measure the height of the Big Dipper in order to determine the latitude of his or her location.[23] The north–east/south–west orientation of the Chinese subcontinent had, in this case, been clearly noted by both the Kangxi emperor and knowledgeable observers. Qu also gave descriptions of the important cultivated trees of the far south: "Numerous sugar palms, fan palms, and silk-cotton trees are planted beside shrines, numerous rosary limes by Buddhist temples, numerous banyans by village community altars, and numerous water-pines and lychee trees beside reservoirs formed by dams."[24] Qu also left detailed accounts of how to grow many of these species.[25] The regional differences between heavenly objects and terrestrial flora, illustrated by Qu's accounts, should remind us that many members of the local elite were aware of the cosmological and environmental differences from locality to locality. Environmental variation and the availability of various herbal therapies from north to south had long been important information for the daily lives of literati and merchants, few of whom would have been easily manipulated by morality-based cosmology or meteorology.

SUCCESSION CRISIS

In 1676 the Kangxi emperor named his second son, Yinreng (1674–1725), heir apparent, in clear disregard of the conventional rule of succession. That rule, based on Confucian thought and established as a political principle in the Zhou polity, stipulated that the first son of the first wife (that is, the empress) was to be the legitimate successor. Kangxi elected to follow a different tradition, the one employed by the Mongolian rulers of the Yuan, and named his most competent son as heir. The Ming dynasty had seen many feeble-minded and incompetent emperors, and the Qing rulers were determined to avoid the fate of the ruling house they had routed.[26] As the Kangxi emperor approached the end of his life, Yinreng's eagerness to take over drove him to a series of rather erratic maneuvers. A confrontation became unavoidable, and in 1708 the emperor announced that his second son would not succeed him after all. Then, after a fair amount of hesitation, he restored him to the position of heir apparent that same year, citing a need for stability.[27]

The question of succession dominated the later years of the Kangxi emperor's rule. After Yinreng was temporarily stripped of his title, competition sprang up among the emperor's wives and their sons. Many courtiers chose a favorite candidate and rallied behind him. Several contenders, including the third prince, Yinzhi (1677–1732), the fourth prince, Yinzhen (1678–1735), the eighth prince, Yinsi (1681–1726), and the fourteenth prince, Yinti (1688–1756), built up vigorous factions.[28] In the winter of 1711, Kangxi again deprived Yinreng of the coveted next-in-line position. Increasingly evident factional strategizing could no longer be tolerated. The emperor assembled his counselors and declared, "There are great officials forming factions around the heir apparent. You are all people I selected for senior office, and have been favored for fifty years. What is the meaning of this cleaving to the heir apparent?"[29] This scolding appears to have had little effect, as more factions subsequently formed around different princes.

To be sure, by the second decade of the eighteenth century the Kangxi emperor was a weak old man.[30] The grand councilor was also feeling the burden of his years quite acutely, and in 1711 Li Guangdi requested permission to retire, citing age and illness (he complained about a "nasty sore"). Perhaps Li wished to avoid becoming drawn into the politics of the succession. The emperor's reply came from the heart: "As I read your memorial, I felt miserable. As far as I can recall, death has already taken most of the officials with whom I have been close. Among the senior officials like you, only one or two remain. Now I too am old. I find myself unable to put into words how I feel. Return to the palace soon, and we shall discuss this face to face."[31]

The emperor could not quell the perpetual bickering over his heir, and his frustration grew. As this divisive maelstrom swept up more and more officials and scholars, factional boundaries became extremely sharp and the atmosphere became poisonous.

The third prince, Yinzhi, who was among the contenders, presided over a project to train Chinese mathematical astronomers at court. Since the Jesuits had disclosed their unwavering loyalty to the pope, the emperor had decided that they could not be relied on for all matters astronomical.[32] He created the Studio for the Cultivation of Youth (Meng Yan Zhai) in 1712 and the Academy of Mathematics (Suanxue Guan) for more advanced students in 1713. Chinese, Manchu, and Mongol officials who had some basic knowledge of mathematical astronomy were commanded to come learn more from Jesuit instructors.[33] The emperor appointed Yinzhi to act as head of both schools.[34] Foucquet, who had reached Beijing in 1711, witnessed this mobilization:

> He [i.e., the Kangxi emperor] had sought everyone skilled in the different fields of mathematics among Chinese and Tartars [i.e., Manchus], in Peking and in the provinces. The viceroys and great mandarins, in order to pay court to him, introduced the elite of great minds, and the most suitable for sciences. They were brought to him from all directions, mainly young men, in order to constitute the academy we are talking about. He chose more than a hundred of them to run it: learned officials who were experts at calculations and geometry, musicians, astronomers, chosen for all these abilities, to say nothing of a great number of workers who make the scientific instruments.[35]

Foucquet described the scale of this mobilization (more than one hundred literati), the siting of the schools in the imperial garden, and the three classes of participants: literati officials, technical specialists, and craftsmen. Foucquet himself conveyed the discussion taking place in France on cosmology and algebra to the Kangxi emperor and wrote extensively on the topic. Foucquet's manuscripts are fascinating, both intellectually and mathematically, as they include the first serious attempt to express symbolic algebra in classical Chinese.[36] But they were never included in any imperial collections and presumably were read only by a very small number of specialists.

Many of those associated with the studio and the academy subsequently formed a faction that worked to advance the candidacy of Yinzhi for the throne. Li Guangdi, whom the emperor had cajoled and teased out of his attempt to retire, had accepted the job of recruiting and training Chinese men likely to excel in mathematical astronomy; he informally patronized some of the teachers at the new schools.[37] In particular, he encouraged Mei Wending, whose lifelong study of mathematical astronomy set him apart. Mei gradually advanced from studying the Ming system to mastering all of Jesuit mathematical astronomy. He wrote a number of treatises in which he incorporated Jesuit mathematical astronomy into the Chinese theory of classical knowledge. It was not the role played by Jesuit sciences in legitimating the Qing imperium that concerned Mei but rather how this new knowledge could enhance its Asian equivalent.

THE SECOND PAPAL LEGATE

Among the Jesuits there was conflict. The Portuguese dominated the China mission, leaving the French, as newcomers, to shift for themselves at different churches within the Forbidden City. They squabbled over imperial favors while reporting to the same hierarchy in Manila and Rome, forming alliances with other court factions, playing the same political games as all other courtiers. Invited to tutor Manchu princes, Jesuits did not hesitate to meddle when a bitter struggle over succession to the throne broke out among their students. In the middle of all this, Pope Clement XI sent his second legate to the Qing court to settle the Rites Controversy once and for all. His timing could not have been worse and soon the Jesuits fell out of imperial favor.

By 1710 the Kangxi emperor had consolidated Qing control over all of China and had launched military campaigns against the leaders of Mongolia. By any measure, he ruled over a rapidly expanding territorial state administered by a sophisticated and centralized bureaucracy. The Jesuits knew very well that they were serving a powerful, prosperous, and cosmopolitan empire; it was precisely the Qing dynasty's cosmopolitanism that allowed them to continue exercising their influence at court. But that influence became a source of tension with the intervention of the first papal legate to China in 1705. It was then that the emperor learned the degree to which the Jesuits served the Roman Catholic Church. Suspicions aroused, he looked askance at Jesuit proposals, and in 1706 he announced to his officials that while Jesuit know-how was valuable he had no use for their evangelical agenda.[38] Five years later he went further, declaring, "What the Jesuits told us now appears quite different from before; we should be careful!"[39]

During this time, the Kangxi emperor attempted to open direct communications with the pope, to persuade him that China would not give up its own customs and rituals, that any foreign system—such as Catholicism—that washed onto the country's shores would have to make compromises with Chinese ways. Although he tolerated conversion, he insisted that those who embraced Christianity still needed to perform indigenous rituals, such as making regular offerings to ancestral spirits. The emperor's resistance to Catholicism stiffened in 1707, when the pope's edict explicitly prohibited worship of ancestors and Confucius in converts' homes. He banished the first papal legate, Tournon, to Macao.

The missionaries at the court had appealed against the papal edict, but in 1710 Pope Clement XI rejected their appeal. Still, until 1715 the Vatican made no effort to see that the edict was enforced. When that changed, the emperor simply disregarded the news. Another papal bull, *Ex illa die* (March 1715), which formalized the pope's decision, arrived in Beijing in 1716. Discountenanced, the Kangxi emperor would not simply ignore it.

The second papal legate, Carlo Ambrogio Mezzabarba (1685–1741), who had been ordained a priest in 1718 and consecrated a bishop in 1719, arrived at Beijing on December 26, 1720. The pope had sent two envoys in advance to prepare the Kangxi emperor for Mezzabarba's arrival.[40] The pope's letter never reached the emperor. When Mezzabarba reached Beijing, the emperor did not receive him with the same pomp and enthusiasm he had shown Tournon. Instead, he summoned to the court the missionaries he trusted and asked them to authenticate Mezzabarba's identity. The arrogance thus far exhibited by papal envoys rankled, and Kangxi wanted to send a clear and powerful message.[41]

When Mezzabarba distributed a pastoral letter to the Jesuits at the Qing court, he requested that no translation be made into Manchu or Chinese. He announced that he had brought with him the papal bull promulgated five years earlier in which the pope upheld the ban on non-Christian worship. While this line was viewed by many of the Jesuits in Beijing as unwise, they were pleased to learn that the pope hoped to negotiate compromises in other areas.

But before he had even been granted an audience with the emperor, Mezzabarba brashly requested that the emperor enforce the pope's ban. On the same day of Mezzabarba's arrival (December 26, 1720), the Kangxi emperor replied:

> The pope's prohibition is entirely at odds with Chinese principles. Therefore, Catholicism cannot be practiced in China and must be banned. Since Catholicism has now been banned in China, you shall take all missionaries back to Europe except those who are expert in valuable arts [such as painting, mathematical astronomy, clockmaking, and so on] and those who are too sick or too old to leave. Your pope can ban only Europeans, not the people of China, from worshiping Confucius and their ancestors. So those [Jesuit missionaries] who stay will obey his wish. In this way,

> your pope's wish will be fulfilled. If this is agreeable with you, then come to see me. If not, there is no need to come.[42]

Mezzabarba did not respond at once to this ultimatum. Instead he appealed to God to help him understand the emperor's will.[43] The next day, the emperor sent another edict, reiterating his refutation of the papal bull and ordering the Jesuits at the court to bid farewell to Mezzabarba. The emperor knew that most of the missionaries had no intention of complying with the pope's demand and would want to remain. Mezzabarba pleaded that he had travelled 90,000 *li* (28,000 miles) from the West Ocean to China and only wanted an audience with the emperor. "Will the emperor have mercy on me?" Mezzabarba pleaded. "My entourage and I are exhausted. Can we wait [to depart] until next spring when the Grand Canal is reopened, so we can travel south to Canton by water?"[44] If he could just postpone his confrontation with the emperor, Mezzabarba thought, he might have time to work something out with the missionaries at the court.

But the emperor then demanded that Mezzabarba extradite from Europe to the Qing court the two main thinkers behind Tournon's prohibition of ancestral and Confucius worship, Bishop Charles Maigrot and Teodorico Pedrini (1671–1746), both of whom had served as missionaries in China. He maintained that the idea of banning China's traditional forms of worship resulted from distorted accounts of customs and rituals. Here was a tactic frequently practiced by Chinese sovereigns, meant to save the pope's face by scapegoating two unfortunate villains: it could have salvaged the diplomatic relationship between Beijing and the Vatican. To that, Mezzabarba once again could not reply. Instead, he changed the topic by asking that two important papal documents be translated and submitted to the emperor. The emperor responded by saying that Chinese principles and rituals were profound and difficult to fathom; Westerners should not pass judgment on them. He ordered his missionaries at the court, collectively, to translate the papal documents so that he might read them. In accordance with this concession, the emperor received Mezzabarba, allowing the papal legate to pay his respects according to Western etiquette.[45] The message could hardly be misunderstood: the emperor could respect Western customs, so Westerners should respect Chinese customs.

During his audience, Mezzabarba reported that as he had been in Europe at the time of the crimes allegedly committed in China by Maigrot and Pedrini, he knew nothing about them. He hoped the emperor would condescend to explain to him what had happened. The emperor ordered a feast in honor of the papal legate and the missionaries at the court, instructing the Europeans to wear their Western garments. During the feast, Mezzabarba submitted the papal documents, along with their translations, to the emperor.[46] The emperor bestowed a quart of wine upon Mezzabarba and engaged him in the following dialogue:

> Emperor: "I have seen many Western paintings in which people are portrayed with wings. Why?"
>
> Mezzabarba: "This is an allegory. It shows that the heavenly spirits can move so quickly that they almost seem to be winged. It does not mean that people really have wings on their backs."
>
> Emperor: "We Chinese do not understand Western words and their meanings. Therefore, we do not discuss and judge Western principles and affairs. You Westerners do not understand Chinese words and their meanings. How can you thoughtlessly discuss and judge our principles and affairs? That is why I asked you about the winged man."[47]

At the feast's conclusion, the emperor showered Mezzabarba with elaborate gifts. Having punished the first legate, the Kangxi emperor had decided on a different course this time: he would use Mezzabarba as a messenger.[48] He dictated three communiqués for the pope, adopting a diplomatic approach, reiterating the cultural relativism he had emphasized throughout his encounter with the legate. Unfortunately, he would never receive a reply.

As to the documents presented by Mezzabarba, the emperor read them and left his thoughts in their margins in imperial vermilion ink:

> Having read these documents, I cannot say anything but this: "How can we speak to the narrow-minded Westerners about the grand principles of our country?" There is no one in the West who can understand Chinese books. Discussions and debates are, therefore, ridiculous. Today I read their papal documents and discovered that the pope's position is similar to the heresies and heterodox beliefs preached here by [Buddhist] monks and Daoists. I am putting an end to all further irrational discussions of

> this kind. From now on, the Westerners may no longer practice their religion in China. Let us ban it and avoid the troubles it brings.[49]

The missionaries at the court, who had to serve as the intermediaries between the emperor and Mezzabarba, recorded the entire episode in detail, including the emperor's harsh denouncement of the papal document. Those who appended their names to that record are listed as follows (table 4.1).[50]

These missionaries, all of whom signed the record of the second papal legation, were not happy about its outcome. They came from Italy, Portugal, France, Bohemia, and elsewhere, and they represented a range of theological and intellectual orientations—what they shared was a determination to protect their livelihood and maintain their political influence at the court. The papal decision, not to mention Kangxi's decision, undermined both aims, destroying the China mission.

In the last few decades of Kangxi's reign, the emperor changed his policy vis-à-vis the Jesuits: having formerly restricted Jesuit astronomy to the court, he called for transmitting it to all scholars with an interest in astronomy. The subsequent emperors—Yongzheng and Qianlong—continued to allow the teaching of European science to Chinese scholars. Their use and development of Jesuit mathematical astronomy thereupon ceased to be relevant to court politics and entered the vast intellectual terrain of classical knowledge. It was channeled through large intellectual projects staffed by accomplished scholars—most of them from the Lower Yangzi Delta. This set the basic political and social parameters for the rise of Dai's classical vision.

Since the time of Schall and Verbiest, Jesuits had pursued the evangelical strategy of sharing technical expertise in order to gain the ears of the powerful and ensure that evangelical activities continued in cities and villages. Then came Mezzabarba. The Jesuits in the Qing court (listed in table 4.1) were unified in opposition to the message conveyed by Mezzabarba; their defiance of the Catholic hierarchy was the flipside of their submission to the Qing dynasty. But they could do nothing about the new status quo: the Qing bureaucracy had received a clear signal from the emperor that Catholicism was no longer protected. Persecution of local Catholic groups soon broke out.

TABLE 4.1. Jesuit missionaries serving at the Qing court in December 1720

Pfister no.	Name	Date	Chinese Name (Pinyin)	Chinese Name (Characters)
171	Bouvet, Joachim	1656–1730	Bai Jin	白晋(進)
293	Castiglione, Giuseppe	1688–1766	Lang Shining	郎世寧
297	Kögler, Ignaz	1680–1746	Dai Jinxian	戴進賢
233	Parrenin, Dominique	1665–1741	Ba Duoming	巴多明
236	Régis, Jean-Baptiste	1663–1738	Lei Xiaosi	雷孝思
269	de Moyriac de Mailla, Joseph-Anne-Marie	1669–1748	Feng Bingzheng	馮秉正
161	Suarez, Joseph	1656–1736	Su Lin	蘇霖
248	Mourao, João	1681–1726	Mu Jingyuan	穆敬遠
280	Stadlin, François-Louis	1658–1740	Lin Jige	林紀格(濟各)
307	Gravereau, Jean-Baptiste	1690–1757	Ni Tianjue	倪天爵
301	Slavíček, Karel	1678–1735	Yan Jiale	嚴嘉樂
	Ripa, Matteo	1682–1746	Ma Guoxian	馬國賢
284	Cardoso, Jean-Francois	1676–1723	Mo (Mai) Dacheng	莫(麥)大成
274	Fridelli, Xavier-Ehrenbert	1673–1743	Fei Yin	費隱
294	de Costa, Jean-Joseph	1679–1747	Luo Huaizhong	羅懷忠
275	de Magalhãens, Antoine	1677–1735	Zhang Anduo	張安多
276	Simoes, Joseph	1676–1750	Li (Mu) Ruose	李(穆)若瑟
305	Simonelli, Jacques-Philippe	1680–1755	Xu Maosheng (Dasheng)	徐茂(大)盛

Names were listed in Chinese in *Qing zhong qian qi xiyang Tianzhujiao zai Hua huodong dang'an shiliao*, 1:47. Latin names and dates are cross-referenced in Louis Pfister's *Notices biographiques et bibliographiques sur les Jésuites de l'ancienne mission de Chine*, except Matteo Ripa (the enemy of the Jesuits).

THE SCIENCE FACTION

After the Kangxi emperor's death in 1722, the Yongzheng emperor continued his father's policy of containing Jesuit mathematical astronomy at the court, but he ratcheted up the suppression of Catholic proselytizing in the provinces. The year 1724 saw the persecution of Dominicans in Fu'an County, Fujian.[51] The two-pronged imperial policy had been fully articulated a few years before the Kangxi emperor's death; the Yongzheng emperor took it further.

Chinese mathematical astronomers at the court were determined, during this period, to repackage Jesuit science. A series of publications had begun to appear before the Yongzheng emperor ascended to the throne in 1723, and it continued without provoking any murmurs from the new ruler. Titles included *The Essence of Mathematical Principles* (Shuli jingyun)—a survey of imperial mathematics issued in 1722—and, issued the following year, *The Compendium of Observational and Computational Astronomy* (Lixiang kaocheng)—a massive compilation of the astronomical learning acquired under the Kangxi emperor.[52] The official adoption of Jesuit mathematical astronomy—a process many historians in China and Japan consider a grand synthesis—formally introduced it to elite society and integrated it into classical knowledge.[53] The scholar officials who carried the torch of Jesuit science, ensuring the institutional continuity of imperial mathematical astronomy, constituted a science faction.

As grand secretary, Li Guangdi stood at the peak of a powerful faction in the Kangxi emperor's court that may be thought of as the science faction, though no member of it ever used such a label. A group of technically savvy scholars and officials, it fragmented and faded when Yinzhen ascended the throne to become the Yongzheng emperor in 1722. In addition to establishing new state rituals to demonstrate the legitimacy of his reign, the emperor took steps to undermine the factions of his former rivals: the eighth prince, Yinsi; the fourteenth prince, Yinti; and even the less aggressive contender, the third prince, Yinzhi.[54] Most of the members of the aggressive Yinsi faction were executed or exiled. In addition to the systematic elimination of his enemies, the new emperor also slowly centralized and strengthened monarchical power.[55]

Most prominent members of the science faction (table 4.2) suffered but survived through the Yongzheng emperor's reign. Their remarkable resilience ensured the institutional continuity of Jesuit mathematical astronomy in the Qing court, though many were targeted by the emperor and sent into virtual exile. Even brilliant men who had studied at the Studio for the Cultivation of Youth and the Academy of Mathematics were given

TABLE 4.2. The leading members of the science faction

Name	Life Span	Affiliation	Career after 1735
Mei Juecheng	1680–1763	Mei Wending's grandson and a studio student	Chaired Institute for the Revision of Mathematical Astronomy in the Temporal Model Calendar (Zengxiu Shixian Suanshu Guan); was most important mathematical astronomer at Qianlong's court from 1737 to 1746
He Guozong	?–1766	Studio student	Participated in numerous large compilation projects; worked with Jesuits
Wei Tingzhen	1669–1756	Mei Wending's student	Appointed Minister of Works (Gongbu Shangshu) in 1738
Wang Lansheng	1679–1737	Mei Wending's student	Served three years in the interim council of the Qianlong reign
Fang Bao	1668–1749	Studio student; received Li Guangdi's patronage and protection at court	Worked in Institute of the Three Ritual Classics (San Li Guan) from 1736 to death

administrative jobs in distant border regions. But others remained at the Qing court, where they were employed to assist the Jesuit astronomers. An example is Ming Antu, whom the Yongzheng emperor named as one of two joint directors of the Astrocalendrical Bureau in 1730. In collaboration with the other director, Ignaz Kögler (1680–1746), and Tomé Pereira (1645–1708), both of whom were Jesuits, Ming helped improve the accuracy of astronomical readings and predictions.[56]

The Jesuits, who had supported the Yinzhi coalition, were entirely stripped of their political power; they retained a place in the Yongzheng emperor's court solely for their expertise as painters, cartographers, clockmakers, and, of course, mathematical astronomers.[57] Expertise so important that it trumped factional considerations: published in 1724, the magnum opus of the Yinzhi group and the science faction, *A Compendium*

Name	Life Span	Affiliation	Career after 1735
Chen Houyao	1648–1722	Studio student; received Li Guangdi's patronage at court	n/a
Ming Antu	n/a	Studio student; Mongolian Bannerman	Worked with the Jesuits; worked on sequel to *Lixiang kaocheng* project
Gu Zong	1685–1755	Studio student; Manchu Bannerman	Governor of Jiangsu Province (Jiangsu Xunfu); worked on sequel to *Lixiang kaocheng* project
Chen Wance	1667–1734	Mei Wending's student	n/a
Li Zhonglun	1663–1706	Li Guangdi's son; Mei Wending's student	n/a
Li Qingzhi	n/a	Li Guangdi's grandson	Imperial diary officer (*Rijiang qijuzhu guan*); compiled Li Guangdi's *Lixiang benyao*

Not all members of this group remained mathematical astronomers. In fact, many refashioned themselves into classical scholars instead of mere technical specialists (see, e.g., Guy, "Fang Pao"). SOURCES: Hummel, *Eminent Chinese of the Ch'ing Period*; *Qing ru zhuanlue*; *Qing shilu*.

of Observational and Computational Astronomy (Lixiang kaocheng), bore the emperor's imprimatur.[58]

The clearest sign of the resurrection of the science faction in 1736, the first year of the Qianlong emperor's reign, was the appointment of Mei Juecheng (1680–1763) to the directorship of the Institute for the Revision of Mathematical Astronomy in the Temporal Model Calendar (Zengxiu Shixing Suanshu Guan). Backed by some of the court's most powerful Manchu princes and by the emperor's trusted councilors, Mei memorialized to alert the emperor that an important trilogy, *The Sources of Mathematical Astronomy and Music* (Lüli yuanyuan), was no longer in print. The first volume, *The Essence of Mathematical Principles* (Shuli jingyun), had been published in 1723. *A Compendium of Observational and Computational Astronomy* was the second part, and the final part was *The*

Correct Meanings of Mathematical Music (Lülü zheng yi). This collection, especially the first volume, has a clear logical structure and rests on the foundation of Confucian cosmology.[59] Mei explained that all of the woodblocks used to print the set were stored in the Ministry of Rites, but they had lain unused since the first edition. Producing another print run would glorify the Kangxi emperor's farsightedness and grand achievement, Mei suggested. If hundreds of copies were distributed to provincial governors, imperial mathematical astronomy would be studied in all corners of the empire. The Qianlong emperor tacitly approved this recommendation but ruled out any commercial circulation: his grandfather's work could not be sold to all and sundry.[60]

Another member of the science faction, Gu Zong (1685–1755), reported in 1737 that only three people in the entire empire could understand and use the formulae and tables that appeared in *A Compendium of Observational and Computational Astronomy*—its authors. More importantly, some data needed to be verified from time to time by observations, then run through the entire astronomical system. It seemed to Gu that the time had come for Kögler and Pereira to reveal their trade secrets. All of the algorithms and procedures crucial to drawing up astronomical tables should also be published, he said. The emperor agreed; he ordered Gu, Mei, and He Guozong (d. 1766)—members in good standing of the science faction—to oversee the compilation of *A Supplement to "A Compendium of Observational and Computational Astronomy"* (Lixiang kaocheng houbian); most of the work on the supplement was performed by Ming, Kögler, and Pereira, and it was published in 1742.[61] By the 1740s, it was quite clear that the key members of the science faction were shaping the direction of imperial mathematical astronomy.

Yet two key works—*A Compendium* and *A Supplement*—still remained unsatisfactory to some scholars, who felt that they failed to address the length of the tropical year and difference between the lengths of consecutive years caused by the precession of the equinoxes. Those who hoped that astronomy stood to contribute a lot to classical knowledge would not rest until such questions were settled. A tension between two intellectual ideas was growing, and it was far from clear whether mathematical astronomy at the Qing court would continue to be driven by the demands for accuracy in the calendar and state rituals or by the longing to integrate the new science into the universal framework of the classical scriptures.

FANG BAO AND THE SCIENCE FACTION

A prominent member of the science faction, Fang Bao (1668–1749) played a decisive role in the group's eclipse under the Yongzheng emperor after nearly losing everything under Kangxi. Fang was born into a landed family in Tongcheng in 1668. His grandfather was a *jinshi* in the Ming dynasty. He passed the provincial civil service examination with highest honors in 1700 then distinguished himself in the metropolitan and palace examinations in 1706. He was a close friend of the renowned essayist Dai Mingshi (1653–1713)—both were from Tongcheng County in the Lower Yangzi Delta. Quite interested in different literary theories, they routinely wrote in the archaic style (*guwen*), quite different from the fashionable "eight-legged style" mandated in examination essays.[62]

In 1702 a disciple of Dai's published his master's collected essays under the title *A Collection from Southern Hill* (Nanshan ji). Fang prefaced the work by suggesting that Dai's true intention had not yet been realized:

> When Dai was young, he was famous for composing examination essays. A book merchant eagerly published a collection of them and made a handsome profit. Everyone considers those examination essays his true writing, but I know that they are not. At a certain age, after Dai had traveled quite a bit, he began to circulate among his friends his travelogue, historical anecdotes, biographies, and many other things he has written. All of his friends consider his essays in the archaic style, collected in the present book, his real writing. But I know that they are not. Dai Mingshi's true writing, at that point in his life, was still concealed within his bosom. Not one bit of it had been revealed![63]

Evidently Fang spoke as a close friend who truly knew Dai's intentions and potentials. It was likely they shared a great deal of despair over how the Qing forces vanquished the Southern Ming state, for which Dai wanted to commemorate many heroic individuals.[64] Fang indicated that he knew exactly what Dai was getting at.

One element of Dai's undisclosed project was inscribed in the chronology used in *A Collection from Southern Hill.* Dai (or perhaps Fang) relied on the politically perilous dating of the Southern Ming—a way of expressing sympathy with the overthrown regime and disloyalty to the

Qing. Hong Taiji (temple name: Taizong) had proclaimed the establishment of the Qing dynasty in 1637, ergo all Qing documents would be dated "Taizong 1" in 1637, "Taizong 2" in 1638, and so on. Any reference to the rulers of the Ming dynasty or the Southern Ming for the purpose of dating events after 1637 was a treasonable offense. When the Kangxi emperor's chief censor discovered that Dai had used the reign names of the Southern Ming dynasty in his book, he brought charges. The case snowballed into the largest literary inquisition of the era.[65]

The prosecutor took Fang's preface as evidence of a shared political vision and issued a summary judgment against him. In the winter of 1712 Fang was arrested in Nanjing and transported to Beijing, where he was imprisoned in the Ministry of Punishment while awaiting execution. Li Guangdi, at the time the grand academician (*daxueshi*) and the emperor's most trusted advisor, pleaded five times for Fang's life, on the grounds that this was a profoundly learned scholar and a brilliant writer, the sort of man rarely found in the Qing dynasty. On April 17, 1713, Fang was provisionally released from prison and ordered to demonstrate his literary talent by composing a stele inscription, a celebratory rhapsody, and a discourse on music and mathematics. When these were accomplished, the Kangxi emperor declared himself very pleased by the performance, pardoned Fang, and assigned him to the Southern Studio (Nan Shu Fang)—the name given to the elite team that served as the emperor's personal secretaries. In October 1713 Li invited Fang to study mathematical astronomy in the Studio for the Cultivation of Youth, which had been established the previous year. At that moment, he became de facto a member of the science faction.

Li and Fang had been friends since 1691. Before he had even passed the provincial examination, Fang had been invited to Li's mansion to discuss philosophy and politics. In Fang's own account, he impressed Li greatly by his display of classical knowledge.[66] After Li rescued him, Fang became a graceful and loyal servant of the new Qing dynasty. He was also inextricably enmeshed in the network of the science faction. Fang transformed from a former sympathizer of the Ming loyalist resistance to a loyal and articulate theorist of the Manchu Qing dynasty. His political conversion had everything to do with his traumatic and near-death experience in jail and his unexpected release from captivity and eventual restoration to privileged position at court.[67]

Fang's membership in the science faction was not limited to those with political aspirations, though. Take, for instance, Fang's obituary for Mei. As a key aspect of his friend's life, Fang cited his intellectual and moral alliance with Wan Sitong. Neither man had managed to pass the civil service examination, and they had sworn to dedicate themselves to scholarly pursuits outside the Qing bureaucracy—they scorned fame. At the same time, they feared that their specialized learning might become extinct, so both lectured to promote their erudite predilections. Were they covertly seeking fame? Fang did not think so. But fame resulted differently in each instance of preserving their learning. Wan became a leading light of literati society in 1700 or so. Over the course of a lecture series given in Beijing, he garnered several hundred followers. In contrast, Mei's astronomical investigations attracted much less attention; his lectures drew only Li Guangdi and a few of his disciples.[68] But after Li submitted Mei's trigonometry monographs for imperial review, Kangxi's praise was generous. All of a sudden, Fang noted, ordinary scholars envied Mei's glory. Wan, who had shouldered so much of the responsibility during the compilation of *The Ming History*, received no recognition from the imperial house. In contrast, Mei was showered with praise and recognition in his old age because he instrumentally assisted the science faction.[69]

THE SPLINTERING AND RESURRECTION OF THE SCIENCE FACTION

It was during their darkest moments that Yang Mingshi (1660–1736), a disciple of Li and an acquaintance of Fang, held the science faction together. He was known for his stoical temperament, determination, and uncompromising punctiliousness. He passed the metropolitan examination in 1691 and became a presented scholar (*jinshi*). In 1720 he was appointed governor of the southwestern province of Yunnan and became arguably the most capable administrator from the science faction. Quite naturally, he became the protector of the science faction after Li's demise. In 1725, the third year of the Yongzheng emperor's reign, Yang was promoted to serve in the Ministry of War for the new administration; he continued his station in Yunnan to assist the preparation of the military campaign against Tibet. That same year, however, the Yongzheng emperor accused him of including secret messages in his regular reports (*tiben*), dismissed him from the central administration, and ordered him to wait in Yunnan for further instructions.[70]

In contrast to Fang's remarkable talent in *belles lettres* and his deep classical knowledge, Yang was an able bureaucrat with no claims to erudition. Under Li, he served as compiler of the *Ming History* project, as an examiner, and as a supervisor of river work in the Lower Yangzi Delta.[71] In 1713 he was recalled to serve with Fang in the Southern Studio. Fifteen years later, Yang clashed with the Yongzheng emperor and was almost executed.[72] Sent to Guizhou, a remote southern province, he decided to stay there, lecturing and tutoring until the emperor died in 1735. According to Fang and Yuan Mei, the Qianlong emperor recalled Yang to the court soon after assuming power; he served as minister of rites and tutor to one of the imperial princes. By this time, Yang was seventy-five years old; he died the following year.[73]

During his last year (in 1735) in high office, Yang promoted seven scholars, all affiliated with the science faction in one way or another, to the Directorate of Education (Guo Zi Jian).[74] One of them, Qin Huitian, was put on the fast track and promoted to the post of junior minister in the Ministry of Rites. He then became the key sponsor of many forms of technical knowledge under the rubric of the rapidly expanding category of *li* (rites, protocols, and institutional designs). More importantly, Qin's efforts had coalesced into the establishment and expansion of the Institute of the Three Ritual Classics.

Qin came from a family of wealthy officials in Wuxi County, in the Lower Yangzi Basin. When the Kangxi emperor toured the area in 1703, he chose to reside at the Qin family mansion. The emperor then invited Qin Daoran (1658–1747), Qin Huitian's father, to accompany him back to Beijing, where he appointed him tutor to the ninth prince, Yintang (1683–1726). Qin Daoran easily passed the metropolitan and palace examinations in 1709, upon which he became an editor at the Hanlin Academy.

In 1726 the Yongzheng emperor expelled his brothers Yinsi and Yintang from the imperial family. As a part of the purely pro forma legal proceedings, Qin Daoran was summoned to testify. The prosecutor was merciless: "You are a Chinese scholar. The Kangxi emperor ordered you to tutor Prince Yintang. Why did you get involved in the domestic affairs of the Yintang household? You conspired with Yintang and encouraged his actions. You told everyone that Yintang resembled an emperor, a crime meriting capital punishment and worse. In your previous deposition, taken in the Lower Yangzi Delta, you were clearly hiding something. Now you had better confess everything, or you will risk death as a follower of Yinsi and Yintang."[75]

Qin confessed to embezzlement and any number of petty misdemeanors. Like many members of the science faction who suffered under the Yongzheng emperor, Qin was jailed and his property confiscated. After this setback, his son, Qin Huitian, kept a low profile until the Qianlong emperor ascended to the throne in 1735. That year, Yang Mingshi invited him to the court and offered him a position. The following year, Qin passed the metropolitan examination. As part of the general rehabilitation of the science faction, he was appointed to the Hanlin Academy, his father was released after having spent fourteen years in prison, and, in 1737, the family fortune and property were restored.[76]

So keen was the Qianlong emperor to undo the damage done to scholarship by his father's witch hunt that he devised a mechanism for reviving the fortunes of those formerly persecuted. He announced an extraordinary examination targeting scholars of "wide learning and illustrious words" (*boxue hongci*); those who distinguished themselves would serve on the ambitious cultural projects he was planning. The special examination committee chose 267 candidates, of whom only 180 appeared. The eccentric writer and author of *An Unofficial History of Scholars* (Rulin waishi), Wu Jingzi (1701–54), for instance, refused to participate. The renowned historian, writer, and evidential scholar Quan Zuwang (1705–55), on the other hand, launched his career on this occasion, becoming a "presented scholar" (*jinshi*) in 1736.

Among the imperial cultural projects of the 1730s were the creation of the Institute of the Three Ritual Classics, the expansion of the Historiography Institute, and the revitalization of the Institute for Ming History. The emperor named Fang to direct the first of these, where Mei Juecheng, already the court's chief astronomer, served as chief consultant for all questions regarding the use of astronomy in the interpretation of *The Rituals of Zhou*. In 1759 Mei edited his grandfather's works, published as *A Collection of the Works of the Mei Lineage* (Meishi congshu jiyao).

Qin, the most successful member of the science faction in the Qianlong emperor's court, initiated *A Comprehensive Investigation of the Five Rites* (Wuli tongkao), an ambitious project on state rituals. From 1724 to 1734 he led a group that applied evidentiary methods to the study of the Three Ritual Classics: *The Rituals of Zhou* (Zhou li), *The Records of Ritual* (Li ji), and *The Ceremonial Rituals* (Yi li).[77] This led, after the Qianlong emperor appointed him a member of the Hanlin Academy in 1737, to the *Comprehensive Investigation* project. By 1745 Qin had risen to the post of minister of rites and chief compiler of *The Comprehensive Gazetteers of the Great Qing Dynasty* (Da Qing yitongzhi). When his father passed away in 1747, Qin went back to Wuxi to mourn. There, he did further work on the Three Ritual Classics, and when he returned to the Qing court in 1748—still minister of rites—he assembled a team to work out the first draft of *A Comprehensive Investigation of the Five Rites*.[78]

Following the organizational structure of *A Comprehensive Study of the Ritual Classics* (Duli tongkao, published in 1696) established by Xu Qianxue, the chief compiler of *The Ming History* under the Kangxi emperor, Qin's team collated an enormous amount of historical accounts and classical commentaries on ritual practices. Since the duties of the patriarch (*dazhongbo*) of spring offices (*chunguan*) described in *The Rituals of Zhou* are conventionally divided into five categories—auspicious rites (*ji li*), rites of misfortune (*xiong li*), guest rites (*bin li*), military rites (*jun li*), and joyous rites (*jia li*)—the same scheme was used to divide up the materials collected.[79]

If we were to try to map onto modern thought the ancient idea to which we attach the word *rites*, we would find that we need several modern concepts to cover it: our anthropological notion of rituals, our legal notion of precedents and customs, and our sociological notion of norms. Literati studied rites for several purposes, such as improving the performance of marriages, funerals, birthday celebrations, and ancestral sacrifices; clarifying imperial and bureaucratic norms; and framing political, social, and cultural policies implemented by the state.[80] Such efforts to revive the Zhou polity belonged to a more general trend: a serious reconstruction of the classical world in eighteenth-century China.

As important, *A Comprehensive Investigation* significantly expanded the scope of ritual. For example, the book categorized all of the astronomical tools given to the emperors by the Jesuits as "etiquette instruments" used in ritual actions. *The Illustrated Compendium of Qing Rituals* (Huangchao liqi tushi), compiled in 1759, drew on this precedent. It relied on six categories: sacrificial instruments (*ji qi*), etiquette instruments (*yi qi*), apparel (*guan fu*), musical instruments (*yueqi*), banners and decorations used in imperial processions (*lubu*), and martial equipment (*wubei*). Astronomical instruments—placed in the second class—were described in detail. The message was loud and clear: astronomical instruments served a ritual purpose.

A Comprehensive Investigation of the Five Rites was also an expansion of the previous compilation of state rituals. According to The *Annotated Catalogue of "The Complete Libraries of the Four Treasuries"* (Siku quanshu zongmu tiyao), Xu Qianxue's *A Comprehensive Study of Ritual Classics* had addressed many of the same questions. The *Annotated Catalogue* explains that *The Rituals of Zhou* had very little to say about the five rites, while Xu broke them down into seventy-five categories. Qin arranged his categories more or less as Xu had, but he included more categories under auspicious rites, such as music, imperial sacrifices, astronomy, and mathematics, and under joyous rites he set historical geography and cartography.[81]

Given the subject matter described and catalogued in *A Comprehensive Investigation*—institutions, technology, rituals, cosmology, music, historical geography, artifacts, and architecture—the debates that the book faithfully recorded usually focused on concrete details, such as how to subdivide the twenty-four divisions of the tropical year or how to determine the number of halls in the ritual building called the Palace of Light. The book's major contribution, in contrast to the official compilation in the Institute of the Three Ritual Classics, was to bring such unresolved debates to the forefront of classical scholarship. It was more concerned with the standardization of contemporary institutions than debates concerning classical knowledge. Therefore, the goal of Qin's *A Comprehensive Investigation* did not deviate from the official compilation in the Institute of the Three Ritual Classics. The difference between them, however, lies in how they each assembled their team to advance and clarify the expansion of imperial institutions in the name of classical scholarship.

Qin's *A Comprehensive Investigation* had two clear advantages in building his team. First, Qin had direct access to the science faction. Second, Qin's team was semi-official, so it could recruit fresh minds with new insights. It was the very nature of Qin's team that had a profound and unintended consequence to the heroes of this book: Dai Zhen and Qian Daxin. It is to this significant connection we now must turn.

CHAPTER 5

AN OUTSIDER ENTERS THE MAINSTREAM

As an eight-year-old child, Dai Zhen appealed to a common-sense understanding of time to challenge interpretations that millions of scholars had accepted for four centuries. The setting is a community school at Xiuning, the year 1732, the scene a tutorial dialogue, the subject a passage from *The Great Learning* (Da xue) as explained by Zhu Xi:

> Dai Zhen: "How does one know that these words recorded by Master Zeng [Zengzi, 505–436 BCE] are the words of Confucius? Moreover, how does one know that these words recorded by Master Zeng's followers are really *his* words?"
>
> Teacher: "A literatus named Zhu Xi said so some time ago in his annotation."
>
> Dai Zhen: "When did Zhu Xi live?"
>
> Teacher: "During the Southern Song [1127–1280]."
>
> Dai Zhen: "When did Confucius and Master Zeng live?"
>
> The teacher replied: "During the Eastern Zhou [770–221 BCE]."
>
> Dai Zhen asked: "How much time separates the Zhou from the Song?"
>
> Reply: "About two thousand years."
>
> Dai Zhen asked: "Then how could Zhu Xi know it was so?"
>
> The teacher could not reply and said: "This is no ordinary child."[1]

Zhu's annotations to *The Great Learning* were never meant to restore the authentic and original text. On the contrary, Zhu meant simply to illustrate his own moral philosophy. Venerated and canonized by early Ming emperors, Zhu's edition of the Four Books was the only edition that mattered in the civil service examination system. Then came Wang Yangming, whose discovery of the Old Text edition of *The Great Learning* upset everything.[2] The very same book provoked Dai's precocious suspicions.

Historians have seen in a childhood anecdote the first sign of Dai's brilliance, a smart-alecky kid who would, decades later, alter the direction of eighteenth-century scholarship.[3] Given its significance, this anecdote merits examination, for what it masks—intentionally or not—is the true process of Dai's intellectual formation, which owed as much to his humble social status as it did to the new academic resources available during his early years. This "anecdote of genius" alone, however, can be misleading, without emphasis on certain other key moments of childhood and adolescence, when his character was in flux and his eventual intellectual destiny was still uncertain.[4]

MERCHANTS AND EXAMINATIONS

Dai's native prefecture, Huizhou, is a geographically isolated basin located along the southwestern edge of the lower Yangzi. It is a disaster-prone, hilly, agricultural upland that has long been densely populated (fig. 5.1). A remarkably complete series of drought and flood records struck five of Huizhou's counties for the period between 1544 and 1800; the suffering of Dai's native county, Xiuning, was well above average.[5] The ratio of merchants to the general population was particularly high in Huizhou during this period, and these merchants traded over both short and long distances.[6] As some of these businessmen grew rich, they invested in the education of their sons, and the prefecture became known for producing outstanding scholars from merchant backgrounds.[7]

Dai was one such scholar. What little is known of his childhood and adolescence offers some clues as to how a boy from a poor merchant family was able to cross the nearly unbridgeable divide of the two-tiered educational system, one tier for the elite (gentry and successful merchants) and a second tier for the lower classes (peasants, craftsmen, and part-time traders).

Only rich families could afford to pay for a good classical education.

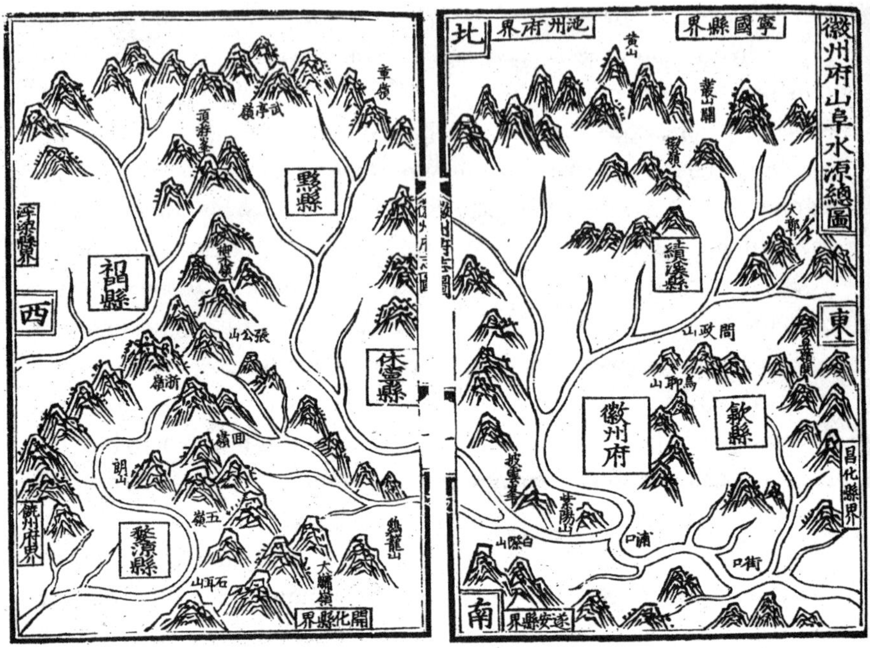

FIGURE 5.1. Mountains and waterways in Huizhou Prefecture. Huizhou is in the center of the right leaf. To the east is She County (also known as Xinan), the seat of the old prefecture and the cradle of Huizhou merchants, such as Wang Daokun (1525-93). To the north is Jixi County, where Hu Shi was born in 1891; north of Jixi is Huangshan, arguably the most admired scenic landscape in Chinese history, where Dai Zhen and Shi Jing met in 1750. On the right side of the left leaf is Xiuning County, where Dai Zhen was born in 1724. In the bottom-left corner is Wuyuan County, where Jiang Yong was born in 1681. North of Wuyuan is Qimen County, where the book collectors, wealthy merchants, and brothers Ma Yueguan (1687–1755) and Ma Yuelu (?–1761) originated. The Ma brothers later resided in Yangzhou and patronized many poets and scholars. From *Huizhou fuzhi*, the earliest extant gazetteer of Huizhou, which was completed in 1502 and became the template of all subsequent gazetteers of Huizhou.

For the children of the poor, a wide variety of charity, community, and small clan schools offered an elementary education that left students who stayed the course with an active vocabulary of about two thousand characters and the rudiments of arithmetic. Second-tier schools varied widely; some provided fragments of the elite curriculum—the Four Books, the Five Classics—and some training in how to compose examination essays.[8]

Few of the children who attended these schools aspired to sitting for the civil service examination; their parents needed them to work, and after a year or two of school they were trained to perform a trade. By contrast, between the ages of two and five the children of literati, officials, and wealthy merchants began learning to read and write at home. A family member (usually their mother) provided informal instruction to both boys and girls. After memorizing roughly two thousand characters, the boys began to memorize the Four Books and one of the Five Classics (until 1786, after which all five had to be memorized). After extensive study of classical scriptures, these adolescent boys learned how to compose examination essays.[9] With notable exceptions, girls in wealthy families seldom progressed beyond basic literacy.

At the age of eight, Dai entered a community school for children of the lower classes. His own recollections of this period are vague, but we can imagine that it was an average community school for sons of petty merchants. Duan Yucai (1735–1815), who compiled Dai's biographical annals (*nianpu*), wrote that his educational progress was dramatic, and he quickly mastered the two thousand basic characters.[10] Later in life, Dai told Duan that he was able to recite most of the classical scriptures and their commentaries but had not been able to memorize the commentaries of later scholars on earlier commentaries; by the standards of the day, he had a poor memory.[11] He shared this limitation with two of his contemporaries: Wang Wufeng (1725–73) and Cheng Yaotian (1725–1814), both from Huizhou. Cheng was especially hindered by a bad memory; after chanting a chapter more than one hundred times, he was still not able to recite it by heart.[12] All three had in common a sense of alienation from the prestigious classical curriculum that relied heavily upon rote learning.

Most community schools in the sixteenth century were established and funded by the Ming state. During the eighteenth century, 60 percent of Lower Yangzi Delta charity schools, the functional equivalents of community schools, were funded by the local elites, while 40 percent were funded by the Qing state.[13] There were also significantly more community and charity schools to provide elementary education than there had been in the sixteenth century.[14] All of these young boys were educated through memorization, the decisive measure of intelligence.[15] Community schools were distinguished from charity schools in Qing China, with the state and local elites jointly overseeing the elementary education provided in both kinds of schools. The clan school, once open to both rich and poor students from the same clan, had become by the eighteenth century a near monopoly of the clan's most powerful branches. Attendance was certainly considered a privilege, and students were given free food, lodging, books, and writing supplies.[16] The most expensive and exclusive way to prepare for the examinations was to hire a private tutor. Many accomplished scholars taught for powerful families in Beijing and the Yangzi Delta.

Once classical excellence was certified by the awarding of a licentiate degree (*shengyuan*)—the lowest rank in the civil service examination system—candidates could take a place, with a stipend, in one of the county schools run by the state, or in one of the private academies run by local elites. The next significant examination students would face was administered at the provincial level; successful candidates would then go on to Beijing for the metropolitan and palace examinations.

Two kinds of elite domination existed in late imperial China—the national and local elites. As the gap between the two groups gradually widened, it reshaped their intellectual and political agendas.[17] National elites maintained their political prominence by using the examinations as a springboard to bureaucracy and political power. They formed a close network and remained focused on issues relevant to the court and the top echelon of the imperial bureaucracy. Local elites, by contrast, were more diverse in geopolitical and social terms. They were members of the gentry, the merchant class, or the military, and dominated local society by controlling material, social, personal, and symbolic resources.[18] As a result, the political agenda of the local elite became increasingly alienated from that of the national elite, who came from wealthy and powerful lineages in large cities.

The gulf separating national and local elites can be further measured by examining the scale and structure of literati access to imperial power. During the Qing dynasty, each of the county licensing and qualifying examinations, which were held twice every three years, saw approximately two million candidates enter examination halls all over China. If a candidate passed, he became a licentiate. Of the entire Chinese population (close to 300 million people, the current population of the United States) only about only 30,000 have been estimated to have achieved licentiate status in the mid-eighteenth century—that is, about 0.01 percent. Licentiates then faced their second crucial hurdle, the provincial examination:

it became the decisive barrier after the 1570s, when only one out of one hundred candidates became a provincial graduate or "recommended man" (*juren*).[19] The next hurdle was the metropolitan examination, held in the capital, which conferred the right to attend the palace examination and a court review. Those who passed were, by definition, members of the national political and intellectual elite, and they had the right to a bureaucratic appointment. Obviously, only a very small number of literati could ascend to this group, and the members of the national elite who did so had a vested interest in perpetuating the status quo and setting the intellectual agenda both inside and outside of the examination system. Peasants had virtually no chance to obtain genteel status. It is this enormous divide between national and local elites that Dai Zhen had to cross at the beginning of his career. The odds looked bad for him.

ACHIEVING CLASSICAL LITERACY

As a boy, Dai admired elite knowledge and status, coded in the word *dao* (way), though they were accessible only to the privileged few and were clearly out of his reach. This scholarly passion, or, as Dai put it, his "will to pursue the way" (*youzhi wendao*), was recalled in a letter he wrote to Duan. He dated the genesis of his determination to pursue the way to the year 1740, when he was sixteen.[20] If we may rely on such reminiscences, he appears to have had no clear idea of what his intellectual goal might be, but he surely understood that scholarly success would bring a change in social status.[21] Moral cultivation was probably never a goal.

A centuries-old Huizhou tradition dictated that when a boy turned sixteen, he would be obliged to leave his native place and learn how to conduct business by serving an apprenticeship in a new place.[22] Dai Zhen's father followed this tradition, and in 1741 he took his son south to Nanfeng County, Jiangxi. Perhaps the apprenticeship helped him to realize that commerce was not for him. He took a job in Fujian—further still from home—tutoring children in a community school.[23] According to Duan's biographical annals, that year Dai "made significant progress in his classical studies."[24] This is not unlikely, as many scholars preparing for the examinations served as teachers, posts that permitted much time to be spent reading and writing. For a literatus in late imperial China, pursuing "the way" usually meant going beyond examination preparation and moral cultivation to attain sagehood. In Dai's case, at the age of sixteen he was intent on advancing his social status. In the fall of 1763 he failed in his first attempt at the metropolitan examination. He tried again in 1766, 1769, 1771 (a special examination), 1772, and 1775—without success.

Although Dai was determined to pursue classical scholarship in 1741, neither his father, his grandfather, nor his great-grandfather had ever held an official position.[25] This suggests that his paternal ancestors were not classically literate, and that he came from a politically weak branch of the Dai clan—he could depend on neither collateral nor affinal relatives to help him achieve his goal.[26] Moreover, his standing was too low to beg for favors (a routine part of intrafamilial politics in the more influential branches) from other more powerful branches of the Dai clan or from other successful clans, such as the Chengs and the Wangs in Xiuning.

Without proper schooling or the assistance of his family, Dai had no choice but to teach himself.[27] In the long history of the civil service examinations, few ever succeeded in such an endeavor. Many young men from wealthier merchant families had failed to make any progress climbing the examination ladder, and most ended up looking after their family's business. Dai's perseverance attested to his confidence and determination to ascend the social hierarchy. In 1748 he married a woman named Zhu, who, according to Duan, "endured bitterness and scanty meals in order to serve her in-laws and Dai. She managed the entire household so that Dai could concentrate on his learning."[28] She was illiterate.

The greatest challenge for those who desired examination success was the acquisition of classical literacy. Dai's friend and admirer Hong Bang (1745–80) provided an interesting account of how this challenge was overcome:

> When Dai read books [in the community school], he wanted to find out the meaning of each word. The teacher would quickly respond with the gloss offered by the official commentary. Dai was never satisfied with that and would not let the matter drop. After a while the teacher became annoyed and fetched him fifteen chapters of a paleographical dictionary, *The Explication of Simple and Compound Characters* [Shuowen jiezi], compiled by Xu Shen [CE 58–147]. Dai loved it. He spent three years studying this paleographical dictionary and then moved on to a classical dictionary, *Approaching Elegance* [Er ya], and an ancient treatise on dialect words,

> *Dialects* [Fangyan], by Yang Xiong [53–18 BCE]. After having studied these dictionaries, Dai went on to collect all available Han dynasty commentaries on the classical scriptures. He compared and cross-referenced these dictionaries and commentaries in order to figure out the meanings of each word by the basic principles of the Chinese writing system known as the six etymologies [*liushu*], as well as to confirm their meaning in each occurrence of each word in the classical scriptures.[29]

Traditional education began with memorization: only once a student was able to recite the entirety of the Four Books and the Five Classics did he make any attempt to understand their meaning. Dai reversed the process: he achieved his goal of attaining classical literacy by taking the highly original route of decoding the classical scriptures. No sensible scholar would have advised a fifteen-year-old, even a talented one, to study this way, simply because the curriculum of the examinations did not encourage it. Such an approach disrupted the accepted practice of progressing from classical literacy to writing examination essays in the approved elegant style. It took a young man with average talent years to learn how to compose such examination essays.[30] Hong described how his friend had proceeded: "Dai's family was poor, so he could not find a way to make a living. Therefore, he began learning how to write examination essays. His style is like something from the remote past, and it is extremely difficult to understand. . . . The meaning of each of his compositions was profound and difficult to grasp, the words far above their everyday use. When a superficial scholar read one of these works, he was usually confused. The superficial scholar then turned to his fellows to mock or denounce Dai."[31]

While this unusual approach could only yield a style far from the current fashion, considering every possible meaning of each word in the classical scriptures—including meanings rarely invoked in the eighteenth century—exposed contradictions and mistakes in the writings of contemporary scholars. Dai preferred to emphasize the correct interpretation in his own jottings, only implicitly pointing to the scholarship he was criticizing. It is therefore difficult—yet essential—to identify his foils. Moreover, his prose was fraught with allusions to difficult and obscure passages in the classical scriptures.

The title he eventually gave to his first collection was *Evidential Analysis of the Classics* (Jing kao).[32] In that early work, Dai meticulously traced the provenance of the classical scriptures down to his times in order to authenticate them. He also ventured into several highly specialized areas, applying technical methodology to analyze "Records of Ancient Technology" (Kaogongji), a chapter in *The Rituals of Zhou*.[33]

Writing about the classic dictionary *Approaching Elegance*, Dai identified it as the starting point of his academic endeavor. Indeed, the book is not merely a dictionary, as it catalogues a range of specialized knowledge about palaces and ritual buildings, cosmography and geography, ritual vessels and astronomical instruments, and insects, fishes, birds, and other animals.[34] Dai proclaimed:

> I think literati should start their classical learning with the classical dictionary *Approaching Elegance*. Ten years ago I fetched a copy and read it, and I have worked on it ever since. . . . From time to time, I find [in it] something important and write it down because I fear that I may forget it as time goes by. I call the collection of those notes *Evidential Analysis of Words in "Approaching Elegance"* ["Er ya" wenzi kao]. This project was for my own reference alone. As for a comprehensive discussion of this dictionary of 10,791 words as well as how it relates to all of the classical scriptures, I must postpone that until a future date.[35]

This kind of etymological and phonological research could yield no benefits for the literary essays and answers to practical policy questions required by the examinations. Dai kept at it, however.

To no one's surprise, in 1749 he failed in his first attempt at the county examination.[36] But among his contemporaries there were some who did pass the county examination early on. Weng Fanggang (1733–1818), for example, earned his reputation as a distinguished scholar in part by becoming a licentiate in 1744—he was ten years old.[37]

Driven by his urge to obtain genteel status, Dai went boldly and somewhat naïvely into uncharted territory. His choices reflect the combination of social and cultural obstacles with which he was forced to cope. As we have seen, he managed to overcome the most difficult status barrier—classical literacy—in an unconventional way. Moreover, he may have been born with a mathematical and a linguistic facility; he was drawn to analyzing patterns and excelled at inferring quantitative and qualitative regularities from discrete and concrete evidence. No doubt his dictionary approach prompted strong deductive and inductive reasoning in his classical research. This unconventionality, nevertheless, cost him dearly when it came to writing competitive examination essays.

ENGAGING DONGLIN SCHOLARSHIP

After a decade of idiosyncratic research, Dai had a brief encounter in 1750 with the famous moralist Shi Jing (1693–1769).[38] Shi was fifty-seven years old, at the peak of his career (fig 5.2); Dai was twenty-six, a nobody.

When Shi showed considerable interest in Dai's research on the classical scriptures, Dai responded that he dared not share *Evidential Analysis of the Classics* with others, since he feared they might be "dazed, confused, or stunned into disbelief."[39] Such readers, he meant, were not up to his level, and he was pleased that Shi—whom he considered a peer—was willing to be frank with him. Afterward, Dai sent Shi a phonological analysis of two names of vassal states in *The Book of Songs*; it so impressed the older scholar that he requested the longer study from which the excerpt was drawn. Politely declining, Dai sent instead a letter outlining how he went about deciphering classical scriptures and offered an evaluation of the current state of classical research.[40]

The biography of Shi Jing suggests why the older scholar proved such a fine fit for the younger. He was born into a wealthy gentry family in Changzhou, one of the metropolitan centers in the Lower Yangzi Basin. His father had married the granddaughter of a high Ming official, so the family enjoyed a prominent position in the politically influential Changzhou network.[41] At the age of twenty-five, after attending a few meetings of the Xishan Association for Moral Cultivation (Xishan Xuehui) in Wuxi County, Shi experienced a sort of conversion and decided to forsake the examination route to a career.[42] Instead he devoted himself to a distinctive combination of moral self-cultivation and classical scholarship that can be traced back to the Donglin Academy.[43]

An important event for early Qing intellectuals was the contest between the Donglin Academy and the nefarious eunuch Wei Zhongxian (1568–1627).[44] The culmination of a long tradition of private academies that extended from the Song, the Donglin Academy was founded in 1111. But not until it was rebuilt in 1604 by Gu Xiancheng (1550–1612) and Gao Panlong (1562–1626), both natives of Wuxi County, did it achieve distinction

FIGURE 5.2. Shi Jing's portrait. The poem in the corner reads: "When does the beginning begin? / When does the end end? / The cosmos is so immense that countless kinds of creatures move around in it. / The ancients [sage and wise men] have long gone, and I bow to them [with respect] from a remote vantage point." Signed by Shi Jing from Jinling (archaic name for Wujin County, Changzhou Prefecture), courtesy name Zhongming. From *Shunshan Shi Zhongming xiansheng nianpu*.

as a center of Confucian study firmly opposed to the radical teachings of Wang Yangming.

Donglin activists espoused the need to limit monarchical power, finding support in the crypto-democratic aspects of local academic communities.[45] The Donglin activists soon clashed with the entrenched court faction headed by Wei. As the quiescent Tianqi emperor slumbered, Wei mounted a reign of terror: six Donglin activists were tortured and executed. After Wei fell from grace in 1627, his faction was dissolved, and progressive literati, many associated with private academies, reclaimed a prominent place in civil society, such as it was. In time, a group calling itself Fushe (meaning "a return to the society [of antiquity]") inherited the Donglin combination of moral philosophy and gentry activism, becoming the leading voice of political dissent as the Ming teetered toward collapse. Such protests came to an abrupt end when the dynasty fell in 1644.[46]

Some celebrated Ming loyalists, such as Huang Zongxi and Gu Yanwu, were painfully aware of the outcome of Donglin activism and haunted by the conviction that limits had to be placed on the despotic nature of imperial institutions.[47] Qing epigones of the Donglin group rejected Wang Yangming's teachings as empty moralism and turned to a variety of practical methods of self-cultivation.

Although their influence gradually declined, groups continued to operate in Xishan, Suzhou, and Huizhou, and the Donglin emphasis on techniques practiced daily was embraced by the Xishan Association for Moral Cultivation.[48] The group's organizer, Gu Pei (dates unknown), advocated the meditation techniques invented by one of the revivers of the Donglin Academy, Gao Panlong, and introduced them to Shi Jing along with other important techniques of moral cultivation, such as keeping a ledger of good and bad deeds, meeting to confess and discuss one's moral failures, chanting *The Classic of Filial Piety* (Xiao jing), and confessing one's bad deeds alone in a dark room.[49]

These techniques, drawn from a variety of traditions, had been woven into the Neo-Confucian texture of the moral cultivation associations (*jianghui*) of the waning Ming.[50] During the late seventeenth and the early eighteenth centuries, the groups were given a new life here and there, but Wuxi County was clearly the center of such activity.

The Xishan group declared unwavering support for the Donglin legacy. Shi joined in 1718, and by 1736, the first year of the Qianlong emperor's reign, he came to feel that a decline had set in: the various groups he knew of had lost their political edge.[51] And the rules prohibiting members from participating in the examination system kept the pool of contributions rather shallow. In a response to what he saw as an institutional crisis—his enthusiastic engagement with several groups kept him well informed—Shi refashioned himself as an orthodox scholar by hosting a series of lectures at the Shunshan Academy. Like other ambitious freelance scholars, he took on disciples. His moral message shifted into alignment with the Confucian orthodoxy as he championed Zhu Xi and denounced Wang Yangming.

Then, in 1740, he orchestrated a moral spectacle that attracted wide attention. His father died that year, so Shi built a small hut adjacent to his father's grave and took up residence there. He summoned his disciples, who came to live in a community shrine not far from the filial hut. When he was not lecturing to his loyal followers, Shi traveled to nearby academies to speak there.[52] At the time, most officials mourned for twenty-seven months. Typically this involved taking an extended leave from one's post and spending the allotted period in one's native place, in the comfort of the family home.[53] Scholars who did not serve in the bureaucracy, like Shi, would take their lead from the official standard. But to display his filial loyalty, after living in his hut for twenty-seven months (1740–42) Shi declared that he would extend the mourning period indefinitely.

News of this remarkable man spread to the capital, and Shi soon counted among his correspondents Chen Shiguan (1680–1758), the minister of works also serving as the grand academician of the grand secretariat (*neige*). Chen had just been dismissed from the Ministry of Works by the young Qianlong emperor for his poor performance managing flood control in North China. In 1744 he requested an honorable retirement, but instead he was summoned back to the court against his will and repeatedly humiliated by the emperor.[54]

It was at this juncture that Shi wrote to Chen, offering advice on how to stay the course of his moral cultivation and avoid the influence of external disturbances.[55] Chen responded gratefully, and Shi wrote again: "Please allow me to explain [the politics to you]. You will not make a difference without investigating the minds of the emperor and your fellow bureaucrats; you will not govern well without correcting the literati ethos and enriching peasants' lives. The reason that we find ourselves unable to challenge the literati ethos is that we only hear what they say instead of observing what they do. We let them produce only useless writings instead of useful learning."[56] Shi's "useful learning," in this case, comprised moral cultivation techniques and regular encouragement. The frustrated official liked what he read and continued to exchange books and long letters with his new guru, engaging in long discussions of "sage learning and mundane politics," a focus of the Donglin legacy.

Meanwhile, Chen's career sank into oblivion. In 1749 the Qianlong emperor stripped him of his post as grand academician on the grounds that he "lacked the ability of a counselor and acted only on lowly and trivial matters." When the emperor accused his former adviser of embezzling land in North China, things looked very grim indeed.[57] This scandal shut down Chen's career. Sent home, he made a special trip to visit Shi in his hut—finally meeting him after six years of correspondence—and paid obeisances at the paternal tomb.[58] After this, several officials of high standing at the court recommended that Shi be presented with an honorary degree so that he could serve the emperor. Shi declined their offer.[59]

Already known to reverent Confucian scholars for his extraordinary devotion to his father, by refusing the degree Shi cemented his reputation. By 1750 Dai Zhen was aware of this controversial figure; he may well have seen him as a kindred spirit. Yet there were others who launched aggressive attacks on Shi's reputation. Some accused him of being a hypocrite; others claimed that he had committed many petty crimes. Ruan Kuisheng (1727–89), for instance, described him as a "cunning and quirky man who does not know anything but pretends to know everything."[60] Dong Chao (1729–64) claimed that Shi's own brother had reported him as a thief. This "thief," Dong went on to say, was known to all in his village as a "licentious and evil monk."[61] And Wu Jingzi, a rather eccentric personality himself, included in *An Unofficial History of Scholars* a grotesquely hypocritical and ridiculous character named Quan Wuyong—a thinly veiled Shi Jing.[62]

Before he met Dai Zhen, Shi got to know Yuan Mei, the most successful writer and poet of eighteenth-century China, known for advocating the liberation of elite women from Confucian norms and for his unapologetic indulgence in sexual pleasure. Dai too was an eloquent advocate of individuality and the primacy of human desire, and the connection between Yuan and Shi may shed light on Dai's own reaction to the "licentious monk."[63]

Over the course of their warm and varied correspondence, they discussed the design of an ancient legendary building—the Palace of Light. In a friendly letter, Shi gently suggested that the only mention of the palace in *The Rituals of Zhou*, a passage that identifies it as the temple devoted to the ruler's ancestors, was a forgery.[64] That comment—really a commonplace—appears to have lit Yuan's very short fuse. Breaking with the faultlessly polite tone the correspondence had maintained until then, Yuan questioned the value of Shi's very existence:

> A gentleman is usually pleased with the fame he receives and yet attentive to criticism, too. If a gentleman wins his reputation by being the best at what everyone does, then it is difficult both to establish that fame and to lay it waste. If a gentleman gains his reputation merely by being different from everyone, then it is easy to establish that fame and to lay it waste. Now, is there anyone who camps by his father's grave beyond the required year? Is there anyone who collects followers [*jiangxue*]? Is there anyone who refuses to sit for the examinations?[65]

Yuan was not recommending a life of conformity—he would have been the last to do so. Rather, he was trying to show Shi that his position was politically and intellectually untenable, based ultimately upon an inherent contradiction. If living in a hut next to your father's grave indicated a commitment to seclusion as a way of life, why then continue to gather disciples and even travel to give lectures? If one lectures as a way of engaging in contemporary politics (that is, in Donglin activism), then one *should* participate in the Qianlong emperor's political system. Among Shi's various activities, Yuan warned, lecturing to associations for moral improvement was the most dangerous and hypocritical. At the end of the letter a final admonition rang out: "All that Donglin activism in the Ming was really boring and useless—and now you follow suit. What a mistake!"[66]

TECHNICAL METHODS AND CLASSICAL SCHOLARSHIP

If Donglin activism was no longer viable in Yuan Mei's eyes, what then did it mean for young Dai Zhen to propose something new to Shi Jing? Many historians consider the letter that Dai sent to Shi instead of his study of *The Books of Songs* an intellectual manifesto. For the first time, he systematically outlined his method and his research agenda. Did he hesitate? There is a tentative feel to the letter, and no doubt had he not been addressing a controversial moralist and member of the national elite, the ideas would have been presented differently. Such considerations are significant, because Dai was very much in the clutches of his pursuit of a degree, and he had spent a decade pursuing classical literacy and then research. By positioning himself as a critic of classical learning, Dai implied that his theory of knowledge, if taken seriously, could change the very nature of how the Confucian scriptures were studied. Dai's posture was to aim high and shoot straight, with consequences not so different from Shi's own career.

The methodological outline he offered turned out to be a summation of his earnest and strenuous exploration of the classical scriptures over the ten previous years. It critically adopted the legacy of pioneering thinkers of an earlier generation—Gu Yanwu and Yan Ruoju—and anticipated the evidential scholarship movement that rose to prominence in the next generation.[67] This text is pivotal in the intellectual development of eighteenth-century China and deserves a close reading.

Dai opened with an anecdote about his first encounter with one of the so-called Six Classics: "As a young man, I could not have a private tutor, since my family was poor. I heard that one of the ancient sages was [called] Confucius and that Confucius had edited the Six Classics to show the way to his disciples and followers. I borrowed one of the classics and read it. But I was bewildered and got nothing out of it."[68] Here was the decisive event that propelled his self-acquisition of classical literacy and his subsequent exploration of the great canonical works of the distant past. After the shakeup, however, came an epiphany: "The way is what the classics get at. What illuminates the way is their words and phrases. If I can comprehend the words and phrases first, then I will *gradually* arrive at the way through words and phrases."[69]

Following this epiphany, Dai began to unveil his method. Deciphering the classics, he explained, took place on three levels. These were lexicology, technical analysis in a broad context, and overall assessment of the quality of classical research.

The first step, lexicology, may sound naïve and tedious at first. Dai claimed that he spent three years studying *The Explication of Simple and Compound Characters* and figured out how to use it to decipher the great works. When encountering a problematic word in one of the classics, Dai first identified it in *The Explication of Simple and Compound Characters.* He was always looking for entries that included a version of the character in seal script; each of these characters had to be old, as seal script was ancient, having evolved from Zhou dynasty (1122–256 BCE) bronze inscriptions to the bamboo-strip writings of the Qin (221–206 BCE). Dai then compared how the word was used by other authors of antiquity. By comparing different usages of the word in different contexts, he eventually became critical of Xu's work and suspected that he might have failed to put to use all of the texts available to him. To second-guess Xu, Dai borrowed from a friend that compilation of orthodox interpretations of the classical scriptures, *Commentaries on the Thirteen Classics* (Shisanjing zhushu), and studied it diligently. Here were many of the texts that Xu would have relied on to provide the historical meanings of Chinese characters. After an intensive examination of these canonical glosses, Dai cross-referenced all occurrences of each word in *The Thirteen Classics.* He concluded that he could settle on the exact meaning of a word only after examining its form, based upon the method of six etymologies, and cross-referencing all of its occurrences in the classical scriptures.

The second step, technical analysis in a broad context, was designed to address the interpretive problems that could not be resolved by his meticulous lexicology and cross-referencing. These problems included issues in astronomy, phonology, institutional analysis, nouns and their referents, mathematics, and geography. Dai spent an entire decade—from 1740 to 1750—developing his technical analysis of these fields. In his letter to Shi, he enumerated six examples that illustrated the indispensability of technical analysis for the investigator of the past (table 5.1).[70] Just as Dai championed the use of specialization and precise scholarship for grasping the meaning of ancient writings, he also advocated the overall integration of specialized knowledge into a single framework. His beloved etymology, paleography, astronomy, phonology, institutional analysis, geography, mathematics, and natural studies should be taken together as a package.[71] Using such methods made possible a more accurate and complete picture of the classical world. Although it was taken for granted that the scriptures gave reliable access to an ideal and remote past, Dai deployed his eclectic erudition to realign, authenticate, reconstruct, and reinterpret the order

TABLE 5.1. Examples of technical analysis

	Classical Source	**Technical Requirement**
1	"The Canon of Yao" (Yao dian), *The Book of Documents* (Shang shu): He [the legendary Yao emperor] commanded the *Xi*s and *He*s, in reverent accordance with their observation of the wide heavens, to calculate and delineate the movements and appearances of the Sun, the Moon, the stars, and the zodiacal spaces, and so to deliver respectfully the seasons to be observed by the people.	Astronomy: One should know the movement and trajectory of the Sun, the Moon, and the five planets (those visible to the naked eye: Jupiter, Venus, Saturn, Mars, and Mercury).
2	"Songs from the South of Zhou" (Zhou nan 周南) and "Songs from the South of Zhao" (Zhao nan 召南), *The Book of Songs*: "Fair, fair," cry the ospreys, On the island in the river. Lovely is this noble lady, Fit bride for our lord.	Phonology: One should know the pronunciation of each word in ancient Chinese (Old Chinese in modern linguistics), and avoid any projection of later voicings onto ancient pronunciation.
3	"Capping Ceremony for Literati" (Shiguanli), *Ceremonial Rituals* (Yi li), one of the Three Ritual Classics	Protocols: One should have a good understanding of architecture, palace floor plans, ancient robes, utensils, etc., in order to know how ancient state and family rituals were properly performed.
4	"Tribute to Yu" (Yu gong), *The Book of Documents*	Geography: One should know the specific location of events and institutions reported in the ancient political system.
5	"Records of Ancient Technology," *The Rites of Zhou*	Mathematics: One should know geometry and algebra in order to calculate correctly the dimensions of ancient Chinese ritual bells, the wheels of a chariot, weapons, sundials, etc.

(continued)

TABLE 5.1. *(continued)*

	Classical Source	Technical Requirement
6	*The Book of Songs* (in general)	Nouns and Their Referents: One should know the "ten thousand things" (*wanwu*) referred to in *The Book of Songs*. That includes the names of birds, animals, insects, fishes, plants, and trees. We may call this "natural studies." Once we understand the exact appearance and features of animals and plants, we can accurately interpret the poetic metaphors that allude to their appearance and features.

and meaning of the sacrosanct texts. He opened a window on a whole new view of classical knowledge.

The straightforward premise on which Dai operated was that the classical world was real and perfect. The knowledge that described various aspects of that world, therefore, had to be seamlessly interconnected. His technical methods were not mere tools but indispensable and integral components of classical knowledge, without which access would be impossible.

Yet if these technical methods were integral, why had so many scholars dispensed with them over the centuries? Dai sought to address the absence of any consideration of technical methods in mainstream exegeses by arguing that they were deliberately excluded. He gave accounts of this process in the fields of phonology, mathematical astronomy, and music.

Why had so many scholars stopped using the phonetic method, or phonological analysis, as part of textual exegesis? Dai said that in the past many scholars considered it foreign—borrowed from translated sutras during the fourth and fifth centuries—and therefore inferior. His reply was that the methods had been invented in China, and he credited Sun Yan (386–534), a disciple of the classical master Zheng Xuan (127–200), with the invention of a powerful method for conveying the pronunciation of Chinese characters (*fanyu*, later called *fanqie*). To convey its sound, a character was designated by two Chinese characters, the first giving consonant formations, the second giving vowel formations. This phonetic method provided the basis for subsequent discussions of poetic rhyme. Dai suggested that, during its later development, the phonetic system was stolen by Buddhists who claimed it as their own invention. Deceived, scholars could no longer recall the system's ancient provenance. Therefore, to use the phonological analysis of the classic scriptures was in fact to restore an ancient Chinese usage.[72]

The case of mathematical astronomy was more profound and urgent because the thieves were Jesuit missionaries, whose influence was still very much alive. Dai recycled the narrative he had applied to phonology, arguing that sages in ancient China had used sundials to calculate the position of the Sun in the sky, measuring the height of a gnomon (*gu*) and relating that to the length of its shadow (*gou*). For Dai, the *gougu* calculation, which has been called the Chinese equivalent of the Pythagorean theorem, was the embryo that eventually evolved into a general theory of plane and spherical trigonometry. Mathematical astronomy had long played an important role in classical knowledge because of the prominence of the cosmos in the scriptures, yet the scholars of the eighteenth century overlooked it. Was this only because it was difficult to master? Dai once again pointed to foreign manipulation. The Jesuit missionaries had imported trigonometry, which by 1750 had completely superseded traditional, primitive trigonometry.[73] As far as Dai could tell, the Jesuits purposefully tried to obscure the true Chinese origin and subsequent development of mathematical astronomy. The remedy he proposed was to insist that this tool, as used in classical exegesis, was not something imposed by the foreign missionaries, but rather a specialized body of knowledge with a long and continuous indigenous history.[74]

The intersection of mathematics and music was even more to the point. Why had scholars become so ignorant about music, an art viewed as a significant dimension of the symbolic universe in ancient China? As with phonology and mathematical astronomy, musical knowledge had been lost over time. Scholars no longer understood the mathematical structure of musical notes and were doomed to fail in their efforts to decipher the classical scriptures.[75] They saw connections where none existed and made assumptions with no basis in fact.[76] Dai strongly urged that the origins of these methods—even if foreign—not be held against them. He advocated methodological neutrality, even objectivity. Far from the first thinker to advocate objectivity, Dai was nevertheless its most coherent and sophisticated apologist to date.

At the end of his letter, Dai offered a general assessment of classical research. Here he revealed his own intellectual mettle, grounded in the accumulated study of ten years, while passing judgment on some towering thinkers and scholars.

How does one know whether one's thinking is sound? It must possess erudition and profundity, perceptiveness and judgment, veracity and accuracy. Would any of the great scholars of the past have passed this tripartite test? Dai's answer was an astonishingly bold "No." Consider two candidates: Zheng Qiao (1104–62), an outstanding historian and erudite classicist of the Song dynasty, and Yang Shen (1488–1599), a talented polymath of the Ming. Both were very productive in their lifetimes. But while their erudition and profundity were indisputable, the veracity and accuracy of their scholarship left something to be desired. Other scholars, such as Lu Jiuyuan and Wang Yangming, insisted on directly approaching the great way by moral intuition and ignored laborious learning. But how could anyone cultivate his morality without first inquiring into the ultimate moral goals?[77] This is precisely what has been dubbed "intellectualism": a new direction that Confucianism took in the eighteenth century.[78] What Dai proposed, however, was not merely intellectualism. In his letter to Shi, Dai outlined a research agenda that implied a grand academic system. If the classical world was to be reconstructed, a special set of skills would be needed.[79] The dream of such objectivity may seem familiar to us, but it occurred to very few in eighteenth-century China.

DAI AND OTHERS

Between 1740 and 1750, Dai mastered the classical canon, struggled to claim gentry status, and set out on a search for the truth in the classical scriptures. At the end of that decade, when he turned twenty-six, Dai had developed his trademark methods for understanding the most revered of classical texts, and in the process proposed a new theory of knowledge.

Dai knew that every examination candidate expended countless hours of mental labor memorizing the commentaries to the classic scriptures and very few hours critically questioning those interpretations. Yet at a young age, if we believe the legend, Dai had been far more interested in challenging the commentaries than swallowing them whole. His background as the son of a lower-class merchant, his unconventional way of achieving classical literacy, his engagement with a controversial moralist, and, most importantly, his intellectual agenda all clearly indicate that he was an outsider in relation to polite society, which was dominated by the classical curriculum and a focus on the perpetuation of status.[80] Dai was also aware, from his study of many historical figures and his familiarity with the case of Shi Jing, that eccentric maneuvers could easily be ignored or dismissed.

Ying-shih Yü, borrowing Isaiah Berlin's famous analogy, described Dai as a hedgehog who knew only one big thing, as opposed to the mainstream scholars who, like the fox, knew many things.[81] Building upon Yü's landmark work, new directions are possible for reflecting upon the formation of this hedgehog.

Dai could hardly overlook the importance of Neo-Confucian thought to both the classical scholarship and the civil service examination system of his day.[82] Yü considered Dai someone who, in his youth anyway, "supported the Cheng-Zhu Confucian orthodoxy and rejected Lu-Wang moral idealism."[83] But Dai did not always and exclusively follow the Cheng-Zhu line. His fascination with a collection of highly technical skills can never be squared with the academic paradigm set out by the Cheng brothers and Zhu Xi. When we fully appreciate the range of disciplines that Dai mastered, it becomes clear that he was blazing a trail. While he incorporated many Western insights into his outlook, he also understood the dilemma and difficulties that confronted predecessors such as Mei Wending.

Dai's method also differed significantly from that of other evidential scholars. Yü argued that an "essential tension" emerged when Dai chose to pursue one big thing—the philosophical foundation of the cosmos and the world—whereas other evidential scholars only hoped to correct the readings of a few antique passages.[84] But when we reflect on his manifesto of 1750, let us not forget that it was a single formulation at a single moment, addressed to an audience of one. A careful reading of that letter allows us to better situate Dai's intellectual and political position. At that moment, when one thinker reached out to another, young Dai and old Shi were both outsiders. Their intersection reveals a multidirectionality and a multitude of layers in the development of Confucian discourses. Shi has long been seen as a despicable character, a buffoon to be ridiculed. The young nobody who reached out to him, however, grew into an intellectual giant who foreshadowed the rise of modern Chinese thought.

CHAPTER 6

HOW TO DETHRONE JESUIT SCIENCE

The arrival of Dai Zhen in Beijing in 1754 changed the trajectory of Chinese intellectual history.[1] That same year, a group of technically proficient scholars passed the civil service examinations, led by another rising academic star, Qian Daxin.[2] All had immersed themselves in mathematical astronomy, analytical phonology, and historical geography. After Dai paid a call on Qian, his name was on everyone's lips. Over the following years, the cohort of 1754 began to engage, absorb, and challenge Dai's synthesis of technical methodology and classical knowledge.[3] And they passed his ideas along through their networks.

Dai's move also intersected with an important scholarly project, the composition of *A Comprehensive Investigation of the Five Rites*, presided over by the minister of rites, Qin Huitian. Dedicated to upholding the political legacy of the science faction, he sought ways to effect a partnership between the Jesuit sciences and the Confucian heritage. Qian and Dai—in the right place at the right time—emerged as the most qualified scholars to manage this challenging task. As this new wave of "technical literati" swept into Beijing, they undermined the putative superiority of European astronomy by championing a grand synthesis of classical knowledge.

THEORIES OF CLASSICAL KNOWLEDGE

From 1740 to 1754 Dai had developed his technical methodology, declaring classical lexicology, analytical phonology, historical geography, and mathematical astronomy the foundation of classical knowledge. As Ruan Yuan was to lament, such disciplines had constituted "practical and substantive learning" (*shixue*) for respectable scholars in the distant past, but had come to be scorned as the specialties of fortune-tellers and technicians. Dai's great contribution, as Ruan saw it, had been to elevate technical methods back to the level of respectable learning.[4]

A century after the Qing conquest, examination candidates considered classical lexicology and analytical phonology a part of the "lesser learning" (*xiaoxue*). Historical geography and mathematical astronomy had long been scholarly taboos, monopolized by the imperial state because of their utility in military tactics and building political legitimacy. Still, some eminent scholars who freed themselves from the single-minded focus on passing tests had taken them up in order to investigate the fall of the Ming. And with the tacit approval of the Kangxi emperor, a few classical scholars had continued to study these disciplines under the rubric of practical and substantive learning. How did Dai, a young man from the lower merchant class, gain access to knowledge so meticulously guarded by the imperial state? The answer can be found by examining the career of his mentor, Jiang Yong (1681–1762).[5]

Almost by chance, after Dai, aged eighteen, returned from his merchant apprenticeship in Shaowu, his family paid a courtesy visit to Cheng Xun (*jinshi* 1724, provincial examiner in 1748), a respectable gentleman and scholarly official living in Xiunin.[6] As it happened, Jiang was residing in Cheng's mansion. Already sixty-one years old, he had spent his entire career as an itinerant scholar who, never having passed the provincial examination, moved from patron to patron. In their first meetings, Dai and Jiang discussed, for the most part, mathematical astronomy and analytical phonology. Jiang encouraged Dai to rethink his approaches to these subjects, providing him with critical new materials and educating him about current scholarly debates. Here was a classic example of the sort of exchanges that characterized "Huizhou fellowship" (*Huixue*), named for their home prefecture.

It was Jiang who acquainted Dai with the full gamut of Jesuit mathematical astronomy and Mei Wending's scholarship. In 1709 Jiang had traveled to Nanjing to sit for the county examination, a requirement for maintaining his licentiate status and continuing to receive the stipend that went with it. He found there a copy of *Mathematical Astronomy of the Chongzhen Reign*, the earliest Chinese-language collection of Jesuit mathematical astronomy, and was soon consumed by it.[7] Jiang mastered every facet of the Western science as transmitted by the Jesuits. During this period of intensive study, he became acquainted with Mei's reputation and happened on some of his incomplete manuscripts in a bookshop. This early work disappointed him: the basic tenets of mathematical astronomy were presented with naïve simplicity. Though he retained some curiosity about this celebrated figure, Jiang abandoned his plan to study with Mei and turned his attention to ritual matters and the phonology of the classical scriptures.[8]

That meant missing the chance to meet Mei, who died in 1721. It was not until some time between 1725 and 1730 that Jiang could finally afford a copy of *The Complete Collection of Mei Wending's Mathematical Astronomy* (Mei Wuan xiansheng lisuan quanshu, 1723), when he realized that he had drastically underestimated the scope and depth of Mei's contribution. Penitently, Jiang devoted himself to a close study of the book.[9] In 1740, after securing Cheng Xun's patronage and moving into his mansion, he began to compose a collection of essays criticizing and correcting Mei's mathematical astronomy, with the goal of preserving and expanding it. He called this collection *Wings of Mei* (Yi mei).[10] The most important issues, according to Jiang, were the length of the tropical year and the annual difference between a sidereal and a tropical year.

Cheng proposed that Jiang travel to Beijing to present his study to Mei Wending's grandson, Mei Juecheng. The latter was the chief astronomical consultant for the Institute of the Three Ritual Classics (San Li Guan), the Historiography Institute (Guo Shi Guan), and the Ming History Institute (Ming Shi Guan), where he devoted himself to applying new tools to unlocking some of the truths in *The Rituals of Zhou* and the mathematical astronomy section of *The Ming History*. Jiang had modest hopes that the editors and compilers would appreciate his work on mathematical astronomy, analytical phonology, or the Three Ritual Classics. But after an initially courteous exchange, Mei politely dismissed the older scholar.[11]

Although most of the classical scholars at the Qing court were vaguely aware of the utility of mathematical astronomy for deciphering classical scriptures, they lacked the technical competence to do anything about it. Mei Juecheng was an exception. But he was not exactly savvy in the profound implications of classical knowledge and tended to treat the Jesuit science as a mere tool. Jiang's detailed exposé was, on the contrary, an attempt to synthesize and validate the foreign system as legitimate, cutting-edge technical knowledge. After Mei sent him away, a despondent Jiang returned to Cheng's mansion in Huizhou.[12]

It was shortly after this setback that Dai and Jiang first met. The older scholar astutely perceived his new friend's raw mathematical talent and introduced him to *The Complete Collection of Mei Wending's Mathematical Astronomy*. This is why, two years later, a degreeless young man who had never been to the capital produced a systematic volume on arithmetic, explaining how to add, subtract, multiply, divide, and find square roots. His tool of choice was not an abacus, the calculating device used by so many merchants, but something called "Napier's bones." This tool had been introduced by the Jesuits, was widely used at the Astrocalendrical Bureau, and was described by Mei Wending.

Within two years, Dai mastered the basics of arithmetic and trigonometry. Between 1742 and 1744 he drafted four mathematical studies, only to graduate to a more ambitious treatise, *Napier's Bones Arithmetic* (Ce suan).[13]

In 1750 Dai enrolled at the private Ziyang Academy in Huizhou prefecture and began to share his technical methods with some of his schoolmates, especially Cheng Yaotian.[14] This move won Dai a membership in the local elite of Huizhou, but he remained an intellectual outsider far from the national elites.

As Cheng soon noticed, Dai's interest in the classical scriptures was quite narrowly focused on the technical (and widely ignored) materials known as "the Taifu rituals" (*Taifu li*). Later scholars would pay homage to this focus by referring to this collection, which provides many links between cosmology and rituals, as "Venerable Dai's ritual records" (*Da Dai liji*).[15] When the term "Heaven" (Tian) occurred in the Taifu rituals, Dai always invoked the Tychonic cosmos to explain its operation, insisting that the Western model provided the key to portions of the classical scriptures that had previously been obscure. In this regard, Dai did not differ much from Jiang, who used advanced mathematical astronomy to assess classical knowledge. Although Cheng praised Dai for his collation of the Taifu rituals, he was really responding to the magisterial application of a technical method that unlocked formerly unreadable passages. Deeply influenced by this encounter, Cheng accumulated many interests himself, including mathematics, historical geography, palace design, family rituals, natural studies, ancient technology, and funeral rites.

Jiang Yong celebrated his seventieth birthday in 1750, and in his honor Dai composed a laudatory letter. Listing prominent scholars who had left their marks on the approved technical fields (lexicology, historical geography, analytical phonology, and mathematical astronomy), he lauded Liu Chang (1019–68), Wang Yinglin (1223–96), Hu Wei (1633–1714), Gu Zuyu (1631–92), Yan Ruoju, Gu Yanwu, and Mei Wending. Such lists, far from unusual, are standard academic practice in building an intellectual lineage. Dai expressed regret that each of these men had mastered only one discipline, whereas Jiang had mastered all four.[16]

It need hardly be mentioned that Dai was also describing his own achievement, the technical foundation on which he planned to reconstruct the classical world. The detailed research notes he had kept during his study of the scriptures and later published—*Evidential Analysis of the Classics* and *A Supplement to "Evidential Analysis of the Classics"* (Jingkao fulu)—trace his incremental progress toward a fully formed method.[17] His goal was ambitious, and his method was unconventional. Rather than reveal prematurely his ideas on the classics, he decided to put his name on the intellectual map with three more focused treatises: *The Illustrated "Records of Ancient Technology," On Base-Altitude and Circle-Division Mathematics* (Gougu geyuanji), and *A Commentary on Qu Yuan's Rhapsody* (Quyuan fu zhu).

All three treatises emphasized the necessity of using specialized techniques to decipher the classical scriptures. Dai prefaced *Illustrated "Records of Ancient Technology"* by noting: "To read this chapter [i.e., "Records of Ancient Technology"] from *The Rites of Zhou*, we should thoroughly understand *The Six Books of Etymology* [Liu shu] and *The Nine Chapters of Mathematics* [Jiu shu]. All of us have worked on the classics since we adopted the queue. Yet our hair will have turned grey without many having heard of applying either [*The Six Books of Etymology* or *Nine Chapters of Mathematics*] to classical studies. Why?"[18] *The Six Books of Etymology*, previously classified as "lesser learning," designated basic philology, including classical lexicology and analytical phonology, while *Nine Chapters of Mathematics* was the oldest mathematical text then known. Together, they provided details needed to reconstruct the classical world. For instance, it was crucial to determine the scale and dimensions of each ancient artifact and practice. Chariot, wheel, bow, sword, arrow, dagger-axe, halberd, bell, drum, sundial, all kinds of bronze drinking vessels and decorative jades for ritual performance, buildings, palaces, well-drilling apparatus, and canals—each had to be precisely comprehended.

The other two treatises were no less technical. *On Base-Altitude and Circle-Division Mathematics* was a full elaboration and synthesis of Mei's spherical trigonometry; *A Commentary on Qu Yuan's Rhapsody* relied heavily on Tychonic cosmology and cosmography as well as analytical phonology.

Dai completed all three treatises in 1752 before taking a job in Xixi, not far from his hometown, as a private tutor. He had been hired by Wang Wufeng (1725–73), who already employed Jiang Yong. Thanks to Wang's patronage and Jiang's intellectual stimulation, Dai recast his three treatises, bringing them to a point that he considered publishable.[19] This is clear from the various manuscripts that have survived to our time, in which we can clearly trace the development of his thinking.

For the next two years, Dai developed his own theory of classical knowledge, gradually diverging from his mentor's philosophy. One could say that Dai shifted his intellectual orientation toward realism, coming to believe that "entities, states, and processes described by correct theories really do exist," rather than a philology that valued classical scriptures as an empirical priority to access the classical world or a nominalism that valued technical methodologies as empirical priorities for accessing the classical world.[20] Granting that Dai was both a realist and a philologist, it was his realism that set him apart from Jiang who was a strict nominalist. These differences were brought out when Dai and Qian came face to face.

DECISIVE ENCOUNTER

In the spring of 1754 Dai abruptly fled Huizhou with the local magistrate on his tail. After a showdown with a member of a branch lineage who was trespassing on his family's burial site, Dai had come out the loser.[21] He was suddenly in serious trouble. The only conceivable option for him was flight. Where would he go?

He chose Beijing. The existing sources shed no light on that decision. Perhaps he was seeking fame, perhaps patronage, perhaps both. There is no question that the capital—the heart of the national examination network—seemed a more suitable place for his work and for the debates with the national elites that could sharpen his arguments. He was looking for someone who could appreciate his work, and the chance of finding that someone was higher in Beijing than in Xixi.

The price was high, however. It was a difficult and expensive journey, and there was no chance of sitting for an examination once he got there.[22] From every conceivable perspective, this unplanned escape was rushed and risky. Still, Dai was intellectually ready.

On arrival, he settled in one of the lodges set up for commercial and scholarly travelers from Huizhou.[23] Dai eagerly approached the many examination candidates who passed through, but he impressed no one. One day, manuscripts in hand, he visited a newly minted *jinshi* named Qian Daxin, perhaps the brightest young scholar to emerge from Suzhou. Years later, Qian vividly depicted this first meeting: "Dai Zhen is a loner. On most social occasions, he appears awkward, with no idea how to behave. He has few friends and does not seem content with his situation. In his thirties, he dragged himself to Beijing. Trapped in a lodge, he could barely figure out how to get his next meal. Everyone saw him as a defiant and unconventional fellow [*kuansheng*]. Yet one day he lugged his manuscripts into my studio. We talked all day long. I gazed at him while bidding farewell and then sighed, 'He is truly a marvelous talent!'"[24]

With Qian's introduction, Dai became a sensation among the scholars from Suzhou. Their applause instantly launched his career as a prominent classical scholar and expert in Jesuit mathematical astronomy. As historian Kondō Mitsuo observed, if we picture the intellectual shape of the Qianlong emperor's reign from 1735 to 1796 as an ellipse, then Dai and Qian were the twin centers around which many significant scholars revolved.[25]

Dai's rapid success simultaneously promoted his method and his scholarship, creating an essential and latent tension with Qian and other Suzhou scholars even as it altered the course of his interests.[26] The tension emerged from their subtle disagreements over what counted as technical methodology and how it might enhance their historical understanding. Shelving their differences, they pressed forward with a shared intellectual agenda.

Qian was born in Jiading County, right at the center of the Lower Yangzi Delta. His father and grandfather were both licentiates but never passed the provincial examination. For two generations, they taught locally.[27] With their encouragement, the third generation satisfied pent-up hopes. A precocious and intelligent boy with a remarkable memory, Qian learned how to read and write at the age of four and how to write examination essays at the age of nine. He attempted his first county examination in 1738, at the age of ten, but failed, as he did on his second attempt in 1740. He passed the county examination on his third attempt, in 1742, becoming a licentiate at the age of fourteen. His reputation quickly spread, attracting the attention of the wealthy Wang family in Jiading; Wang Mingsheng (1722–97) had earned a licentiate's degree in the same year, at age twenty.[28]

The patriarch of the Wang family, Wang Erda, had a brilliant daughter named Wang Shunyin (1728–67). Her father adored her, had seen to it that she received a good education, and had already refused several marriage proposals. Delighted by his son's success in 1742, he also spotted the academic potential of Qian, who came from a family of lesser wealth and prestige. Wang Erda thought that he might tempt Qian with a uxorilocal marriage, meaning that Shunyin and her new husband would remain in the Wang home after they wed.[29] But some members of the Wang family, according to Qian's own account, had reservations about the match. The notional groom came from a family with no land, and no matter what the young scholars of Suzhou were saying, Qian's bone structure and facial features showed that he would not succeed in life. Everyone, he recalled many years later, from various relatives of the Wang family to the servants in the house, wondered why the patriarch "had chosen such a poor and ugly scholar for his favorite daughter." In spite of all the carping gossip, the engagement took place.[30]

Now Qian and his future brother-in-law had to confront the challenge considered the toughest in the examination hierarchy. Neither Qian's father nor his grandfather had ever surmounted the provincial examination. In fact, about 98.5 percent of the licentiate population never passed. Those who did earned membership in the national elite. Qian and Wang failed in 1744. When they both tried again in 1747, the older man passed.

This certainly put competitive pressure on Qian, but in the fall of 1750 he failed once again. That winter, he married into the Wang family, ratcheting up the pressure to succeed.[31]

During this period, both Qian and Wang Mingsheng enrolled at the Ziyang Academy in Suzhou, expanding their intellectual horizons into classical research. This academy was far more resourceful and prestigious than the identically named one in Huizhou, where Dai was a student. Suzhou was the largest and wealthiest academic center besides Beijing. The academy's patrons came from the city's elite circles; some, like the Hui family, had distinguished themselves by their scholarship over four generations. Many Ziyang academicians were also members of the network of national elites. Some were leaders in belles lettres, examination success, and classical research.

Qian enjoyed the intellectual companionship of his student cohort as well as the mentorship he received from scholars such as Hui Dong (1697–1758) and Shen Tong (1688–1752), both of whom adamantly championed "ancient learning" (*guxue*), a radical rejection of the Song Confucian ontology in favor of the scholastic tradition of Han-era classicists. Other academies, including the Ziyang Academy in Huizhou, maintained a critical stance toward the state orthodoxy, but Hui and Shen insisted that it was only one mode of Confucianism among many, and believed that it failed to offer an interpretation of the entire classical canon. The young band of Ziyang academicians encouraged each other to study ancient learning, reinventing many exegetical tools deployed by the Han classicists. Mathematical astronomy and historical chronology, for example, were included in their toolbox. This was a critical reorientation for Qian and Wang. From 1749 to 1751, both absorbed the Suzhou version of ancient learning and slowly developed their own research agendas.[32]

In 1751 a special provincial examination was held in Jiangsu to commemorate the Qianlong emperor's visit, part of his southern tour. Qian submitted a rhapsodic poem (*fu*) to the governor of Jiangsu, Zhuang Yougong (1713–67). Zhuang selected it as one of a batch of qualified pieces he presented to the emperor in his temporary dwelling place in Nanjing. Three examiners approved the poems of a few young and poetically talented men, awarded them the provincial degree, and instructed them to head to Beijing for the metropolitan examination in the spring of 1752.[33] Qian was among them.

Upon his arrival in Beijing, Qian took a room in a house with Chu Yinliang (1715–90), one of his fellows from the special examination. In the fall, both failed in their first attempt at the metropolitan examination. Since his honorific degree in the provincial examination still qualified him for a government position, Qian assumed a prestigious yet not overly burdensome staff position at the Grand Council while he waited for the next metropolitan examination, which was two years away.[34]

It was then that Qian, possibly stimulated by Suzhou-style evidential analysis, turned his attention to mathematical astronomy for the first time. Together with Chu and Wu Lang (1719–71?), he purchased a copy of *The Complete Collection of Mei Wending's Mathematical Astronomy*, the same work that Jiang Yong had shared with Dai Zhen in 1742.

According to his own account, so fascinating did Qian find the book that he "almost forgot to eat or sleep" during the next two years.[35] He then applied its lessons to historical chronology and began his research on the first complete system of ancient mathematical astronomy, the Triple Concordance System (*santong li*) drawn up by Liu Xin (50 BCE–23 CE). Liu had offered a theory of solar and lunar motion sufficient to run an efficient lunisolar calendar. In other words, this system was basically a prototype of early mathematical astronomy, in that it followed the average lunar cycle closely and inserted intercalary months as needed. Qian found it fascinating to trace each dynasty's astronomical system back to the Triple Concordance System. This titanic effort made it possible for him to corroborate several textual sources and detect many chronological mistakes in the standard histories. Many technical conundrums, however, also began to emerge.[36] Moreover, Qian grew all the more acutely aware of the problems with the old systems as he confronted the successes the Jesuits had had in measuring the length of the tropical year.

This was precisely why Qian immediately recognized and embraced Dai's project. He and his Suzhou mentors had long promoted several technical disciplines, including classical lexicology, analytical phonology, and historical geography. Fundamentally a historical philologist, Qian paid meticulous attention to historical sources and would not be persuaded to discredit them without compelling evidence. Mathematical astronomy was a new tool, allowing him to appreciate fully Dai's remarkable talent. Still, whereas Dai, a realist, hoped to reconstruct the classical world, Qian was most concerned about verifying textual sources.

UNGRACEFUL ASCENDANCE

In spite of the warm reception his ideas received, Dai proved too "defiant and unconventional" (Qian's words) to charm Beijing's polite society.[37] He found it as difficult to accustom himself to high society's manners as elite scholars did to accommodate his intellectual agenda. To give some idea of the cultural gap, the man never composed a poem in his life! His ungraceful forays into the salons of both Beijing and the Lower Yangzi Delta were agonies for him.

In the scholarly zone as well, he had much to learn. During the summer of 1755, Dai first met Ji Yun, who had passed the metropolitan examination in 1754 and impressed him with his manuscript, *The Illustrated "Records of Ancient Technology."* According to Ji, when he first saw the manuscript, it comprised only a description of each technological artifact and the titular illustrations.[38] In other words, the manuscript neither cited nor quoted from *The Rites of Zhou* chapter that it inventoried. Such contemptuous treatment of a sacred text was unacceptable. As Ji wanted to publish the study, he asked Dai to conform to academic norms by working both the text of "Records of Ancient Technology" and the official commentary by Zheng Xuan, an exegete of the Eastern Han dynasty, into the book. Dai's commentary would appear after Zheng's. Grateful for Ji's patronage, Dai willingly complied, and the book was printed at Ji's publishing house in 1758.[39]

The poor merchant's son had begun to attract admirers. Yao Nai (1732–1815), the renowned essayist from Tongcheng County, crossed Dai's path in 1754. Yao recalled that he and Dai had both been living in Ji's Beijing mansion when he came to admire this eccentric scholar that everybody was talking about. Summoning his nerve, he asked to be taken on as a disciple, only to meet a polite but firm refusal.[40] Still, Yao did not take the rejection well and later wrote a short essay to refute a detail in Dai's book.[41] Such vaunting claims did Dai make for the explanatory power of his technical methodology that most Suzhou scholars, not just those suffering from sour grapes, wondered whether he might have overstated his case.

Wang Mingsheng, Qian's brother-in-law, had a critical exchange with Dai in 1755. As he was winding up *A Late Investigation of "The Book of Documents"* (Shangshu hou'an), Wang showed Dai his exegesis of "The Canon of Yao" (Yao dian), a chapter in *The Book of Documents*. Dai conveyed his reactions in a lengthy letter. After the usual expressions of courtesy, he disputed a gloss of a seemingly unproblematic word, *guang* (in modern Chinese, "luminosity"), in the sentence "*Guang bei si biao.*"[42] Wang, following the same Zheng Xuan whose commentary Dai had included in his book after prodding from Ji, glossed *guang* as a verb meaning "to fulfill" (*chong*). Dai disagreed. From an extensive comparison of similar passages in the classical scriptures, Dai had concluded that the character had appeared in the sentence as a mistake; the original must have been *heng* (horizontal). At the end of the letter, Dai delivered an impromptu lecture:

> It has been two thousand years since *The Six Books of Etymology* was abandoned and classical knowledge declined into error and viciousness. So, after two thousand years here you are; you love antiquity and are insightful about the sources. So I assume that it is not a matter of lodging a claim to antiquity to elevate your status. I am an eccentric man, and my view is probably narrow. I believe that those who blindly place their faith in antiquity are stupid, only slightly better than those who do not know what they are writing about. We should infer from, rather than dwell on, evidence. Take, for instance, the single character *guang* in *The Book of Documents*. In a debate over its meaning, every relevant ancient source could be cited by both the critics and the defenders of any specific reading.[43]

After this exchange, Dai obtained more evidence to support his reading from Qian, Yao, and his cousin Dai Shoutang (dates unknown), refining his evidential analysis of the sentence. Wang stuck to his own reading and, obviously annoyed by his friend's criticism and arrogance, remarked a few years after Dai's death, "If Dai Zhen were still alive and had heard of my interpretation [of a different passage], he would not have been convinced at all. I daresay Dai Zhen considered himself superior to everyone else in history."[44] The intellectual difference between them was real and profound. What Wang despised, and his brother-in-law tolerated, was the very fact that Dai was an arrogant straight shooter.

The meteoric rise of this unpolished outsider, who found a patron in 1756 when Wang Anguo (d. 1757) hired him to tutor his son, resulted not from his brilliant classical scholarship but, as we shall see, from his astonishing competence in mathematical astronomy. But neither Wang nor any other fan of this unpolished prodigy could have anticipated the thread connecting Dai's predilections: his grand vision of classical knowledge, his uncompromising argumentation, and his occasional intrusions into the realm of moral philosophy—all were linked to his method. The ranking members of the national elite had attempted to promote a technical expert and got a visionary classical scholar instead. As a result, the repercussions of his scholarship were felt in a range of disciplines embracing technical methods and in the overall direction of classical knowledge.

When Qian came to Beijing in 1752, he was invited to consult on a number of slippery questions bedeviling a new project that would be published as *A Comprehensive Investigation of the Five Rites.* The team led by the minister of rites, Qin Huitian, had already completed a first draft. Qian saw mathematical astronomy as the project's weakest element, but he could not decide how to proceed. Therefore, when a defiant and unconventional expert showed up on his doorstep in 1754, he understandably concluded that his quandary had been resolved. The next day, Qin and Qian and their entourage rode to Dai's humble room at the lodge and invited him to join the *Comprehensive Investigation* task force. Rising to the task, Dai helped create a new section, entitled "Observing the Celestial Patterns and Bestowing the Timetable for Farming" (Guanxiang shoushi). The section was fleshed out with his *On Base-Altitude and Circle-Division Mathematics* and Jiang's *Explanation of Algorithms in Astronomy* (Tuibu fajie), loyally preserved in Dai's bookcase, both included verbatim. Almost overnight, Dai became known as a profound classical scholar who could apply mathematical astronomy to explain technical passages in the Three Ritual Classics.

TECHNICAL VOCABULARY

Cast in different languages, *On Base-Altitude and Circle-Division Mathematics* and *Explanation of Algorithms in Astronomy* suggest the two extremes of the use of mathematical astronomy for ritual purposes. For the study of algorithms, Jiang adopted the translated vocabulary of Jesuit mathematical astronomy with few reservations, whereas Dai borrowed the language of the ancient treatise *The Gnomon of the Zhou Polity: An Arithmetical Classic* (Zhoubi suanjing) to describe his spherical trigonometry. By incorporating these diverse texts under one heading, Dai assumed a position marked by compromise, something his master, Jiang Yong, had failed to do.

But taking one's terminology from *The Gnomon of the Zhou Polity* does not fuel popularity. The obscurity of this language was brought home in 1755 when Dai found a sponsor from Huizhou prefecture—Wu Sixiao (dates unknown)—to publish *On Base-Altitude and Circle-Division Mathematics* independently. In the preface, Wu cautioned the reader about Dai's terminology in 1758:

> Early this summer, Dai showed me *On Base-Altitude and Circle-Division Mathematics*. When I read it I came across expressions that had not been used since the Han dynasty! It is indeed marvelous that he uses roughly two thousand words to explain ancient and modern mathematics in their entirety. Dai derives his vocabulary from the first chapter of *The Gnomon of the Zhou Polity* and consistently uses the ancient terms for his methods and mathematical subjects. . . . I am concerned that those who have learned the newest sort [i.e., the Jesuit sort] of mathematics will not be able to comprehend the ancient terms used in this work. I have therefore supplied the terms currently used in plane and spherical trigonometry in a smaller print right under the ancient terms used by Dai.[45]

Perhaps Wu cautioned his readers in order to protect his investment in publishing this book. The work was challenging because spherical trigonometry was transposed onto classical knowledge, and making out the meaning demanded a high level of classical literacy, something few mathematical astronomers possessed. It was exhilarating because mathematical astronomy was seamlessly integrated into ancient thought, raising the possibility it might become respectable among members of the national elite.

Owing to the science faction, the Jesuits' astronomical vocabulary had swept the field in the early part of the Qianlong emperor's reign. The appearance of Dai's technical vocabulary inspired the Suzhou literati to contrast his "Chinese method" (*Zhong fa*) with the "Western method" (*Xi fa*) of Jiang. Seizing the home-grown terms, they rejected the imperially authorized Jesuit vocabulary.[46] Officially they could not attack the science endorsed by the throne, but they could reject Jiang, whom they accused of bending all too willingly to Western ways.

It is important that we recognize the ethnocentric strategy of driving a wedge between Us (the Chinese) and Them (the Jesuits), because many historians have uncritically described the distinction and applied it to the historical narratives of many members of the science faction. The Suzhou scholars, Qian in particular, fabricated a distinction between the so-called Chinese method and Western method in order to discourage scholars like Jiang who adopted the language of imperially sanctioned mathematical astronomy. They wished to promote the technical language pioneered by Dai to reinforce the official astronomical discourse and undermine the long-standing assumption of Jesuit technical superiority.

COSMOLOGICAL REALISM AND TEXTUAL EMPIRICISM

The conjoining of works by Dai and Jiang in *A Comprehensive Investigation of the Five Rites* reminded many scholars at the Institute of the Three Ritual Classics—especially Mei Juecheng—why they had disliked Jiang's mathematical astronomy. As these feelings were inspired by more than language, an analysis of the differences among the theories of Jiang, Qian, and Dai is warranted.

Qian had just completed his study of Mei Wending's work when Dai offered a glowing account of his mentor's insights. Following this recommendation, Qian quickly obtained a copy of Jiang's *Wings of Mei*. But after careful examination, Qian concluded that the book lacked substance. He could not hide his disappointment, and in the emotional letter he wrote to Dai he expressed misgivings about his correspondent's motives.[47]

A problem, Qian said, arose when Jiang compared the recorded length of the tropical year in the standard histories with his own calculations. Qian argued eloquently that Jiang's technical vocabulary and his numbers should both be rejected. Accusing Jiang of blindly following the "European doctrine" (*Ouluoba zhishuo*), Qian defended Mei's achievement.[48]

Moreover, Qian argued that the Jesuits were not trustworthy: had they not changed their cosmological models more than once since the beginning of the seventeenth century? At the Ming court, they had introduced a model built from the basic insights of Ptolemy (90–168); then, at the Qing court, they presented a Tychonic cosmos. As the prediction of solar and lunar eclipses proved challenging, the Jesuits replaced the perfect circles of celestial movement with the ellipses of Johannes Kepler (1571–1630). To yield more accurate computations, they also changed the number of deferents and epicycles. In the 1750s they even transmitted the Copernican heliocentric model, though the Catholic ban on Copernicanism was not rescinded until 1757.[49] Couldn't the Europeans make up their minds? As Qian saw it, these unreliable scholars merely threw everything against the wall to see what would stick. Evidently they had never developed and would never come up with a universal technical standard.[50]

Jiang acknowledged that the length of the tropical year had been widely disputed. Mei, he said, had basically followed the system designed and implemented under the Mongols in the thirteenth century; it was known as "granting the seasons" (*shou shi*). Historian of science Shigeru Nakayama has reconstructed the eight-hundred-year-old formula for predicting the minute decrease in the length of the tropical year where T is the length of the tropical year and t is the number of years subsequent to the epochal year 1282. By the time Jiang wrote his treatise, in 1750, $t =$ 468, and the decrease in the length of the tropical year would have been 0.000936. Jiang rounded it up to 0.001. Therefore, the length of the tropical year should decrease about 0.01 days every 4,700 years.[51]

According to Jiang, Mei had been skeptical about the estimate of a 0.01 annual decrease, recorded in the Yuan dynasty and presented as a sudden fluctuation rather as the gradual change brought about by the Earth's annual motion. More importantly, the figure given for the length of the tropical year in the temporal model calendar was 365.2425 days, slightly *larger* than the 365.2415 days used in granting the seasons in the thirteenth century. How could that be? Although the framers of the temporal model calendar had relied on a different epochal year and had therefore changed the starting point of the calculation, a longer tropical year was unthinkable. Something had gone awry in their measurements, Jiang argued.[52]

Citing difficulties in working with sundials during the cold season when the sun was low, Jiang proposed the spring equinox, instead of the winter solstice, as the starting point for measuring the tropical year. Identifying the winter solstice was notoriously difficult because, while measuring the length of shadow on the sundials in the winter, the length is rather short and it changes very little from day to day. The ancient Chinese chose the winter solstice as their starting point because the cosmic energy's lowest point seemed like a natural point of origin. While Jiang cited technical reasons for making a change, this meant a break with millennia of tradition. And in fact most literati perceived his true motive: he was following the Jesuit model.[53] But shifting the standard had powerful ritual and political implications.

Jiang also proposed a new cosmological theory to account for the minute variation in the length of the tropical year. He suggested treating the length of the tropical year as a constant, which he called the "mean length." But had not measurements taken in both China and Europe shown without a doubt that the length of the year varied? Jiang explained away this phenomenon by describing the solar orbit as a Keplerian ellipse, as had been done in the imperially sanctioned *Supplement to "A Compendium of Observational and Computational Astronomy,"* published in 1742. According to Nathan Sivin, the *Supplement* was a remarkable achievement: it incorporated new and improved constants and tables based on the observations of J. D. Cassini (1625–1725) at the Paris Observatory and John Flamsteed (1646–1719), "the King's Astronomical Observator" at the Royal Greenwich Observatory. More intriguingly, the *Supplement* introduced the Keplerian ellipse to the Qing court. Kepler was a devoted Copernican and rejected the Tychonic cosmos; his cosmology was entirely elided by the Jesuits, however. The *Supplement* informed the Chinese reader only of the mathematical advantage of using an ellipse to describe the celestial orbits: because the numerous epicycles were unsatisfactory for computing the solar anomaly and retrograde movement of Mars, Kepler posited that the deferent (the larger circle on which epicycles are centered) should be an ellipse, and took equal areas of the ellipse in order to derive the daily mean motion. The compilers of the *Supplement* stated elsewhere that "the ellipse method precisely reconciles eccentrics, epicycles, and equants."[54] The *Supplement* adopted the Keplerian ellipse because it was a more effective mathematical tool to improve astronomical accuracy. Historian Shi Yunli argued that the solar theory that the authors adopted from Newtonian lunar theory dramatically improved predictions of solar eclipses. While the imperial authorship of the preceding *Compendium* meant that the authors of the *Supplement* had to maintain the appearance of continuity between two works, the two works differed in kind rather than degree.[55]

Jiang embraced the novelties in the *Supplement*. He used the apogee and perigee of the Sun as well as the differences accumulated by epicycles to explain the minute variations in the length of the tropical year.[56] More importantly, he used the Newtonian lunar theory, the Keplerian ellipse, and other innovations to assess the recorded length of the tropical year in the standard histories.[57]

For Qian, it made no sense to correct the canonical records of Chinese history by means of the Jesuit model he deemed so ephemeral and unreliable. So he and Dai set out to compile a list of the recorded lengths of the tropical year in Chinese history and to enumerate the pros and cons of each dynasty's astronomical system. According to Mao Yuesheng (1791–1841), their project petered out quickly.[58]

Jiang's use of the tools described in the *Supplement* was equivalent to using NASA's computers to estimate the features of a solar eclipse that took place, say, one thousand years ago. This kind of "cosmological realism," pioneered by Jiang and championed by the young Dai, stressed current observations and new methods over the textual evidence in the classical scriptures. Qian disputed that epistemological priority because he had misgivings about the methods borrowed from the Jesuits.[59]

Historical records of astronomical phenomena, by contrast, consisted of first-hand observations. There was no reason to be skeptical about the empirical validity of such records. Using demonstrably unreliable tools fresh from the Jesuit workshops to discredit the reliability of historical records struck Qian as utterly wrongheaded.[60]

To comprehend the operation of something as vast as the cosmos, said Qian, required the accumulation of knowledge of many generations of mathematical astronomers. When China's early mathematical astronomers realized that the length of the tropical year and even the length of the sidereal year varied, they devised all sorts of techniques to calculate those cycles. All of these—including the work of Guo Shoujing in the Yuan dynasty and the configuration of orbits, deferents, and epicycles imported by the Jesuits—were mere geometrical auxiliarics (*qiaosuan*), far from true descriptions of cosmic reality.[61] They could be improved over time to increase their accuracy, but cosmic reality could only be revealed by the constant modification of computational techniques. Since Dai also made a clear distinction between the cosmological model and the physical sky it represented, Qian's adamant philosophical premise, as outlined in this letter, was—coincidentally—consistent with Dai's. Qian did not reject the use of Jesuit mathematical astronomy, but his textual empiricism prevented him from accepting it as a corrective for the classical record, and he disputed any form of cosmological realism in Jiang's argument.

CONCLUDING REMARKS

Chinese, Japanese, and Western scholars have long ranked Dai Zhen among the most important thinkers in modern Chinese history. Yet no previous studies, whether brief and focused or long and comprehensive, have paid much attention to how his social ambitions affected his scholarly idiosyncrasies. As he moved from place to place, seeking examination success, companionship, and patronage, he found inspiration outside of the intellectual mainstream, and by the force of his own intellect he drew others along the same path. At a crucial moment, an acquaintance attacked Dai's mentor for adopting Jesuit methods, providing a decisive impetus for a rupture between mentor and follower.[62]

For Dai, the goal that trumped all other considerations was the overall reconstruction of the classical world—a perfect realm that served as his moral foundation. If only he could figure out every single detail, retrieving the precise scale, design, and shape of everything recorded in the great books of antiquity, his mission would be complete. Mathematical astronomy would retrieve only some details. Such a vision of classical knowledge distinguished him from his mentor, Jiang Yong. But his insistence on historicizing European knowledge would have greater repercussions than he could have imaged.

CHAPTER 7

BRINGING IT HOME TO THE PALACE OF LIGHT

When Dai Zhen joined the team preparing *A Comprehensive Investigation of the Five Rites* in 1754, he set about preaching the virtues of his technical methods. However, his audience did not always greet his new ideas with enthusiasm and support. After three years had passed, Dai had a chance to visit Yangzhou and Suzhou. Qian Daxin urged him to pay homage to Hui Dong and to seek support from the respectable leader, who had established himself as a master of "Han learning." Dai arrived at Yangzhou and met Hui in 1757. Their unique meeting (Hui died in 1758) revealed a subtle yet profound disagreement on a subject of great interest to both scholars. Their disagreement also shed new light on the intellectual landscape of the eighteenth century.

Heavenly principles, they agreed, were insufficient. Instead of the abstractions beloved of Zhu Xi, Hui and Dai located the most reliable cosmological model for political action in a legendary edifice—the Palace of Light. Where Neo-Confucians spoke of "political mission," "moral responsibility," and "self-cultivation," these men demanded something practical—a blueprint for institution building and political organization. Their new orientation in speaking of concrete models of political action has long been misconceived as "apolitical" or "scholastic" by modern historians.

Hui and Dai differed on what this blueprint should look like, however.

The difference shows that young Dai had already developed his own classical vision before he met the older and more established Hui. An in-depth analysis of contemporaneous debates around the Palace of Light reveals the divergence between their visions and political orientations. The historic significance of the Palace of Light and the three ways it was imagined—as a cosmological palace, an ancestral temple, and a policy hall—are central to the debates concerning it. The key difference between Hui's and Dai's classical visions must be cast in the contexts in which their predecessors and contemporaries had extensively debated in terms of those three ways of imagining, arguing, and challenging the Palace of Light.

HISTORICAL SIGNIFICANCE

The historical significance of the Palace of Light cannot be overemphasized. For millennia, the ruling elite viewed it as a mysterious site of power, an enigmatic structure that embodied the mystery and sanctity of imperial legitimation.[1] The first references to the building link it to the Zhou polity. Members of the Zhou dynastic house (1045–256 BCE) regarded a personified Heaven as a universal and supreme deity, and they instituted, as a check on earthly rule, the notion of a mandate of Heaven.[2] King Wen and his son, King Wu, the founders of the Zhou, believed that Heaven had compelled them to overturn the Shang ruling house. The origin of the Heavenly mandate was therefore intertwined with the origin of the Palace of Light.

After King Wu died, his son was deemed too young to rule unassisted. As an array of ambitious contenders geared up to seize power, the Duke of Zhou, Wu's brother, was chosen to serve as regent. According to *The Grand Scribe's Records* (Shi ji), written by Sima Qian (BCE 139–86), the duke assembled all the vassals of the realm in the Palace of Light, a defining moment for the new dynasty. The duke defeated the remaining contenders for the throne and established a stable polity, a flourishing culture, and an orderly society. All literati in times to come would see this as the pinnacle of their civilization. For those who, like Confucius, looked upon the Zhou and in particular the Duke of Zhou as the model of enlightened rule, this stands among the finest moments in a golden age.[3]

But what did the Palace of Light look like? No ancient record clearly and unambiguously indicated its location or its appearance. Its purpose, for that matter, was also vague. A clue from the medieval period underlined the palace's symbolic power: "[Mulan] returned home and had an audience with the Son of Heaven; / The Son of Heaven sat in the Palace of Light [to receive her]."[4] This couplet is from "The Ballad of Mulan" (Mulan ci, fifth century), the first documented account of the legendary woman who dressed as a man to fight for China's Tuoba Xianbei rulers. The Xianbei underwent rapid Sinicization and, as is evident in the ballad, erected a Palace of Light to claim a connection to the Zhou legacy.

Two centuries later, during the Tang dynasty, Empress Wu Zetian (r. 690–705) built a grand Palace of Light to glorify her rule.[5] And almost a millennium later, during the 1660s, an unknown Jesuit missionary mentioned the palace in a poem included in a "traditional history of the Chinese script": "In the beginning of the winter, the Yellow Emperor rules the Palace of Light. / According to the imperial calendar, a letter is issued from the Highest Office."[6] This foreign report corroborates the pervasive association of the Palace of Light with imperial rule in early modern China.

As frequently as it had been evoked, the building inspired few scholars to investigate its concrete reality. The first serious attempt did not occur until the eighteenth century, when scholars reopened debates on ancient institutions and rituals, attempting to reconstruct a golden age that, by that time, was nearly three millennia old.[7] Its name consists of a pair of characters that revealed its cosmological significance. The Chinese character *ming* 明 is built out of the characters for Sun 日, on the left, and Moon 月, on the right, and its meaning is *bright*, *brightness*, or simply *light*. It suggests that the Sun and the Moon, inhabitants of the skies of day and night, bless the polity. The character *tang* can mean *hall*, *building*, or *palace*.[8] *Ming tang* has been translated as "Hall of Light" by William Edward Soothill in 1934 and "Spirit Hall" by Michael Nylan and "Bright Hall" by Wu Hung in recent scholarship.[9] Unlike these predecessors, I choose to render it as *Palace of Light* to stress the distinct meanings it had acquired in the eighteenth century and to suggest the architectural grandiosity that eighteenth-century scholars ascribed to it.

In Dai Zhen's day, those who scrutinized elements of the Zhou world, such as the Palace of Light, hoped to reveal the role of human interest in distorting the hallowed texts, but above all they sought to recreate the universal standard that had existed in the Zhou, providing a concrete and objective model to follow. The Institute of the Three Ritual Classics at court was the center where the classical scholars hoped to standardize the classical model. Some of the scholars with the most penetrating insights into the historical vicissitudes of the Palace of Light, however, would not have been welcome at the institute. Fang Wuru (*jinshi* 1706), for instance, was politically marginalized from the institute. In 1751, when Dai was studying at the Ziyang Academy in Huizhou, he met Fang, a senior lecturer who had studied with Mao Qiling and He Zhuo (1661–1722), both of whom were very critical of the work done at the Institute of the Three Ritual Classics.[10] In contrast, Jiang Yong, having produced a large number of monographs on ritual matters, had hopes of finding work in the capital on one of the big imperially sponsored projects. His patron, Cheng Xun, a ritual official at the Institute of the Three Ritual Classics, invited him to Beijing as a consultant. The debates Jiang launched into with senior colleagues on the project, including Fang Bao and Mei Juecheng, earned him dismissal. Dai Zhen, meanwhile, offered his service not to the institute but to Qin Huitian's semiofficial team in compiling *A Comprehensive Investigation of the Five Rites* in 1754. Dai subsequently revised his short treatise on the Palace of Light for publication.

Meanwhile, the Palace of Light had been invented in three ways: it was the cosmological place to which all vassals were summoned so that the new ruler's possession of the mandate of Heaven might be declared; the ancestral temple where the Son of Heaven performed sacrifices to the imperial lineage and the cosmos; and the policy hall where the king managed his vessels and deliberate the political issues.

Each way of imagining the Palace of Light was, firstly, linked to a controversial precedent in the imperial past and, secondly, challenged on the grounds of inconsistency or contradiction as recorded in the Confucian canon. Both aspects in each imagination were intimately related because debating certain idioms, passages, or expressions of the classical scriptures in each historical precedent had constituted the battlefield on which fierce political struggles unfolded. All three historical precedents—the founding of the Xin dynasty (CE 9–23) by Wang Mang (45 BCE–25 CE), the reforms of Wang Anshi (1019-86), and the Great Rites Controversy (1520–38)—were all fought out by debating the classical language regarding the Palace of Light. The first two illustrated how *The Rituals of Zhou* could be distorted for political ends. And the story of the succession of the fourteen-year-old Jiajing emperor (r. 1521–67), with its bitter political debate about the orthodox transmission of the imperial lineage, perfectly suited a revival of the Palace of Light. All three cases were parts of the backdrop against which the Qianlong emperor and his officials at the Institute of the Three Ritual Classics went about imagining a replica of the mythical building.

THE COSMOLOGICAL PALACE

The first conception of the Palace of Light—a cosmological palace to which all vassals were summoned—comes from a chapter in *The Records of Ritual* entitled "Hierarchy in the Palace of Light" (Ming Tang wei):

> In the past, King Zhou of Yin caused chaos under Heaven. He killed one of his vassal lords, a man named Guihou, and served his flesh at a feast for other vassal lords. Therefore, the Duke of Zhou helped King Wu conquer Yin. King Wu died. His son, King Cheng, was still young and weak. The Duke of Zhou stepped into the position of Son of Heaven and governed for six years. The Duke of Zhou summoned all vassals to the Palace of Light so that they could assume their positions according to the hierarchy [implied by] the Palace of Light. He called for rituals and music and announced the standard measures of length, weight, and currency. Everyone under Heaven obeyed him. In the seventh year, the Duke of Zhou transferred his authority to King Cheng.[11]

The events surrounding the duke's regency were seized by the founder of the Xin dynasty to justify usurping the Han. The striking, complex parallels between the Duke of Zhou, a loyal regent, and Wang Mang (c. 45 BCE–23 CE, reigned 9–23 CE) caused some consternation among Confucian literati. Wang wanted to convince the bureaucrats and aristocrats of his time that he was merely following in the footsteps of the Duke of Zhou and, according to a number of Qing classical scholars, he was not above tampering with the classical scriptures to justify his claims.

From 1736 to 1748, the Qianlong emperor and his ritual officials mobilized hundreds of scholars to compile the Three Ritual Classics, namely, *The Rituals of Zhou*, *The Records of Ritual*, and *The Ceremonial Rituals*. Several princes and high councilors were nominally in charge of this task force, which was set up in the Qing court. The main compilers at the Institute of the Three Ritual Classics included Yang Mingshi, Wang Lansheng (1679–1737), Li Fu (1673–1750), Zhu Shi (1665–1736), Li Qingzhi (1690–1744), and Fang Bao. All had served prominently at the court of the Kangxi emperor, due partly to their involvement in the science faction led by Li Guangdi. All had been demoted or sent to distant posts by the Yongzheng emperor. All had been recalled to the Qianlong emperor's court to guide the Three Rituals project. One of their thorny missions was to determine the authenticity of various chapters and passages in *The Rituals of Zhou*, especially "Hierarchy in the Palace of Light" and "Records of Ancient Technology" (Kaogongji).

Fang Bao, the chief compiler of *The Rituals of Zhou* for more than five years, tirelessly defended it against a series of attacks.[12] After he survived the literary inquisition case of Dai Mingshi (see chapter 4), Fang began studying harmonic mathematics (*yuelü*) in the Studio for the Cultivation of Youth and in 1723 became known as a fine classical scholar and devoted counselor to the Qing imperial family.[13] A chief compiler at the Institute of the Three Ritual Classics, Fang passionately defended *The Rituals of Zhou*, for him a sacred text.

Of the classical scriptures, *The Rituals of Zhou* has provoked some of the most charged debates. Fang argued that the passage that presented the Duke of Zhou "stepping into the position of the Son of Heaven" was fabricated by Wang Mang to sanction the steps he had taken to seize power. The rituals that took place during Wang's self-crowning, he said, had inspired the language slipped into *The Rituals of Zhou*.[14]

Wang's top advisor, Liu Xin (d. 23 CE), erected a Palace of Light while expanding the theory of dynastic cycles to justify his master's grip on power.[15] Some have argued that Liu had enormous influence over both the sciences of his day and the classification of ancient knowledge, with his Triple Concordance System (which soon became essential to making the imperial calendar), the cosmology of the five phases and five powers, a musical theory, and a system of mensuration.[16] Liu's systematic reworking of the theory of dynastic cycles, based on "Heavenly learning," also had a remarkable legacy.[17] The Triple Concordance System was a hybrid of mathematical astronomy and political astrology, with a good dose of musical harmony and mensuration helping it along. This synthesis of classical knowledge and specialized methodologies had a profound impact on imperial China; the astronomical sections of the standard histories drawn up by each dynastic house were deeply indebted to the Triple Concordance Calendar; this was true until 1911.

Despite their titanic efforts, however, Wang and Liu failed to maintain their grip on power. Liu Xiu (5 BCE–57 CE) overthrew the Xin dynasty and restored the imperial lineage of the Han. The new emperor of the Later Han dynasty declared that he restored the Heavenly mandate of the Former Han dynasty and condemned Wang Mang as an evil usurper who disrupted the mandate of the Han dynasty. Ironically, Emperor Guangwu (Liu Xiu) drew his entire theory of dynastic cycles from Heavenly learning and the Triple Concordance System furnished by Wang Mang's advisor Liu Xin.

In his work on the Three Ritual Classics, Fang took a cynical approach to Heavenly learning, rejecting some aspects of it. He went directly to the history of the Qin and Han empires offered by the historian Sima Qian, drawing on descriptions of specific events to dispute several misconceptions. For example, contrary to the claim that the design and details of the Palace of Light had survived to the Han, he argued that by 221 BCE its original features had been lost forever. A hundred years later, according to Sima Qian, Emperor Wu (BCE 141–87) performed the *feng* and *shan* sacrifices to Heaven and Earth then retired to the Palace of Light to receive congratulations from his vassals and councilors. According to Confucius, Fang pointed out, the *feng* and *shan* sacrifices did not exist. Fang mocked the statement that more than seventy kings had performed these sacrifices. Who could these kings have been? A performance of the *feng* and *shan* sacrifices implied the appearance of a new king to replace the old. Only a virtuous king could effect so profound a change in the mandate of Heaven. Where were these seventy virtuous kings? No records of them could be found in the ancient texts.[18]

In Sima's "On the Feng and Shan Rituals" (Feng shan shu), the historian poked fun at many aspects of the theory of dynastic cycles. He regarded cosmic signs, such as the discovery of a cauldron or a unicorn, as strange occurrences (*guaiwu*). For Fang, Sima was mocking the links between such aberrations and the rituals performed by the ancient rulers to communicate with the spirit world. Fang dismissed the words of magicians (*fangshi*) who encouraged the ancient rulers to see cosmic signs everywhere. He denied that any spiritual communication was involved in the *feng* and *shan* rituals, arguing that Confucius had never mentioned any of the seventy-two rulers who, according to Sima's narrative, performed those rituals. One might bring about spiritual communication, but such conversations with Heaven ought to take place in the Palace of Light. This argument permitted Fang to avoid all of the controversies regarding the meanings and political implications of the Palace of Light.[19]

A number of skeptics considered *The Rituals of Zhou* a forgery, such as He Xiu (129–182) in the Han dynasty, and Ouyang Xiu (1007–73) and Hu Hong (1105–1161) in the Song.[20] Fang argued that He, an opponent of Zheng Xuan, had cast doubt on the work's authenticity because he had seen Wang Mang use it to justify his overthrow of the Han. Similarly, Ouyang and Hu believed that the book had been used for immoral ends. Wang Anshi (1021–86) had said that *The Rituals of Zhou* was "the best source for the policies that can be applied to later ages" and built his reform policies on it.[21] Instead of arguing against Wang's policies, Ouyang and Hu dismissed *The Rituals of Zhou* as a forgery. Even famous scholars could not tell the difference between Wang's misguided reforms and the wisdom embodied in *The Rituals of Zhou*. Fang averred that while some passages of *The Rituals of Zhou* had been tampered with by Wang and his chief councilor, the work's overall quality was definitively at the level of the other classical scriptures. Fang's learned opinion opened the door to numerous debates in the years that followed.

Another chapter of *The Rituals of Zhou*—"Records of Ancient Technology"—includes details on the subjects of palace design, ritual clothes and utensils, and mourning rituals. Most of these subjects were intertwined with state rituals. The search for inconsistencies led some to question the true nature and reliability of "Records of Ancient Technology," which had been disputed sporadically since the eleventh century.[22] "Records of Ancient Technology" was widely known as a work written during the Han dynasty rather than in the high antiquity (Zhou). *The Rituals of Zhou* consisted of supposedly six records of offices in the Zhou system: Heaven, Earth, Spring, Summer, Autumn, and Winter. The sixth record, the "Winter Office," was lost in the Han dynasty and replaced by "Records of Ancient Technology." The Palace of Light in the "Records of Ancient Technology" contradicts other classical works. Historically, however, three Palaces of Light had been built in the Han and Tang dynasties.[23] Previous construction of this mythical building had basically ignored the inconsistency and contradiction surrounding "Records of Ancient Technology." For the Qianlong emperor, settling the many debates the elusive building had inspired over the millennia would set him above the rulers of old. It would be an indication that he had transcended the Han and Tang dynasties, and shifted the Qing decisively closer to the Zhou.

Once again, skepticism had prevailed before. Some classical scholars in the seventeenth century, such as Wan Sida (1633–83), Yao Jiheng (1647–1715) and Mao Qiling (1623–1716), all challenged the authenticity of parts or the entirety of *The Rituals of Zhou*. They all acknowledged that "Records of Ancient Technology" was written in the Han dynasty and was not a reliable record of the ancient Zhou system. Fang Bao, as the chief compiler of *Commentary and Meaning of The Rituals of Zhou* (Zhouli yishu) at the Institute of the Three Ritual Classics, had to confront the challenges from Wan, Yao, and Mao on the one hand and gave a special treatment to "Records of Ancient Technology" on the other. "Records of Ancient Technology" was written from the perspective of craftsmen in the Han dynasty, more than one thousand years after the first Palace of Light was built in the Zhou system. It consisted of many dimensions and details of the making of artifacts, buildings, weapons, and clothes. Not many classical texts provided these kinds of details. Could they be correct?[24]

Jiang Yong best articulated this skepticism. In his treatise *Commentary on the Chapter of Explanation of Buildings in The Ceremonial Rituals* (Shigong zengzhu), Jiang mistakenly attributed this work to the Neo-Confucian master Zhu Xi and consequently raised the book to a nearly canonical status.[25] By following this work as a masterpiece, Jiang Yong denied that the Palace of Light was the blueprint for a cosmological palace. Like Xu Qianxue before him, Jiang argued that neither "Hierarchy in the Palace of Light" nor "Records of Ancient Technology" are reliable. On this evidentiary basis, Jiang separated the cosmological Palace of Light from the sacrificial Palace of Light (or the Ancestral Temple), a view widely shared by Neo-Confucian scholars in the Lower Yangzi Valley.[26] It was at this specific issue that Dai Zhen departed decisively from Jiang Yong. Far from sharing Jiang's skepticism, Dai believed that the Palace of Light was exclusively a cosmological palace.

THE ANCESTRAL TEMPLE

During the Ming and Qing dynasties, the Palace of Light was identified with a building in the Forbidden City, the Ancestral Temple (Tai Miao).

Its presence in the heart of the capital gave the Manchu conquerors, the alien Aisin Gioro clan, a new imperial identity, a ritual practice familiar to the Han, and a cosmic justification for their rule.[27] The Manchu regent Dorgon (1612-50) commanded his army to occupy Beijing in 1644 after the last emperor of the Ming dynasty—the Chongzhen emperor (r. 1627-44)—committed suicide. Upon declaration of the Qing dynasty, Dorgon immediately buried the former Ming emperor with utmost respect and enlisted him in the Ancestral Temple of the Ming imperial lineage. In so doing, Dorgon simultaneously preserved the ritual continuity and integrity from Ming to Qing and highlighted the significance of the Ancestral Temple for the new dynasty.

By the 1740s, countless officials had eulogized the Qianlong emperor for his efforts to "investigate antiquity and promote culture" (*jigu youwen*). Like the Yongle emperor in the Ming dynasty and his own grandfather, the Kangxi emperor, he aspired to be a great patron of literati culture.[28] Identifying the Ancestral Temple with the Palace of Light quickly became the focus of the Institute of the Three Ritual Classics. A work by the Eastern Han scholar Cai Yong (132–192), *An Excursus on the Palace of Light and Monthly Observances* (Ming Tang Yue Ling lun), proved illuminating:

> The Palace of Light is the Ancestral Temple [Tai Miao] of the Son of Heaven, wherein the emperor sacrifices to his ancestors in the company of the Supernal Lord [Shangdi]. Members of the Xia lineage called this place the Hall of Multiple Generations [Shi Shi]; the people of the Shang called it the Multistoried House [Chong Wu]; and the people of the Zhou called it the Palace of Light. The eastern [hall] is called Green Yang [Qing Yang]; the southern is called the Hall of Light [Ming Tang]; the western is called the Assembly of Emblems [Zong Zhang]; the northern is called Somber Hall [Xuan Tang]; and the central one is called the Grand Hall [Tai Shi].[29]

Monthly Observances (Yue Ling) was an ancient text evolving before the unification of Qin in 220 BCE. It was included, in slight variation, in both *Master Lü's Spring and Autumn Annals* (Lüshi chunqiu), by Lü Buwei (251-235 BCE), and *The Records of Ritual* (Li ji) in the Han dynasty. Cai was perhaps the first exegete who systematically connected the Palace of Light with the cosmological and ritual temporality in *Monthly Observances*. The Palace of Light was, in Cai's account, a temple where sacrifices were made to the imperial patriline and to the supreme deity. It was not new to the Zhou but a practice continued from the Xia and Shang. Furthermore, the sacrificial practice was simultaneously cosmological and mundane. The five halls at the center of the temple and the four directions were symbolic, signifying the rotation of the five phases and five powers in the theory of dynastic cycles, and the progression of the ruler from one residence to another, according to the season.

Cai's next utterances proved both confusing and controversial for the court officials of the Ming and Qing:

> [The Palace of Light is] the source from which spring governance and instruction, the origin of all change and transformation, manifesting unity. Therefore, it is said that the affairs of the Palace of Light are great and its meaning profound. If one invokes its pure aspect, it is called Pure Temple [Qing Miao]; if one invokes its aspect as the hall of governance, it is called the Ancestral Temple; if one invokes its venerable aspect, it is called Grand Hall [Tai Shi]; if one invokes its aspect of facing toward the light, it is called the Palace of Light; if one invokes the aspect of the schools maintained at the four gates, it is called the Great Academy [Daxue]; if one invokes the aspect of being surrounded on the four sides by [a body of] water, round like a jade *bi*, it is called the circular moat [*bi yong*]. They are all different names for the same structure—it is a single thing.[30]

The imperial locations Cai listed—the Ancestral Temple, the Pure Temple, the Great Hall, the Great Academy, and the moat—corresponded more or less with places in and around the Forbidden City. The passage's concluding declaration, the unity behind all of the apparent differences, was generally taken symbolically. When ritual officials and classical scholars like Fang and his colleagues began to investigate the institutional design and regulatory structure of the Zhou, the new issues they looked at involved the architecture and location of the Palace of Light, rather than its mythical and cosmological implications. Among these rather vague one-to-one associations, identification of the Ancestral Temple in the Forbidden City as the Palace of Light in high antiquity clearly had the most profound implication politically.

Yuan Mei was among those who doubted the widely accepted identification of the Palace of Light with the Ancestral Temple. He claimed that the identification was fabricated during the Han dynasty; in fact, he blamed Cai Yong. While a number of ritual officials considered the "hierarchy in the Palace of Light" passage a forgery, for Yuan it was a reliable source—partly because of what it did not say. It did not identify the Palace of Light with the Ancestral Temple. Whereas King Cheng of Zhou established his Ancestral Temple in Gaojing (near today's Xi'an), the Duke of Zhou had built the Palace of Light in Luoyang. If the Palace of Light and the Ancestral Temple were indeed identical, why would King Cheng have worshiped his ancestors in two different locations, one of them hundreds of miles away from the capital in which he resided? Yuan went on to argue that the minister of music (*da si yue*) is described as looking after the music for the sacrifices to Heaven and to Earth; he was also in charge of the nine melodies played at the sacrifices to ancestors. If the Palace of Light and the Ancestral Temple were one and the same, surely such an official would have been closely associated with it, yet there is no mention of the Palace of Light in the passage in question.[31]

Yuan also set out to show that the alleged evidence for the identity of the two buildings had been misconstrued. In the Han dynasty the commentators Mao Heng and Mao Chang had written a long preface for *The Book of Songs* and a short preface for each of its chapters. In one of those short prefaces, they wrote, "The Duke of Zhou sacrificed to King Wen [father of the Duke of Zhou and King Wu] in the Palace of Light." And in a commentary to *The Classic of Filial Piety*, they wrote, "The Duke of Zhou worshiped King Wen in the company of the Supernal Lord [Shangdi, i.e., Heaven] in the Palace of Light." These sentences appear to identify the Palace of Light as the location where rulers worshiped their ancestors. Yuan denied the authenticity of both sentences, suggesting that they had both been composed after 221 BCE—still more Han forgeries.[32] This was a barb directed at the growing number of literati who labeled themselves specialists in "Han learning," a label referring to the classical knowledge developed in Suzhou and modeled after the exegetes of the Han dynasty.[33] In other words, Yuan Mei was subtly arguing against the mainstream "Han learning" scholars, such as Hui Dong.

It cannot be said that Yuan Mei's arguments swayed all those who heard them, though if they had the Palace of Light would have ceased to cast so long a shadow over Qing political machinations. This is because its identification with the Ancestral Temple linked it to the succession struggles, so often nasty, among imperial siblings or lineages. The Qing emperors were particularly mindful of the agnatic succession that had occurred under the Jiajing emperor of the Ming, triggering the reconstruction of the imperial lineage.

The trouble began in 1521, when the Zhengde emperor (1491–1521), Wuzong, died without an heir. After thirty-seven days of suspense, his cousin, Zhu Houcong (1507–67), was enthroned as the Jiajing emperor (r. 1521–67). The highest court officials commanded the young emperor to offer sacrifices to his imperial predecessor at the Ancestral Temple and to call his biological father "Royal Uncle."

Zhang Cong (1475–1539), a junior official, submitted a memorial in which he found fault with the misrepresentation of family relations and called on the Jiajing emperor to acknowledge his own father in the sacrifices. A bitter political struggle ensued. The emperor sided with Zhang and mobilized literati support to solve the ritually tricky question of how to honor his father. This Great Rites Controversy (Da Li Yi) and its resolution would set a precedent for future succession crises.[34]

The solution came in 1538, when Feng Fang (1493–1566), a high official known at the court for his book collection and his fine calligraphy, recommended that the emperor "build a Palace of Light and worship his father in the company of the Supernal Lord."[35] The Jiajing emperor liked the sound of this and he promptly rewrote history, removing his uncle's tablet from the Ancestral Temple (treated at the time as the Palace of Light) and installing his father's tablet there instead.[36] The Palace of Light became closely associated with the Great Rites Controversy and an emblem for imperial succession.

As a foreign group, the Qing emperors felt obligated to exhibit greater punctilio than their Ming predecessors in managing the grand sacrifices. Over the first century of their rule, they gave increasing attention to the Palace of Light, identified as the Ancestral Temple at the court. In 1657 the Shunzhi emperor declared that his father, Hong Taiji, and his grandfather, Nurhaci, would both be worshiped in the Grand Sacrifice to Heaven to take place at the winter solstice. The Kangxi emperor added the tablet of his father, Shunzhi, in the Ancestral Temple in 1667, and the Yongzheng emperor placed the tablet of his father, Kangxi, in the temple in 1724.

When the Qianlong emperor ascended to the throne in 1735, he immediately ordered his ritual officials to study whether the grand sacrifices to Heaven, Earth, and his father—duties required of a new emperor—should be held at the summer or the winter solstice. (A precedent in the Song dynasty was found that permitted him to avoid this scheduling constraint completely.)[37] In addition, he needed them to determine how long his mourning period should be and when the harvest sacrifices were to be performed. The emperor's requests were meticulously considered and debated by his ritual officials. Two options were presented, approved, and then carried out by the minister of rites. Another instance of ritual deliberation occurred in 1751, when one of the imperial rituals dubbed "Praying for a Good Harvest" (*qi gu*) was described in a memorial as "not in accord with the design of the Palace of Light." (Many imperial rituals were performed annually, and this one was not significant. The reason it was singled out for change may be that by then a consensus had emerged on the design of the Palace of Light.) The ritual officials recommended it be renamed "Praying for a Good Year" (*qi nian*). The emperor approved the change.[38]

At roughly the same period (1736–48), the Institute of the Three Ritual Classics proceeded with systematic cleansing of the Three Ritual Classics from any misleading implications of historical precedents. These actions were taken after the relevant research at the Institute of the Three Ritual Classics had been carefully deliberated. The cultural project of the Three Ritual Classics, undertaken by the Qing imperial state, should first eliminate uncertain and ambiguous interpretations of the Three Ritual Classics to avoid disruption in the transmission of political legitimacy. The Manchu emperors dictated that the historical precedent of the Jiajing emperor in the Ming dynasty should never happen again. No one would know this better than the Yongzheng emperor, who instituted the secret succession policy for his posterior. At the same time, a more positive note on the cultural project of the Three Ritual Classics should be added. The Ministry of Rites could consequently build and furnish a machinery of state rituals that would operate routinely without regular imperial approval or intervention. Therefore, the classical debates in the Institute of the Three Ritual Classics provided the raison d'être for any general or ad hoc adjustment of the state rituals: the seemingly unsavory details in our modern eyes were in fact crucial components of the ritual machinery that could keep the imperial will in check. The imperial house could rarely step outside the bounds of the ritual machinery because the price was way too high. In other words, the controversy over the Palace of Light was an indirect commentary on their contemporary politics. They were debating how much leeway should be allowed for the imperial house to maneuver within the bounds of legitimate state rituals.

THE POLICY HALL

By the end of the seventeenth century, many thought of the Palace of Light as a locus of government. According to the court diary, during the 1670s the Kangxi emperor ordered his councilors to teach him the Confucian canon; he hoped to learn how to run his empire wisely and to speak the language of the thousands of literati who staffed his bureaucracy. The imperial tutors knew precisely what was at stake and took their job seriously. One of them, Lu Longqi (1630-93), routinely promoted the thinking of Zhu Xi over his Ming critic Wang Yangming.

This passage is from *Mencius,* one of the works Lu explained to the emperor:

> King Xuan of Qi said, "I have been advised to tear down the Palace of Light. Should I do so or not?"
>
> Mencius replied, "The Palace of Light is the seat of kingly governance. If Your Majesty wishes to practice the governance of a true king, you should not tear it down."
>
> "May I learn more of the governance of a true king?"
>
> Mencius replied, "In past times, when King Wen ruled at the city of Qi, he took only one part in nine as a tax on those who tilled the land, and those who served his government inherited their stipends. At the border, goods in trade were inspected, but no fees were levied, no restrictions were placed on the use of fish traps installed near dams and weirs, and penalties for those convicted of crimes never involved their wives and children. Widows and widowers, orphans and the aged without children to support them—these classes of people without means or other people to turn to were always given priority in the proclamations through which King Wen enacted his humane [*ren*] policies."[39]

The Neo-Confucian master Zhu Xi commented on this passage: "The Palace of Light was where the king lived and enacted his policies."[40] To all appearances, the Palace of Light was merely a location suitable to good government. But Lu informed the Kangxi emperor that it was in reality something different. In Lu's view, Mencius was using the discussion of the Palace of Light to persuade the king to govern in an enlightened fashion. In the Neo-Confucian orthodoxy, "humane and enlightened policy" was a direct result of moral cultivation of the ruler. According to Lu, King Xuan of Qi must have admired the Palace of Light as the cosmological palace to summon all other vassals. To destroy the symbolic power of the Palace of Light would have significantly decreased King Xuan of Qi's chance to become the hegemon over other kings. King Xuan of Qi knew this and asked Mencius for a political strategy. Instead of advising King Xuan of Qi with a pragmatic political solution, Mencius attempted to persuade the king to rule in a true "humane and enlightened" manner by way of the Palace of Light. The location of the palace was of little concern.

The classical scholar and polemicist Mao Qiling (1623–1716) sharply disagreed with Lu Longqi. According to Mao, the location of the Palace of Light was exactly the problem. After stating, "Everything in Zhu Xi's commentary on the Four Books is wrong," the polemicist Mao Qiling enumerated 452 mistakes, including a contradiction between the historical record and the alleged function, just quoted, of the Palace of Light:

> [When the Son of Heaven] enacted his policies, he must have done it in his own palace [in Gaojing], where he got up in the morning and went to bed at night. Why then did the King of Zhou travel east to Qi [in today's Shandong province, where King Xuan of Qi reigned and the Palace of Light was supposed to be] in order to enact his policies? If the Palace of Light was designed to be [moved from place to place], a temporary court [for the Son of Heaven] when he toured his territories, then why was there no Palace of Light in the western part or the southern part of his territory?[41]

Mao challenged Zhu Xi by asking: why would a king have lived in a palace so far away from the palace where he would enact his policy? Using the ancient mode of transportation, this was virtually impossible. Mao's deconstruction of the location of the Palace of Light accorded with Yuan Mei's critique. Known for taking provocative stands, he routinely commented on a wide range of topics: funeral rites, lineage laws (*zongfa*), imperial sacrifices to Heaven and Earth, sacrifices to the imperial lineage, and ancient education. He loved a rousing debate, and his belligerent style certainly made him more enemies than friends. Yet after his death his work on the Palace of Light began to be admired and cited by literati society.[42] Mao called into question, for example, when exactly the Palace was built, what function it had, and why so many controversies had grown up around it. More importantly, why had the palace's floor plan become the focus of debate?[43] The analysis he offered in *Inquiries Regarding the Palace of Light* (Ming tang wen) made the debate over the floor plan *the* key to all of the building's other problems—political, ritual, and cosmological.

Why did the floor plan matter so much? Zhu Xi wrote a comment about the Palace of Light that would have appealed to those with an interest in the design and appearance of the building. One of his disciples indicated that there were two different descriptions of the floor plan of the Palace of Light in the classical scriptures. One described five halls (*shi*); the other said there were nine. The political and ritual implications of the two numbers differed radically. Zhu also wrote, "The Palace of Light should have nine halls, just like the nine units in the well field [*jingtian*] system."[44] In this statement, the Palace of Light figures as a political and cosmological principle rather than an actual building at a specific location. What exactly did the Neo-Confucian grandmaster imply?

THE GRAND WAY OF THE PALACE OF LIGHT

The Palace of Light encountered a radical new philosophical orientation from Hui Dong, a paragon of classical scholarship from Suzhou. Like most literati, Hui subscribed to the standard narrative of how the classical scriptures had survived to the present: the First Emperor of Qin gathered and burned most of the Six Classics, works that Confucius had labored over because of a belief that they captured the essence of the Zhou system. One hundred years later, Emperor Wu of the Han dynasty had praised Confucianism and established the Office of Learning (Xue Guan) where the surviving copies and fragments of the classics were gathered for attentive study. A small number of scholarly families dominated the office, each family specializing in a single book and developing a distinctive interpretation of it. Under this system, the classical scriptures were largely restored, and the Zhou system reconstructed from them.

Hui's radical view was that only in the Han dynasty did the classical scriptures exist in their genuine form. Post-Han commentators, however sincere their motivation, tampered with the scriptures in one way or another. Worst of all were the Song commentators (that is, the group whose interpretation of the classics had to be memorized by every civil examination candidate since the Ming), who randomly tailored the classical scriptures to suit their philosophy.[45]

Hui's study of the Palace of Light exemplified his radical new approach. His *Records of the Palace of Light and Its Grand Way* (Ming tang da dao lu) was a byproduct of his lifelong devotion to *The Book of Changes*; in *Records* he wove together virtually all of the exegeses of this seminal text that had been written in the Han dynasty. According to Hui, *The Book of Changes* was basically a catalogue of cosmological principles and operations, and the Palace of Light was an effort to symbolize those principles and operations.[46] The design of the Palace of Light was the culminating act of the Zhou, he argued, and had been recorded in the "Winter Office" section of *The Rituals of Zhou*. Tragically, this text was lost and replaced later with "Records of Ancient Technology," the final chapter of *The Rituals of Zhou* in its final form. Several ritual officials were skeptical about the authenticity of "Records"; Hui simply shoved it to one side. Then he reconstructed the Palace of Light by systematically examining all textual references to it in the Han editions of the Six Classics. Anything later he fiercely rejected.[47]

A proper understanding of the Palace of Light—its modality (*fa*), to use Hui's term—would yield an outline of the way the ancient sages ordered the world.[48] The Grand Way (*Da Dao*) discussed in *The Records of Rituals* as the constant and incessant transformation of the world, had lodged itself in the physical form of the Palace of Light. In the place of the Heavenly principles presented in Neo-Confucian ontology, Hui offered a Palace of Light refracted through his study of *The Book of Changes*, which gives the symbolic and numerological pattern of the transformation of the world.

Furthermore, by recasting the palace as a set of political rules and rituals, Hui shifted the focus away from the problems regarding its location. Yet, he still had to cope with the floor plan, which functioned essentially as the spatial embodiment of the Grand Way. Here, he took his lead from Cai's *Excursus on the Palace of Light and Monthly Observances*, which envisaged the palace as having nine halls and twelve chambers. Hui Dong had based his entire interpretation upon this physical layout. Naturally he dismissed both "Records of Ancient Technology" and Zheng's commentary on *The Rituals of Zhou*—they assigned the building only five halls.[49] The sharp difference between five halls (based upon Zheng Xuan's commentary) and nine halls (based upon Master Zhu Xi's interpretation) became increasingly difficult to reconcile as more and more debates around this deceptively simple issue of the floor plan perpetuated.

The floor plan difference is the evidence necessary to understand Dai Zhen's subtle disagreement with Hui Dong. Intellectual historians have long stressed the difference between Hui and Dai but failed to see what they actually shared. Both thinkers were convinced that the Grand Way was to be found in the material culture of the ancients, not in abstract terms proposed by the Neo-Confucians. Dai had, it will be remembered, written an impressive study of "Records of Ancient Technology" of *The Rituals of Zhou* that Hui banished. The two scholars were fated to come down on different sides of the floor plan debate. Their differences were decisively revealed at their first meeting, in 1757. In the winter of that year, Dai had left Beijing for Yangzhou. He wanted to take advantage of his rapid success in Beijing by exploiting his connections in the south. His patron in Beijing, Ji Yun, had persuaded Dai to join the group of scholars fed and housed through the generosity of Lu Jianzeng (1690–1768), a rich salt commissioner in Yangzhou.[50] Through Ji's introduction, Dai was made welcome at Lu's mansion, and there he met Hui for the first time. Since Qian had urged him to contact Hui, Dai passed along his impressions in a letter written immediately after that first meeting. The two men had discussed the Palace of Light.[51]

For both Dai and Hui, the Palace of Light was the location and symbol of political power. They disagreed on the nature of the political power as well as how it was to be symbolically demonstrated. Unlike Hui, Dai's classical knowledge drew virtually from every scholar whose work he studied, and this is especially true of his contact with seventeenth-century evidential scholarship such as the works of Gu Yanwu and Yan Ruoju. He did not privilege certain kinds of sources—Old Text or New Text, for instance. There was, for him, as much wisdom in an astronomical rhapsody by Qu Yuan (340–278 BCE) as there was in a phonological treatise on local dialects by Yang Xiong. What unified his tireless and eclectic research was his devotion to method. And there he differed from Hui.[52] That remarkable

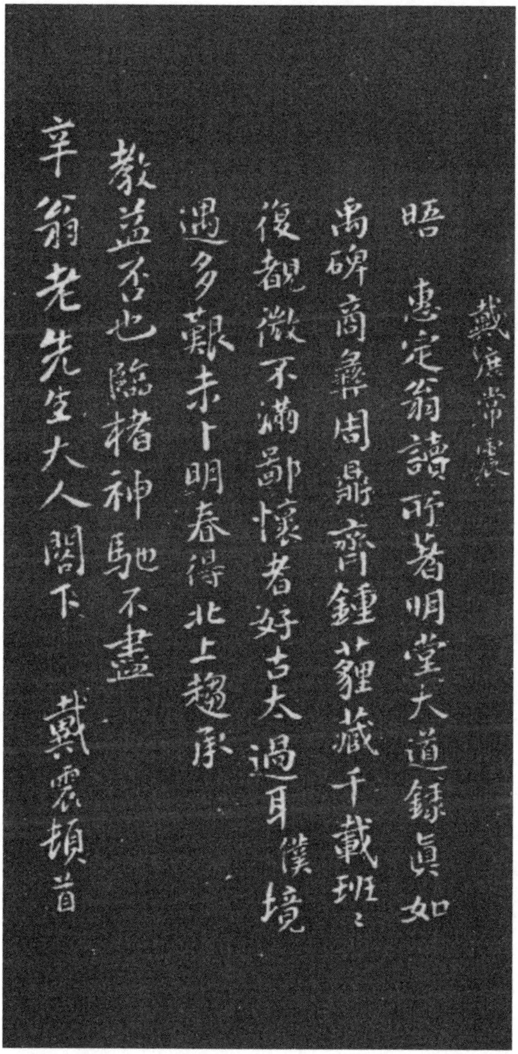

戴庶常震
晤惠定翁讀所著明堂大道錄眞如
禹碑商彝周鼎齊鍾蘿藏千載班班
復覩微不滿鄙懷者好古太過耳僕境
遇多艱未卜明春得北上趨承
教益否也臨楮神馳不盡
辛翁老先生大人閣下 戴震頓首

FIGURE 7.1. Dai Zhen's handwritten letter to Qian Daxin. From *Zhaodai mingren chidu xiaozhuan*, 31-511.

meeting in a Yangzhou mansion provides the most convincing evidence to Dai's intellectual and political orientation.

Their meeting ended not with a conversion but with a polite rejection, based on methods. Dai's exact words to his friend (see fig. 7.1) are as follows:

> I met the old gentleman, Hui Dong, and read his *Records of the Palace of Light and Its Grand Way*. He explains the Palace of Light as though

> he were revealing Xia dynasty steles, Zhou dynasty wine vessels, Zhou dynasty cauldrons, and musical bells [cast at the command] of the King of Qi during the Warring States period. It is as if, after they had been buried for thousands of years, we were now seeing every detail of those steles, vessels, cauldrons, and bells, with all their stains and spots. The only thing with which I was humbly discontented was Hui Dong's excessive pursuit of antiquity [*hao gu tai guo*].[53]

In this private communication—the only letter from Dai to Qian that has survived—Dai discreetly but keenly criticized the older scholar, using the coded expression "excessive pursuit of antiquity."[54] Whereas a mention of his "pursuit of antiquity" would have been words of praise for a classical scholar, the pungency of "excessive" (*tai guo*) is quite devastating.

Unlike Hui's embellished styles in studying the rituals of the Palace of Light, Dai had devoted much labor to the technical specifications in preparing *The Illustrated "Records of Ancient Technology,"* a work completed when he was still a young man. While Hui had nothing to say about "Records of Ancient Technology," Dai focused on it exhaustively. Hui gave equal weight to most of the exegetes of the Han dynasty, while Dai heeded only Zheng Xuan. In his preface to the published version of his study, Dai raised no doubts regarding the authenticity of the "Records"; Hui and many ritual officials at the Institute of the Three Ritual Classics took the opposite approach. For Dai, much of the appeal of the text was its obscurity; he argued that illustrations and descriptions were two sides of the same key to understanding ancient technology. The subject he had chosen had little intrinsic appeal for classical scholars with high social standing, while illustrators seldom appreciated the nuanced details of ancient technology. Thoroughly neglected for centuries, "Records" no longer existed in a form that did it justice. Nie Chongyi had produced the only surviving versions of the diagrams in *Illustrations of the Three Ritual Classics* (San li tu, 1175); according to Dai, most of them were inaccurate.

Basing his argument entirely on the controversial chapter "Records of Ancient Technology," Dai then went straight to the heart of the floor plan controversy: the Palace of Light had five halls and twelve chambers, as Zheng had said. Rejecting Cai's hypothesis also meant rejecting Zheng's challenger, Wang Su (195–296), whose re-creation of the Palace of Light had been adopted in the Tang dynasty.[55] The bulk of Dai's analysis was given over to a resuscitation of Zheng Xuan and rejection of Wang Su. Dai Zhen's distinct route of technical methodology was also a political choice. Instead of defending the authenticity of "Records of Ancient Technology," Dai redirected the issue to its technical difficulty. Once again, he advocated his technical methods as the best way to unravel the Gordian knots in classical knowledge.

MODALITY OF HEAVEN

Hui and Dai disagreed on the floor plan precisely because they agreed so much upon the significance of Grand Way embodied by the Palace of Light. Different floor plans, it seemed to them, exhibit different Grand Ways. Both scholars used exactly the same cosmological claim, alleging that its very form symbolized the "modality of Heaven" (*fa Tian* or *Tian fa*). They disagreed, however, on what the modality should be. This suggests that the two scholars shared a sense of the grand purposes of the Palace of Light. Both scholars imagined the Palace of Light without magic, charms, omens, immortals, or supernatural interventions. They each attributed specific political functions and rituals to it. The inclusion and exclusion of political functions and rituals manifested their distinct political philosophy, following the pattern of cosmological order. The symbolic embodiment of Heaven also depended on how Hui and Dai understood Heaven. After all, they were using the Grand Way of the Palace of Light to refute the ontology of Neo-Confucianism.

The Grand Way attributed to the Palace of Light was therefore an emerging new political philosophy, decisively replacing the centrality of "Heavenly principle" in Neo-Confucianism. Their debates focused on the floor plan because they agreed that what the Palace of Light embodied should be considered the most essential political functions and rituals in antiquity. Each function or ritual requires to be mapped out in specific details and proceedings from one chamber to the next. Thus the floor plan gave a specific arrangement for political functions and rituals. Instead of debating relentlessly over the correct rituals or political functions revolving around the Palace of Light, Dai cut directly to the physical structure of the floor plan to show that many previous debates were moot if the floor plan could be figured out first.

Hui Dong, on the contrary, considered the floor plan as a minor detail.

For him, the Palace of Light was nothing but a physical manifestation of profound meanings and implications in *The Book of Changes*. The inconsistency of some physical details should not curtail his pursuit of the Grand Way in ancient China. Since he aimed high and thought hard about how to refute the Neo-Confucian Heavenly principles, Hui had devoted his life studying *The Book of Changes* and followed the hallmark of the great teacher Confucius's claim by only "narrating" or "following" (*shu*) the classical texts instead of "creating" (*zuo*) his own work. In so doing, Hui placed himself in exactly the same historical aptitude in preserving the true meanings and transmission of the classical texts as Confucius.

Hui's ambition got the best of him. His publication plan, titled *Narrating The Book of Changes* (Zhouyi shu), was supposed to have forty chapters and show the true cosmological claims in *The Book of Changes*. He completed only twenty-three of them before he died in 1758. When Dai met the old gentleman in 1756, Hui was just finishing his *Records of the Palace of Light and Its Grand Way*, intended to be part of his *Narrating The Book of Changes*. Hui called the Palace of Light the "grand temple for the son of Heaven" and provided a long and elaborated analysis of all grand rituals at the Palace of Light (table 7.1).[56] Because of these embellished functions and rituals, the Palace of Light must have been a grandiose edifice that extended to four directions with an elevated square at the center to perform the dual rituals: imperial ancestral sacrifice and sacrifice to Heaven. The entire palace was designed by the first known ruler, Shennong ("divine farmer"), and based upon *The Book of Changes*, which was attributed to Fuxi, the legendary inventor of writing, fishing, and trapping animals. All legendary kings in ancient China (from the Yellow Emperor to Yao and Shun) and rulers from the three ruling houses (Xia, Shang, and Zhou) continued to perform these key functions and rituals in the Palace of Light.[57] This is why, according to Hui, it was a palace of "grand culture." If there were any "ritualism" in Qing China, then Hui Dong would definitely be its champion, and he would also argue that the grand rituals performed in the Palace of Light were the emblematic manifestation of cosmological order.

Dai Zhen rejected Hui's utopian ritualism and directly linked the Palace of Light to the actual operation of the cosmos. Elegant simplicity of the cosmos was Dai's guide to the political philosophy of interlocking individuality or intersubjectivity (*yi qing qie qing*), which was systematically

TABLE 7.1. Major rituals in Hui Dong's *Records of the Palace of Light and Its Grand Way*

禘祭 Diji	Special ancestral sacrifice designed only for the king's family, whose origin could be traced back to Heaven (divine source of political legitimacy).
宗祀 Zongsi	Ancestral sacrifice for aristocrats.
朝覲 Chaojin	Elaborate ceremonies and etiquettes for aristocrats to perform when they have an audience with the king.
耕籍 Gengji	Annual ritual performed by the king at the very beginning of the spring to initiate the year's growing season, similar to the ceremonial first pitch in a baseball game.
養老 Yanglao	Ceremonial feast for the elders in the realm, hosted by the king.
尊賢 Zunxian	Ceremonial feast for the virtuous in the realm, hosted by the king.
鄉射 Xiangshe	Archery competition ceremony, the king awarded the winner with a drink.
獻俘 Xianfu	Ritual in which victorious army commanders presented captives to the king.
望氣 Wangqi	Observing the operation of *qi* in the cosmos to tell auspicious from inauspicious events.
告朔 Gaoshuo	Annual ritual at the end of each year to announce the first day (*shuo*) of the next year.
行政 Xingzheng	Managing state affairs.

laid out in *An Evidential Analysis of the Meaning of Terms in the Mencius* in 1776. Dai's political philosophy resulted from his lifelong devotion to develop a detailed depiction of the cosmos, called *xiang* (literally "imagery," a word taken directly from *The Book of Changes*), which, in Dai's careful elaboration in mathematical astronomy and cosmology, was to be rendered as a mathematical modeling of the cosmos. When Dai began working in the imperial library project, he already furnished his political philosophy based upon the mathematical cosmology he called the "true mathematical model of the cosmos" (*yuanxiang*). It was this cosmic model that could become fully accountable in the language of spherical

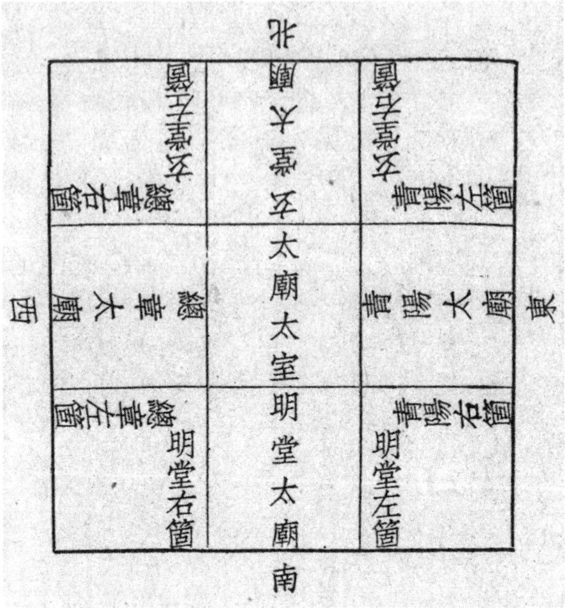

FIGURE 7.2. Zhu Xi's floor plan for the Palace of Light. From Zhu xi, *Zhuzi daquan*, 1218.

trigonometry with predictability and regularity. Dai placed the Palace of Light—allegorically depicted as terrestrial square (*difang*)—in the center of the cosmos, or celestial sphere (*tianyuan*).

The floor plan of the Palace of Light resembled how the cosmic model works. Dai began by separating the palace floor plan into two design elements. We see a Grand Hall in the center and an in-between hall (*jia shi*) at each of the four compass points. The first and relatively uncontroversial element was the cross in the middle (consistent with Zhu Xi's model). The second design element—the undesignated corner blocks—was controversial. Dai Zhen designated them to the "twelve chambers," which were frequently referenced in the relevant literature on the Palace of Light.

For example, Cai claimed that the Palace of Light had nine halls and twelve chambers. Both Zhu Xi and Hui Dong addressed this by making each of the nine blocks a hall, but they could not account for the chambers. Dai saw each of the corners as an in-between hall flanked by two chambers (front and back), giving a total of eight chambers in the corners. As a result, the total number of halls (*shi*) was five: one large one in the center and four in-between halls in four diagonal directions. As for the chambers, there were four large ones (two of which are *tang*, here also translated as "hall" but different from the five main *shi*) and eight lesser ones. The front of each corner chamber was also called a "chest" (*xiang*) or "box" (*ge*). Dai believed that the chambers of ancient palaces usually contained chests, boxes, and smaller rooms (*fang*), which were usually for women. (The Hall of Light in the south, he added, had no rooms because it was designed for ritual purposes only. Women were not allowed to take part in the state rituals.)[58]

So the floor plan of the Palace of Light would look similar to figure 7.3. Once he had figured out the arrangement of the five halls (*wushi*) and twelve chambers (*shi'ertang*), Dai briefly commented on the building's evolution. A long, long time ago, he said, there was a ruler who built a Palace of Light. It gradually evolved from a "Hall of Multiple Generations" (Shi shi) in the Xia to a "Multistoried House" (Chong wu) in the Shang and finally the Palace of Light (Ming tang) in the Zhou. There the Five Emperors were worshiped, the heavenly bodies were observed, and the vassals gathered for important events of state. He left out the palace's magical aspects, as well as its sacrificial functions. His scholarly predecessors, he concluded, had failed to distinguish halls from in-between halls and chambers from corner chambers. It was their neglect of this technical dimension, he explained, that had produced so much confusion in the other areas.[59]

What was its cosmological and political implication? Dai immediately gave a long list, as dictated in the ancient cosmology and ritual temporality in *Monthly Observances* (Yue Ling). With the arrangement of five halls and twelve chambers, each chamber was a room where the Son of the Heaven was to reside in each specific month (table 7.2).[60]

Dai had called out the names of each chamber, corresponding to each month of the year. Why? Perhaps Qian Daxin could give us a clue. In Qian's funeral biography for Dai, he also copied the entire list of correspondence to show its significance. Qian noted the opening sentence of Dai's treatise: "The Palace of Light followed the modality of Heaven. It had five halls and twelve chambers. Therefore [the modality of Heaven] was called the Palace of Light and Monthly Observances."[61] Dai opened his treatise with a direct, provocative, and almost encrypted way of stating the significance of the Palace of Light. Most modern historians had missed it, but Qian Daxin had not. Dai Zhen was fully aware of his own contribution—identification of the floor plan as the missing link between

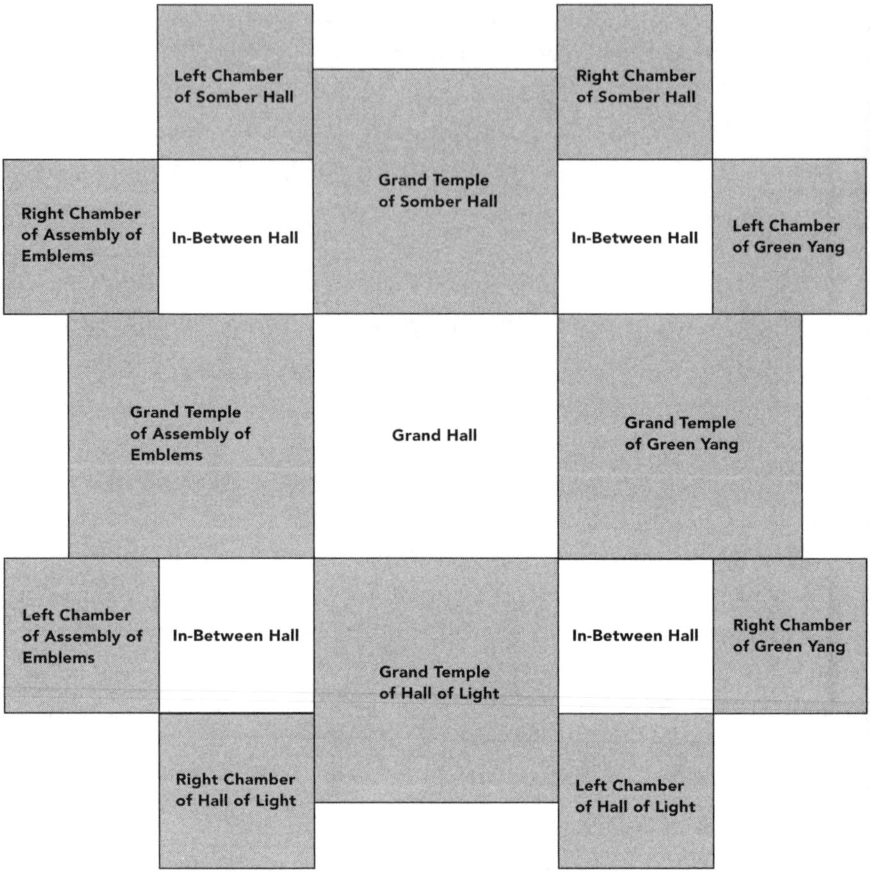

FIGURE 7.3. Dai Zhen's floor plan for the Palace of Light, based on the illustration in Dai Zhen's *Kaogongji tu*. The floor plan is supposed to sit on a ceremonial and elevated platform, similar to the pre-Columbian Maya temple platform at Palenque, with stairs on each side. Each side has a general chamber in the middle facing outward and separating the adjacent left and right chambers. All twelve chambers correspond to twelve lunar months in the classical text *Monthly Observances* (see table 7.2). For Dai Zhen, the Palace of Light was a grandiose architecture designed for specific cosmological rituals.

the Palace and *Monthly Observances*. Dai further explained why this missing link was not discovered before: "Most classical scholars who studied the Palace of Light focused only on five halls but not on chambers. For those who focused on chambers, they could not distinguish box [*ge*] from room [*jia*] in a chamber. That is why the real floor plan had been lost

TABLE 7.2. Correspondence between cosmic rhythm in *Monthly Observances* and the rotation from chamber to chamber in the Palace of Light

	Lunar Months	Chambers in the Palace of Light
1	First month (孟春)	Left Chamber of Green Yang (青陽左個)
2	Second month (仲春)	Grand Temple of Green Yang (青陽大廟)
3	Third month (季春)	Right Chamber of Green Yang (青陽右個)
4	Fourth month (孟夏)	Left Chamber of Hall of Light (明堂左個)
5	Fifth month (仲夏)	Grand Temple of Hall of Light (明堂大廟)
6	Sixth month (季夏)	Right Chamber of Hall of Light (明堂右個)
7	Seventh month (孟秋)	Left Chamber of Assembly of Emblems (總章左個)
8	Eighth month (仲秋)	Grand Temple of Assembly of Emblems (總章大廟)
9	Ninth month (季秋)	Right Chamber of Assembly of Emblems (總章右個)
10	Tenth month (孟冬)	Left Chamber of Somber Hall (玄堂左個)
11	Eleventh month (仲冬)	Grand Temple of Somber Hall (玄堂大廟)
12	Twelfth month (季冬)	Right Chamber of Somber Hall (玄堂右個)

for so long."[62] The way that Dai solved the puzzle of the Palace of Light may sound similar to how Sherlock Holmes solved his detective cases or to modern forensic science. A minor, easily overlooked detail served as the key to unravel the entire plot. "From a drop of water," Holmes once declared, "a logician could infer the possibility of an Atlantic or a Niagara without having seen or heard of one or the other."[63] Dai Zhen was not quite Sherlock Holmes yet, but it seems to me that Dai was claiming a very similar method.

Dai took the technical dimension of ancient artifacts seriously. This was precisely what he said in a letter to Shi Jing in 1750.[64] He had challenged the Jesuits' claim to superiority in mathematical astronomy and changed the terms of intellectual engagement for virtually all members of the national elite. At the same time, everyone disputed the grand purposes and multiple functions of the Palace of Light and regarded the floor plan as a trivial issue. Using deductive logic to examine the pieces of textual and sometimes nontextual evidence, Dai shifted the focus of the discussion. His interests took him to the technical details of the floor plan, following the descriptions of craftsmen to figure out the actual scale of Zhou architecture. While Hui ambitiously fit every single piece of the Palace of Light puzzle (so long as it did not bear a date after 220 CE) into an elaborate, sophisticated, and extended system of political functions and rituals he called the Grand Way, Dai applied a more straightforward and technical approach. More importantly, Dai Zhen never strayed from the path he set out to study mathematical cosmology and astronomy in ancient China. The interconnectedness of the Palace of Light and *Monthly Observances* was a clear demonstration of Dai's commitment to why the cosmos matters to his political philosophy. *Monthly Observances* was one of the ancient texts Dai took seriously to rebuild the cosmos. Dai had studied mathematical astronomy since his teenage years and never ceased incorporating his mathematical, astronomical, and cosmological knowledge into his classical research until very end of his life. Shortly before his death in 1777 while Dai was working in the imperial library project, he had an interesting exchange with his colleague Ding Jie (1738-1807) on a Song dynasty text, *Yi xuanxing jipu*, by Zhao Shuozhi (1059-1129). In their exchange, Dai continued to expound the significance of spherical trigonometry, mathematical cosmology, and astronomy.[65]

Finally, the floor plan issue separated Dai Zhen from Hui Dong, but the impact on their followers and admirers was mixed. Historians have previously presumed a fault line between the Hui Dong circle, Hui's friends and students, and the Ruan Yuan circle, mostly Dai's followers and admirers. The floor plan issue, however, cut across the presumed fault line. For instance, Wang Zhong (1745–94) and Jiang Fan (1761-1831), who were generally regarded as Hui's followers and admirers, differed quite drastically from each other on this issue.[66] Jiang Fan, a disciple of Hui Dong's, confronted his master with a detailed evidential analysis of the floor plan that supported Dai's conclusion.[67] Wang Zhong, on the contrary, used Dai Zhen's systematic connection from the Palace of Light to *Monthly Observances* to dismiss *Master Lü's Spring and Autumn Annals*.[68] Ruan Yuan extended Dai's work in a study of the Zhou palace complex.[69] Sporadic research continued but no one could match the paradigmatic influences pioneered by either Dai Zhen or Hui Dong. Few evidential scholars blindly followed their research as the last words, either. Over the course of the nineteenth century, the topic drew fewer scholars. From time to time, the imperial court invoked the Palace of Light to justify reforms of court rituals. The minister of rites, Li Hongzhang (1823–1901), for example, was asked to settle a ritual dispute over the reconfiguration of the Ancestral Temple in 1877.[70] Concerns about social unrest and the mismanagement of the imperial state gradually replaced debates about ancient institutions in the nineteenth century.

MODERN HINDSIGHT

Dai Zhen's intervention, as important as it was, did not put an end to the many controversies surrounding the Palace of Light. "Among those controversial matters in the study of ancient institutions, nothing has been more unsettled than the debates regarding the Palace of Light," proclaimed one of the most learned classical scholars, Wang Guowei (1877–1927), during the transition from imperial to republican China. With his famous essay "A Comprehensive Investigation of the Palace of Light and the Imperial Court" (Ming tang miao qin tongkao), Wang intended to lay the debate to rest once and for all.[71] Wang could finally draw a conclusion on these long-lasting debates because Qing scholars had exhausted all angles on all sources regarding the Palace of Light. In this sense Wang summed up his intellectual legacy on this issue and critically examined it from a postimperial vantage point.

One could also argue that Wang analyzed the Palace of Light from a modern perspective. It was modern in the sense that Wang experienced the collapse of the Qing dynasty and was compelled to evaluate the intellectual legacy of his Qing predecessors in light of other contemporary Western theories. Wang conducted a review of Qing literature on the Palace of Light and reached his own conclusion. Historian Wang Fansen once remarked on Wang Guowei's similar treatise on ancient institutions of Yin and Zhou (*Yinzhou zhidu lun*): "Under his Qing scholar's hat and long queue, Wang's brain was in fact filled with new and exceptional ideas. He unraveled the intellectual tradition of Yin and Zhou institutions from a geographical perspective and mapped them out in order to enable multiple interpretations of the ancient institutions."[72] A similar remark can be made about Wang's "A Comprehensive Investigation of the Palace of Light and the Imperial Court." With his refreshing perspective,

Wang selectively disregarded Dai Zhen's contribution. In the course of evaluating Dai's contribution to the debate, Wang examined his *Commentary on the Water Classics* (Shui jing zhu), concluding that the study had included portions plagiarized from Zhao Yiqing (1711–64). On the basis of this charge, Wang launched a full-blown attack on Dai's academic integrity.[73] He appears to have decided that his work on the Palace of Light was trivial. Ironically, Wang's own discussion of the floor plan was remarkably similar to Dai's.[74]

New scientific research also confirmed Dai Zhen's old findings. The architectural archaeologist Yang Hongxun is an authority on the construction of ancient Chinese buildings and palaces. He has singled out "Records of Ancient Technology" as not only reliable, but the critical source for reconstructing the Palace of Light as it existed in the Zhou. Even more remarkable is that Yang has essentially proven that Dai was right about the palace's corner chambers (fig 7.4).[75]

The hindsight here is not intended to demonstrate that the Qing scholars' reconstruction of the Palace of Light was the predecessor of contemporary architectural archaeology. What it shows is the differentiation and specialization of academic disciplines as well as their new impartiality in early modern China.[76] These new disciplines, such as historical geography, analytical phonology, historical etymology, and mathematical astronomy, emerged as a specialized methodology for classical research and gradually gained autonomy as academic fields of inquiry. Their common goal was to establish the objective reality of the ancient institutions. Historian and critic Wang Hui calls it the "new cultural discourse of ancient institutions" (*xin liyue lun*) and argues that the new cultural discourse was inevitably in essential tension with the existing political hierarchy and organizations of the Qing state. He also argues that the new cultural discourse manifested a certain process of disenchantment. What Wang Hui calls disenchantment in the new cultural discourse had at least two meanings. The classical scholars of the eighteenth century, such as Fang Bao, Jiang Fan, and Dai Zhen, all avoided further discussion or elaboration on prognostication, strange occurrences, and magical discourses in ancient records. But more importantly, classical knowledge had become alienated from the scholar's immediate concerns in statecraft or self-cultivation and had turned into a study of impersonal and isolated facts in the classical world. This disenchanted move in classical

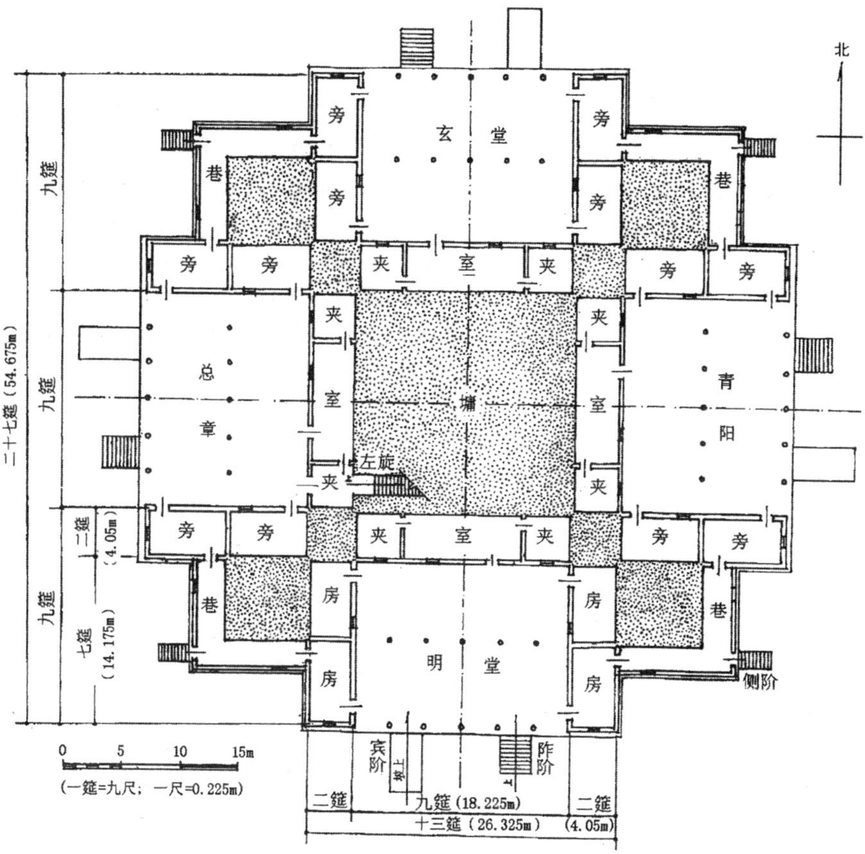

FIGURE 7.4. Yang Hongxun's floor plan of the Palace of Light. Yang was a disciple and assistant of Liang Sicheng (1901–72), Liang Qichao's son and arguably the best and most influential architectural historian of his generation in China. It was not coincidental that Yang would continue unraveling the issues related to the Palace of Light, to which he devoted much research, becoming the leading authority on its archaeological reconstruction and history. From Yang Hongxun, *Gongdian kaogu tonglun*, 110.

knowledge was unprecedented and simultaneously uneasy with the existing political and social hierarchy.[77]

Dai Zhen anticipated the archaeological reconstructions of the turn of the twenty-first century. Although Wang Guowei despised Dai, he came up with the same configuration as Dai for the corner chambers in the Palace of Light. Dai's *polis*, like his *cosmos*, was orderly, mathematical, and predictable. The ancient "cosmopolis," first envisioned by Dai and more fully developed by his followers, was an important intellectual construct in eighteenth-century China, but modern intellectuals and classical scholars alike have dismissed it. Thus contemporary scientific archaeologists, such as Yang, are unaware that the reconstruction of corner chambers had already been foreseen, and historians of Chinese science and technology who seek to compare China to ancient Greece and medieval Europe often neglect to look at the past through the eyes of its inhabitants. The Palace of Light is an intriguing reminder of the limits of hindsight.

CHAPTER 8

LEGIBILITY OF VISIONARY SCHOLARS

A distinction is made in this book between two visions: the pragmatic vision of the imperial state and Dai Zhen's vision of the perfect cosmological and political order of the classical world.

Contemporaneous with the Islamic dynasties (Ottoman, Safavid, and Mughal), Russian Empire, European monarchies (Britain, Bourbon, and Habsburgs), colonies in the Americas, and the African slave trade, High Qing China has recently been recast as an early modern Eurasian empire, coexisting with numerous points of origin of the modern world and paving its way for the emergence of the modern Chinese state. What inspired and transformed the political elites in the Qing? Did they share similar ideological, political, or military features that made them early modern? What kinds of sacrosanct codes constrained them? How did they interpret or even manipulate these sacrosanct codes to facilitate institutional change?[1]

After the Kangxi emperor's consolidation of the Qing state's control over China, a set of early modern conditions resumed courses that had started in the sixteenth century: population growth, commercial expansion, urbanization and the rise of urban commercial class, increasing contact and exchange among religious doctrines, reconstruction and unrest of local society, and the decline of nomadic population.[2] In addition to these early modern conditions, the Qing empire doubled its territory and population over the course of the eighteenth century. The elite Chinese scholars may not have understood those conditions as we term them. They were, however, aware of many new issues that required pragmatic solutions in a diverse and expanding population.

The ambitious Kangxi emperor initiated the compilation of *The Comprehensive Gazetteers of the Great Qing Dynasty* (Da Qing yitongzhi), numerous projects on mathematical astronomy, historical cartography, analytical phonology and rhyme books, *The Book of Changes, Ritual Classics*, the Four Books, and many others.[3] In *Seeing Like a State,* James Scott noted that "certain forms of knowledge and control require a narrowing of vision. . . . The great advantage of such tunnel vision is that it brings into sharp focus certain limited aspects of an otherwise far more complex and unwieldy reality."[4] As the elite Chinese scholar-bureaucrats who ran the largest empire of the day collected information from Central Asia, Southeast Asia, and Western Europe, they gradually developed a global yet incomplete view of the world that bore traces of contacts with peoples near and far. They also developed new intellectual tools. Incorporating the methods of mathematical astronomy, historical geography, analytical phonology, and historical etymology, these learned officials assembled a new array of techniques to handle early modern issues and conditions. Despite the expansion of their geographical and astronomical knowledge, the Qing exclusively focused on the narrow vision of technical utility and political control, expressed in the language of Neo-Confucian orthodoxy. The tunnel vision of the Qing state also attempted to limit any potential subversion of the expanding Qing state.

The second vision, cohesively initiated by Dai Zhen and subsequently elaborated by members of Ruan Yuan's circle, was derived from the rejuvenated debates on political principles and technical knowledge.[5] These debates were usually driven by a multivalent and complex discourse called *li* (ritual, order, norm and custom), which was supposed to offer principles and guidance for Chinese politics, regulations, rituals, legal systems, and social order. This discourse about *li* was anchored in the long commentarial tradition of classical knowledge and was manifested in the imperial orthodoxy and political ideology of the Qing Empire. Moreover, the single guiding system of *li* made sense only if it was derived from the patterns and principles of Heaven and Earth. In Qing China, it drew a considerably large amount of cosmological and geographical knowledge—specialized and technical knowledge in mathematical astronomy, historical geography, and material cultures—into the discussion of *li*. How and why did the single guiding system of *li* flourish particularly after 1736—the year when the Qianlong emperor ascended to the throne and the Institute of the Three Ritual Classics was initiated at court?

The vibrant debates over the multivalent and complex *li* discourse were infused with a body of unexpected and technical materials from the European missionaries. Though Jesuit missionaries were kept on a short leash, Chinese scholars borrowed freely from foreign technical offerings. How did they think and talk about such acquisitions? Were those who embraced the Jesuit legacy wont to adopt a specific political posture? To what did they link the foreign ideas? The technical know-how that appears in the Three Ritual Classics fit the bill, as did some pieces in *The Book of Documents*, *The Book of Songs*, and *The Book of Changes*. Historian Kai-wing Chow wrote of "the profound commitment of the Confucian literati to a *specific* vision of Confucian society—a vision of a new social order based on pure Confucian rituals and doctrines. . . . The articulated vision and the practice of cultural reform were inextricably intertwined. . . . The discourse on ethics, Classical learning, and lineage . . . constitute a major part of the reform movement."[6] The "rise of Confucian ritualism," associated strongly with increasing lineage solidarity of local society, was dated to the seventeenth and eighteenth centuries. In a rather different study of the same period, Chang So-an has painted a picture of great intellectual change and turmoil driven by the acquisition of new technical know-how. According to Chang, classical scholars of the eighteenth century fashioned much of their social world from mundane details painstakingly recovered from the classical era. Her many detailed case studies show Qing scholars attempting to order their lives by ritual studies.[7]

Technical know-how not only served as the template permitting classical scholars to insert the Jesuit sciences into their understanding of the universe but also became an important component in deciphering ancient texts. The frequent debates that broke out over seemingly trivial details from the classical scriptures concerned, more often than not, imperial politics. The intellectual orientation of the seventeenth and eighteenth centuries has been characterized in this book as a pursuit of how to build a vision of external order based upon ancient exemplars instead of how to cultivate internal transcendence to accomplish moral perfection, with the transition to Manchu rule leading many to imagine various possibilities for a new political and social order.[8] In particular, the classical vision in Dai's formulation had an affinity for mapping the totality of the ancient "cosmopolis" systematically. These visionary scholars were certainly not revolutionaries. They were a new breed of Confucians who sought to explain, and hoped to sustain, the integration of imperial authority with provincial and local societies. They came from and represented the provincial and local interests, and hoped to increase their participation in imperial politics without disrupting the political hierarchy. Their classical knowledge could potentially become more legible to the imperial orthodoxy.

The success of Dai Zhen's new classical vision lies in how it was partially selected into the Qing state's tunnel vision. The classical vision itself took a life of its own in the course of intellectual history. But some of it was also absorbed into the Qing's political ideology. To further demonstrate how new classical vision partially merged with the Qing's political ideology from 1750 to 1800 and beyond, a brief survey of the development of specialized and technical knowledge embedded in the Confucian canon will help. The tools or skills—what we may call science and technology today—circled around the understanding of heavenly patterns, earthly geography, and human behavior, and were translated into the classical language in mathematics for calendar calculation, hydraulic engineering and project management, historical geography and map-making, as well as intricacies of ritual bearing.[9]

The Qing state selectively adopted the new classical vision to substantiate three main sources of political-cum-scientific legitimation: cosmological, geographical, and social. Each of these sources demonstrates the pressing concerns of the then expanding Qing empire. Each of them also demonstrates how different branches of classical knowledge were mobilized and certain aspects of the classical world were invented to bolster the political culture of early modern China. The Qing defined and discussed knowledge and action in terms of heaven (*tian*), earth (*di*), and men (*ren*). This tripartite structure relates to three knowledge fields, corresponding to the three sources of legitimation: (1) ethical decision making as well as mathematical and cosmological speculation; (2) topographical formation of the Qing empire's territory and its manipulation; (3) social order, rites (*li*), and the expression of the social order in clothing rites, which had an important material and materiality aspect. The comprehensive analysis of classical knowledge (*tongjing*) in the new classical vision therefore implied a direct applicability to how the Qing ordered its empire.

Put differently, Dai Zhen and other visionary scholars supplied a wide variety of abundant and legible classical knowledge for the Qing state. Three text corpora, each designed to show how the new classical vision became legible to the Qing, will be briefly assessed to conclude this book. Each text corpus is selected to address cosmological, geographical, and social sources of legitimation in the nexus between political thinking and scientific knowledge. Together they constitute what some sociologists call ideological power in preindustrial societies. The ideological power offers a *transcendent* vision of social authority. It unites the political elites by claiming that they possess ultimately meaningful, often divinely granted, common qualities. The ideological power also offers the political group *immanence*—strengthening the normative solidarity by granting the group a sense of ultimate significance and meaning in the cosmos—achieved by giving the group common ritual, knowledge, and aesthetic practices that distinguished them from others.[10] Dai and his followers provided not only the transcendent inspirations for the pragmatics of early modern China but also the critical immanence for the science faction to operate in the Qing bureaucracy.

MATHEMATICAL COSMOLOGY

The Gnomon of the Zhou Polity: An Arithmetical Classic (Zhoubi suanjing) was the most revealing ancient text on the relationship between political ideology, mathematical astronomy, and classical knowledge. *The Gnomon of the Zhou Polity* is a collection of texts assembled in the first century BCE, most of which address how to compute the dimensions of the cosmos from the shadow cast by a single vertical pole known as a gnomon. While *The Gnomon of the Zhou Polity* was overlooked in Ming China, by the eighteenth century it stood at the center of an intellectual movement dedicated to retrieving missing cosmology from the Confucian canon. By looking at how it was rediscovered, read, and used for specific political purposes in the eighteenth century, it becomes clear how important it was to the cosmological justification of the Qing empire.

The Gnomon of the Zhou Polity, an ancient work in ten sections, was first printed in 1084 as part of *Ten Volumes of Mathematical Classics* (Suanjing shishu), compiled for the newly established Institute of Mathematics (Suanxue Guan) in the court of Emperor Shenzong (r. 1067–85) of the Northern Song (960–1127). No single author drew up this work; it probably evolved gradually, as many hands wrote and revised various parts. Most later editions were variants of an important edition published in 1213. The book's contents stabilized after the thirteenth century, when the ten segments were shuffled into two chapters.[11]

The imperial Song publication is now lost. What has survived is a later edition produced by Bao Huanzhi, a literatus and mathematical astronomer of the Southern Song (1127–1279). In 1200, not long after the Mongol conquest of northern China, Bao obtained a copy of *Ten Volumes of Mathematical Classics* in Hangzhou, the capital of the Southern Song. He republished the collection in 1213 after assuming an official post in Dingzhou, located in the western part of Fujian province on China's southeast coast. Only two versions of *Ten Volumes of Mathematical Classics*, independent from each other, survived to seventeenth-century China. The first version was a fragment of the Bao edition: six out of the titular ten works survived. The second version was a copy of *Ten Volumes of Mathematical Classics*, preserved at the imperial court in the early Ming dynasty (1368–1644). It is unlikely that this edition had the same genealogy as Bao editions and their offspring. The Ming imperial edition was divided into many subsections that were subsequently rearranged as a collection of rhymed entries in *The Great Compendium of the Yongle Era* (Yongle dadian), a huge encyclopedia compiled under the direction of the Yongle emperor (r. 1403–25). The fragmentation of *Ten Volumes of Mathematical Classics* made it implausible even for literati in the Ming court to access and study *The Gnomon of the Zhou Polity*.[12]

Only a few scholars could access the Bao edition in the Ming dynasty. A prominent publisher and connoisseur of rare books named Mao Jin (1599–1659) amassed a large private library housed in his Pavilion Inspired from Antiquity (Jiguge). Mao published some six hundred titles, including complete sets of *Thirteen Classics, Seventeen Dynastic Histories,* and *Rare Books of Qian Qianyi* (Jindai bishu).[13] These publications circulated widely in China proper.[14] Mao Jin's son, Mao Yi (1640-?), inherited his father's library. He made facsimiles of prized Song editions by tracing every feature of the rare books he borrowed from other collectors. While the painstaking technique was used to produce single copies for sale as collectables, a facsimile of a work with popular commercial potential sometimes served as the basis for woodblocks from which many copies were made.[15] This process, known as "tracing Song manuscripts" (*ying Song chao*), preserved the exact features of many Yuan and Song dynasty works that have since been lost. In 1684 Mao Yi borrowed a copy of the six-volume Bao edition from the book collector Huang Yuji (1629–91), whose private library in Nanjing was called the Hall of Vastness (Qianqing Tang). Mao hired artisans to make a number of facsimiles of Huang's book; the result was his own Jiguge version of *Ten Volumes of Mathematical Classics*, published in 1684. The Jiguge version of *The Gnomon of the Zhou Polity* was not only hand-copied but also printed with limited circulation. Both the hand-copied and printed versions survive today.

The printed Jiguge version of *The Gnomon of the Zhou Polity* includes a brief yet telling introduction (*tici*) by the book connoisseur Hu Zhenheng (1569–1645), who passed the provincial examination in 1597 and became a magistrate in Hefei in the late Ming. Hu recalled that on his first encounter with this ancient treatise, he felt intimidated by its difficulty and strangeness. But a second look revealed to him that it was comprised of nothing but arithmetical procedures for determining the dimensions of the cosmos. Hu noted the discrepancy between the dimensions of the cosmos recorded in *The Gnomon of the Zhou Polity* and those measured by the mathematical astronomer Li Chunfeng (602-70). Ancient scholars like Cai Yong (132-192) had already disputed the accuracy of the dimensions of the cosmos in *The Gnomon of the Zhou Polity*. The cosmography in *The Gnomon of the Zhou Polity* describes the Heaven as like the large umbrella-like canopy over an ancient Chinese chariot, which symbolizes the Earth. Cai Yong suggested that the canopy heaven cosmography was responsible for the mismeasures of the cosmos. The rival cosmography was the spherical heaven cosmography, which describes the Heaven as a chicken egg. Wang Chong (27-97?), for example, supported the canopy heaven cosmography and refuted the spherical heaven cosmography. Huan Tan (23-50), on the contrary, supported the spherical heaven cosmography and refuted the canopy heaven cosmography. Yang Xiong (53 BCE-18) famously crafted "eight challenges" (*banan*) to the canopy heaven cosmography and the debate continued. Hu not only mentioned that ancient cosmographical controversies remained unsettled but also pointed his readers to the fact that even the title *The Gnomon of the Zhou Polity* was ambiguous. Hu's contemporary, Shen Shilong (*juren* 1597), told him that the word *zhou* in the title referred to the ancient Zhou house (1046-256 BCE). Hu, however, found that Yu Xi (281-356) glossed *zhou* as the perimeter of the canopy heaven. Therefore, the title, contents, and measures in *The Gnomon of the Zhou Polity* were all called into question in Hu's introduction.[16]

By the beginning of the seventeenth century, the circulation of rare and technical books like *The Gnomon of the Zhou Polity* was quite limited, and only high-minded book collectors like the Maos and the Huangs indulged in this costly hobby. They amassed so many volumes that it was out of the question to study them closely. Sometimes they would allow friends to look through their collections, lending items for the sake of authentication or promotion. It is safe to assume that many scholars shared Hu Zhenheng's skepticism toward *The Gnomon of the Zhou Polity* and did not find commentary in the Bao edition helpful in picturing the cosmos and measuring its dimensions. By the eighteenth century, scholarly fashions had changed, and a number of scholars in different parts of the country had devoted considerable efforts to mathematical astronomy. Many viewed it as the key to unlocking the puzzles of ancient cosmology. Dai Zhen was the pivotal figure in this process.

Dai revamped the entire cosmology and cosmography in *The Gnomon of the Zhou* based upon the spherical trigonometry and Tychonic cosmography of the second half of the eighteenth century. He basically created a new cosmos in the name of *The Gnomon of the Zhou*. His mathematical cosmology, however, failed to disseminate widely in the literati society of the Lower Yangzi Valley. The following anecdote will illustrate the limited influence of his cosmology: a young man named Hu Yichang (1743–83) from the southern province of Guangdong had just passed the provincial examination in 1771 and arrived in Beijing in 1772 to take the metropolitan examination for the first time. Dai Zhen made his fourth attempt in the same year. They both failed to pass the highest hurdle and shared a boat on their way south along the Grand Canal. Dai Zhen was then forty-eight years old and desperately attempting to pass the metropolitan examination in order to secure the status that should have come with his intellectual prowess and fame. During their travels, Dai managed to impress young Hu, who subsequently became an admirer of Dai's work and hand-copied Dai's manuscripts—including, probably, *The Gnomon of the Zhou Polity*—with the promise he would publish them later. Unfortunately Hu died in Nanjing in the following year and never published Dai's manuscript.[17]

Meanwhile, for the imperial state it was an entirely different story. In the summer of 2011, I visited the Palace Museum's rare book room and admired the refined artisanship of the facsimile of the Bao edition. To my surprise, I discovered that on the book's first page someone had written a brief remark, a date, and left the mark of a famous seal. Here is what the Qianlong emperor himself wrote in the summer of 1783: "My grandfather [the Kangxi emperor] possessed a good understanding of trigonometry. Unfortunately I have not studied that subject. By the time I was a child, trigonometry had already become essential to our mathematical astronomy. [When we] investigate ancient texts, however, we may not be able to avoid mistakes in our interpretation of them. . . . It may be easy to observe the sky and study trigonometry, but now I am too old to learn such things. Whenever I have reflected on this, remembering my grandfather, I have always felt ashamed."[18] Here the Qianlong emperor rehearsed a widely held view on *The Gnomon of the Zhou Polity* and reduced it into a simplistic and narrow vision. This vision, dreamed up by the Kangxi emperor and his advisors and then circulated widely among members of the political elites, has it that a series of secret messages concealed in *The Gnomon of the Zhou Polity* were revealed by the Tychonic cosmography Jesuits imported to Qing China.

Dai Zhen's reinvention of *The Gnomon of the Zhou Polity* provided the most systematic and succinct expression of the new mathematical cosmology. How and why it was incorporated into the imperial ideology with such an astonishing expediency and efficacy should be viewed in contrast with how it largely failed to impress other classical scholars of his times. The official recognition of Dai's mathematical cosmology, especially articulated in *The Annotated Catalogue of "The Complete Works of the Four Treasuries"* and appreciated by many around 1800, boosted Dai's posthumous fame and increased the circulation of his mathematical cosmology. It was often the legibility of these visionary scholars that brought them official recognition and, in turn, wide dissemination and debates about their works in the literati society. The inherent transcendent and immanent forces of their classical vision served as a competing alternative to the Neo-Confucian orthodoxy. In the eyes of Dai's critics and admirers, it was precisely the legibility of Dai's classical vision that could be conceived as a convincing alternative for the imperial ideology. On the same grounds, Fang Dongshu (1772-1851) later identified Dai's vision as a threat to the Neo-Confucian orthodoxy and attempted to deconstruct its coherence.[19]

HISTORICAL GEOGRAPHY

The so-called three rivers (san jiang) in *The Book of Documents* had long been a conundrum for policy makers, scholars, and hydraulic engineers.[20] The three rivers were believed to be around the Lake Tai region in the Lower Yangzi Valley. Historical geography and precedents in hydraulic histories mattered in their policy debates. These policy makers and hydraulic engineers, sharing similar concerns with statecraft thinkers and evidential scholars of their time, were keenly aware of the political and economic problems of the Lake Tai region while devising their engineering solutions.

In the wake of the collapse of the Ming dynasty and the subsequent duration of the Manchu conquest in the seventeenth century, Gu Yanwu (1613-82) and Gu Zuyu (1631-92) each traveled widely in China proper and began to verify historical records with their observations in each locality. Gu Yanwu published his results and reflections in *Advantages and Disadvantages of the Provinces and Prefectures of the Empire* (Tianxia junguo libing shu), in which he particularly emphasized the effect of topography on political and economic development in imperial China. His study of historical and natural geography and local conditions throughout China proper was frequently cited and widely read. Like Gu Yanwu, Gu Zuyu was a less profound thinker but studiously verified historical names of more than 30,000 places. His empirical contribution was duly noted, and he was later recruited to work on *The Comprehensive Gazetteers of the Great Qing Dynasty*. Like Gu Yanwu and Gu Zuyu, empirically minded scholars also traveled to verify the cultural and physical geography recorded in local sources. As a result, Gu Yanwu and Gu Zuyu were retroactively regarded as the founders of a legitimate branch of learning called historical geography.[21]

The imperial Qing state had mapped its territory differently. Hu Wei (1633–1714), for instance, joined the official team to compile *The Comprehensive Gazetteers of the Great Qing Dynasty* in 1690. In addition to his contribution to *The Comprehensive Gazetteers*, he published his forty-six-chapter magna opus *My Narrow Vision of the Imperial Territory* (Yugong zuizhi) with abundant illustrative maps on grids and to scale. In this work, Hu Wei provided a grand synthesis of a long classical tradition of cartography, local histories, and contemporary survey of the imperial territory commissioned by the throne for compiling *The Comprehensive Gazetteers*. This work quickly became the go-to text from which later controversies on historical geography and hydraulic policies were derived. They identified the separation of the North China Plain and the Lower Yangzi Basin as one of the most crucial problems.

As far as the Qing state was concerned, the crucial mechanism for an agrarian empire—hydraulics—differed in the North China Plain and the Lower Yangzi Basin. The hydraulic systems of both regions can be traced in classical allusion in *The Book of Documents*, especially in the chapter "Yugong." In the North China Plain, the problem was that the Yellow River kept changing course to enter the ocean, flooding large areas every time this occurred. Hydraulic engineers focused on how to prevent the Yellow River from changing course along with other flood-control measures. The flooding of the Yellow River also disrupted political and commercial connections between the political center (Beijing) in the north and the cultural center (urban centers in the Lower Yangzi Basin) in the south. Every time the Yellow River changed course, the subsequent flood also had devastating consequences for the Grand Canal. Much of China's tax revenue was shipped to the capital via the Grand Canal, one of the empire's major inland transport routes. Therefore, flood control in the North China Plain was essential to the stability of the Qing Empire—it prevented peasant rebellion.

In the Lower Yangzi Basin, the main concern was the Lake Tai region, adjacent to the estuary of the Yangzi River. The hydraulic engineers in the south focused on the instability of the waterway system surrounding Lake Tai. Because of general wear and sedimentation, the maintenance of the Lake Tai region required either dredging channels or raising dike-walls as the riverbeds rose. Maintaining this region was essential because of its importance to the Lower Yangzi Basin, China's wealthiest region. Its irrigation waters supported cotton plantations and its waters fostered river transport. However, reasons for stewardship were more about the imperial coffer and how to make the region more productive and wealthier.

Farmers in the North China plain feared the Yellow River because floods destroyed their livelihood. Farmers in the lower Yangzi basin, in contrast, regarded waterways as essential for their labor-intensive rice patty farming. Long before the Manchu conquest of China proper, the Song imperial state initiated more than thirty hydraulic projects in the lower Yangzi; the Ming imperial state also supervised more than thirty hydraulic projects in the region. During the eighteenth century, as Qian Yong (1759-1844) testified, it took almost two years to fix a riverbank, three to four years to dredge a waterway, five to six years to dredge a confluence of two rivers, and seven to eight years to dredge a confluence of three rivers.[22] Any hydraulic projects of this sort meant serious mobilization of local labor and money as well as a tremendous disturbance to local communities.

Qian Yong also observed that the resistance to major hydraulic projects could be classified into six categories: (1) It might cost too much for a specific county office to afford. Sometimes several counties needed to coordinate in order to contribute to the same hydraulic fund. (2) Landlords in County A might not want the harvest to be interrupted in order to fix hydraulic problems in County B. (3) Certain literati or merchant households did not want hydraulic projects that forced them to move family funeral sites. (4) Harsh work sometimes caused laborers to revolt. (5) Runners and clerks in the county office sometimes created unnecessary obstacles or delays for the hydraulic project to begin or continue. (6) Certain literati or merchant households complained of the disruption of their livelihood during the long course of hydraulic projects.[23] In short, hydraulic projects were usually time-consuming, required the collaboration of different levels of bureaucracy, necessitated a lot of money and laborers, and caused grievances in local communities. Why then did the imperial state initiate and supervise so many hydraulic projects in the lower Yangzi?

Unlike the North China Plain, where the main problem was to regulate the sediment-carrying capacity and the depth of the channel of the Yellow River, the situation in the Lower Yangzi Basin was more complex. The level of Lake Tai, for example, was sometimes higher than the surrounding farmland, and the level of the farmland was higher than both the Yangzi River and sea level. Elevation differences among the Lake Tai, farmland, the Yangzi River, and sea level directly affected irrigation technology. Irrigation of the farmland went both ways. When the water level in Lake Tai was high, farmers could allow water to flow down into the rice paddies. When the water level in Lake Tai was low, the farmers could then channel water from the rice paddies back into the lake. Hence, irrigation depended on the water level in Lake Tai. This was what scholars and policy makers called "water benefits" (*shuili*), by which they meant how to maximize the economic outcome and agricultural productivity of the land by managing irrigation and flood control of a given region. Therefore, the imperial state required a good overview of the pros and cons of how to suppress certain local resistance and weigh different local interests in order to promote the overall productivity of the Lake Tai region.

But for such policy deliberation in the Lake Tai region, why did historical geography matter at all? Qian Yong provided an important observation:

> In the recent past, those who discuss "water benefits" in the Lower Yangzi Basin always cite "Three rivers flow in, and Lake Tai is stable" from *The Book of Documents*. By citing the locus classicus, they mean to urge the priority of dredging three rivers. But they only know that waterways flow out of Lake Tai, and remain quite oblivious to what waterways flow into Lake Tai. In ancient China, the land was vast and population scant. Their priority was to prevent flooding. Now [that] our tax and financial burden is heavy, our top priority is to manage [the] agricultural output of our land. Therefore if we do not measure the distance, elevation, and depth of our rivers as well as the origins of our waterways meticulously, how can we increase "water benefits"?[24]

What Qian Yong meant by measuring "the distance, elevation, and depth of our rivers" was hydraulic planning and engineering. What he meant by "the origins of our waterways" unambiguously referred to historical geography. To make his point more bluntly, Qian Yong named a mistake that previous hydraulic officials made by abandoning the Wuyan dyke that controlled inflow into Lake Tai and redirected the Xuanshe River into Lake Tai. This decision resulted in more sedimentation in the lake. Qian Yong lamented that this policy failure resulted from the hydraulic officials' ignorance of historical geography; lessons in the past had already demonstrated the capacity of Lake Tai and showed that only certain waterways could flow into it.

In terms of both transcendent and immanent sources of imperial ideology, these kinds of policy decisions were quickly translated into disputes on classical knowledge. Namely, what exactly were the three rivers in *The Book of Documents*? How many rivers flowed into Lake Tai before? Which ones were they? What kind of irrigation and sedimentation problems did Lake Tai have before? Were they related to the confluence of rivers into Lake Tai? In Hu Wei's *My Narrow Vision of the Imperial Territory*, the three rivers, Lake Tai, and the Lower Yangzi Basin were all carefully discussed to address these concerns. In response to Hu Wei's work, Dai Zhen returned to the oldest text of historical geography—*The Guide to Waterways with Commentary* (Shuijing zhu) to expose Hu's "presentism" and pressing concerns regarding the imperial policy on irrigation, sedimentation, and waterways. It was likely Dai began his work of reconstructing *The Guide to Waterways with Commentary* in 1765, but he had not settled on a final draft until he could access all editions of the text in the imperial library project in 1775. In this final draft, Dai clearly set himself apart from Hu's "tunnel vision" historical geography, commissioned and driven by the Qing state.[25]

The classical debates over the three rivers had an actual effect on policy deliberations when the policy makers became involved. For instance, Cheng Yaotian (1725-1814), Dai Zhen's admirer and a well-known evidential scholar and polymath, took on this issue and published a book called *An Evidential Inquiry into the Three Rivers in the Yugong Chapter* (Yugong sanjian kao), in which he forcefully argued that the three rivers in fact referred to only one waterway. Cheng Yaotian cross-referenced and cross-examined almost all relevant passages in *The Book of Documents*. He also used all his methods from his philological toolbox to support his argument. He concluded from his classical research that only one waterway flowed into Lake Tai. The three rivers in *The Book of Documents* refers to the confluence of the Yangzi River, the Han River, and the waterway from Yuzhang. The joining of three major rivers would definitely result in serious flooding, Cheng imagined. So Cheng assumed that the legendary emperor Yu "created" a large basin to buffer both the speed and volume of the confluence of three major rivers into Lake Boyang. Geologists now describe the formation of Lake Boyang as the result of tectonic movement along a north-south axis about one million years ago. We may no longer believe in the origin myth of the legendary emperor Yu, but it served Cheng's argument well. In addition to the origin myth, however, Cheng Yaotian successfully explained the interdependence of the Lake Boyang and Lake Tai regions, and argued that the Lake Tai region cannot be viewed as an isolated system of waterways.[26] Quan Zuwang (1705-55), a renowned historian and evidential scholar, criticized Cheng's interpretation of *The Book of Documents*. Critical exchanges ensued, and Cheng's view emerged as an acceptable one among many scholars and policy makers.[27]

This seemingly trivial classical debate quickly found its policy avenue. Cheng Yaotian later spoke to the right audiences regarding his historical geography and classical research.[28] Ruan Yuan, Dai's admirer and Cheng's friend, assumed the position as the education commissioner of Zhejiang in 1797. He consulted Cheng before he published a semiofficial work called *An Inquiry of the Illustrations of Zhejiang Province* (Zhejiang tukao), which maps all waterways in Zhejiang onto the geographical terms of the classical world. The mapping itself was an extensive philological exercise corroborated with empirical data from an actual survey. Ruan Yuan's publication was clearly directed toward policy deliberations on the increasing troubles in sedimentation, flooding, and irrigation in the Lower Yangzi Basin. Cheng Yaotian's classical research provided significant feedback to the basic source for policy deliberation in hydraulic engineering, especially on the interdependence of the Lake Boyang and Lake Tai regions.[29] The specific hydraulic condition of Lake Tai certainly demanded a specific and pragmatic solution. Ruan and Cheng not only provided a transcendental source of their policy deliberation but also used classical language to disguise and obscure the actual decision-making process, which was driven by the pragmatic concerns of the Qing imperial state.

SOCIAL ORDER

The scholar's robe or *shenyi* (lit., "the long garment") was an important symbol for certain scholars in early modern China. Both its material form and political function changed from the eleventh to the end of the eighteenth century, when Manchu rulers had outlawed it and made it the subject of abstract debates among classical scholars. For Dai Zhen, the ritual apparel system, like the Palace of Light, reflected the pattern of cosmological order in the political and social world. More importantly, Dai studied the ritual apparel system to support his philosophical notion of individuality, which, according to him, is how the cosmos would allow the biological and economic needs of each member of the society to be fulfilled.

"The classical scriptures make the style and function of the long garment a puzzle," noted the editors of *The Annotated Catalogue of "The Complete Works of the Four Treasuries"* (Siku quanshu zongmu tiyao); "many debates have arisen."[30] The scholar's robe is an item of ritual apparel described in the eponymous thirty-ninth chapter of *The Ritual Records* (Li ji). Besides this short chapter, it is also mentioned in the "Crown Jade and Ornament" (Yu zao) chapter of *The Ritual Records*—a rather long chapter that describes the garb and rituals appropriate for a king. These two passages served as the baseline for the commentaries that reinvented the long garment over two millennia.

Classical symbols, such as this garment, became a useful distinction of scholars' identity. "It was not until the Song dynasty that scholars began to restore the style and function of the long garment!" Huang Zongxi (1610–95) proclaimed, "It was not until Zhu Xi that the corrective effort was made!"[31] According to historian Yü Ying-shih, Song Neo-Confucians like Zhu formulated a new political subjectivity.[32] Did the long garment have something to do with this new sense of shared identity and responsibility? The historian and encyclopedist Ma Duanlin (1254–1323), who was born in the Southern Song and served as a minister under the Mongol Yuan dynasty, was just the sort of man Yü had in mind. In *Comprehensive Studies of Documents* (Wenxian tongkao), published in 1319, Ma included a section on clothes and offered a telling anecdote about what it meant in his day when one put on the garment. He began by citing the philosopher and classical scholar Lu Dalin (1044–91) on the ancient period: "As for the long garment, superior and inferior did not mind wearing the same kind; neither did they mind wearing it during auspicious or inauspicious rituals; men and women could also both use the long garment. Noblemen [*zhuhou*] wore official clothes [*chaofu*] during the day and the long garment in the evening, while scholars [*shidaifu*] wore ritual clothing [*yuanduan*] during the day and the long garment in the evening. The long garment was the ritual garb of commoners."[33]

According to Lu, informality and convenience defined the use of the long garment for noble men and scholars, while commoners used it as formal dress. Court officials, noblemen and noblewomen, palace ladies, scholars and their wives, artisans, merchants, and farmers all wore it. It was the default casual apparel for members of the ancient elite. In a reversal of this logic, the long garment became the formal attire of commoners.

By the Song, few wore the long garment. It had come to signify antiquarianism, unconventionality, and nonconformity; only a few members of the emergent literati society favored it. Ma Duanlin provided a nuanced description of the social implications of donning this old-fashioned robe:

> The uniforms mandated by the imperial state are known as "ritual clothes" [*yuanduan*]. The long garment, on the other hand, is the apparel that was designed by the ancient sages. Although ritual clothes were used as the uniforms of the imperial state, they did not convey rank or status. If the emperor donned them he did not appear humble, and if scholars put them on they did not appear to be overstepping their authority. The long garment was made according to universal principles and models. Therefore the worthless could put it on, and so could the respectable. One could wear it at the court or at home. The aged emperor could wear it while taking his ease, noblemen could wear it to attend the sacrificial feast, scholars could wear it in the evening, and commoners could wear it to funerals. The long garment simply did not imply any sort of rank or status in the ancient world.[34]

The long garment, in other words, represented for ancient China the very antithesis of the hat and dress regulation later mandated by the imperial state. It did not facilitate discrimination, serving instead like a sort of equalizing disguise. This was incorporated by the sages into the garment's style, always admired for its elegance and simplicity. Zhu Xi promoted the new political subjectivity embodied in the long garment—a deliberate decision meant to create a value outside the imperial state. This is why Huang Zongxi credited him with "rectifying" the style and function of the long garment.

But even Zhu did not dare wear the long garment in public—the social stigma was too great. Anyone who dared to fashion such a robe, taking his lead from the description in *The Ritual Records*, and wear it in public, would invite derision. We know this because the virtuous scholar Shao Yong (1011–77) earned just that reaction when he tried out the long garment in public.[35]

Those who believed that by discreetly adopting a practice of the ancients they would achieve a measure of inner tranquility and moral transcendence did better. Sima Guang (1019–86), for instance, was wont to put on the long garment in the comfort of his private garden.[36] And Zhu Xi and his fellow Dao learning master Lu Xizhe (1036–1114) both planned to wear the long garment in order to demonstrate their newfound moral philosophy.[37] But both believed that this private practice could only be taken up after they retired from officialdom. So long as those three gentlemen—Sima, Zhu, and Lu—were in their government posts and pursuing their political agenda, Ma Duanlin believed, they could not put on the long garment: the symbolic act would puzzle and anger those who regarded wearers of the long garment as misfits and lunatics.[38]

In 1368, the inaugural year of the Ming dynasty, Zhu Yuanzhang (1328–98) declared that the time had come to replace the foreign symbols used by the Mongol rulers with figures deployed during the Tang dynasty (618–907). Members of the Chinese political elites, who had supported the prolonged military campaign against the Mongols, applauded the founding emperor's initiative. According to the cosmological theory of the five phases (*wuxing*) and their eternal cycle of mutual generation and conflict, the Ming ruling house was obliged to apply to the rituals of state a new color scheme, new names, and new clothes. Once these symbolic changes had been made, the new mandate of Heaven would be suspended over the new dynasty.[39] The Ming imperial house, energized by ethnic pride, drew up a new set of rules for ritual hats and dresses (*yi guan zhidu*), a demonstration that the Han people had reclaimed political power. Restoring the Tang system would require careful study of a period that predated the foreign domination of North China by outsiders—the Khitans, Jurgens, and Mongols. This turned out to be a much more complicated project than anticipated.

Like the Palace of Light (Ming Tang), the long garment is not a single, unchanging thing in the classical scriptures. The Neo-Confucian master Zhu Xi (1130–1200) and his followers had identified it as a type of ritual clothing, yet controversies regarding the exact style and function of the long garment persisted throughout late imperial China. In the middle of the sixteenth century, for instance, Wang Yingdian (fl. 1540) gave a detailed description of the long garment in his *Illustrations and Explana-*

tions of "The Rituals of Zhou" (Zhouli tushuo). General features include a collar decorated with a band of fabric that extends from the neck down both sides of the robe, large and wide sleeves, and decorative stitching extending down the lower half of the robe. In spite of having provided an illustration, Wang acknowledged that scholars were generally confused about what the long garment should look like since the classics were filled with contradictory depictions.[40]

The shifting associations of the venerable robe persisted after the Mongol conquest, the Yuan-Ming transition, and the reconstruction of the state under the Ming dynasty. During the Ming dynasty, the long garment had become the de facto scholar's robe and served a number of social functions (fig 8.1). First, those who wore it were thereby distinguished from farmers, craftsmen, peddlers, and, most significantly, merchants. When the semiotics of rank and status functioned properly, the distinctions among officials, merchants, craftsmen, and farmers were correctly and unambiguously perceived. Should class markers fade, a sense of social dysfunction would prevail, as occurred with the extensive commercialization of the sixteenth and seventeenth centuries that led to a blurring of categories as wealth was quickly translated into cultural capital.[41] As long as markers of rank remained apparent, individuals were able to negotiate social conventions.[42] Quite importantly, the long garment was a style devoid of foreign influence—it was seen as purely Chinese. Confucius had once praised an ancient statesman, declaring, "Were it not for Guan Zhong I might now be wearing my hair loose and folding my garments [*ren*] to the left."[43] In this passage, the key word *ren* was rendered as "garments" by Arthur Waley. Just what article of clothing the word *ren* refers to became the focus of debates among many classical scholars in the eighteenth century. The debates went directly to the heart of the question: to fold one's garment on top of the left was to wear the long garment—the quintessential Chinese item of clothing; to fold it on top of the right was the act of a barbarian. Consequently the long garment served as both a status marker and a cultural marker distinguishing domestic from foreign.

A remarkable anecdote from the sixteenth century will show how charged the garment remained in the dawn of early modern commercialization. In *The Records of Ming Scholars* (Ming Ru xue'an), Huang Zongxi told the story:

> One night, Wang Gen [1483-1541] dreamed that the heavens were falling. As thousands of people ran, screaming for mercy, he raised his arms and lifted the heavens. Then, noting that the sun, moon, and other heavenly bodies had strayed from their orbits, he raised his hands again and restored all to their proper positions. When he woke, sweat pouring off him like rain, Wang was struck by a deep insight into his mind-in-itself [*xinti*]. . . . Having carefully studied *The Ritual Records*, he now set about applying its prescriptions to fashioning a hat, a long garment, a sash, and tablets. Putting them on, he said, "How can I speak the words of [the sage] Yao and act as Yao did without also wearing his clothes?"[44]

In a comment on this passage, the historian Shimada Kenji praised the symbolic act as the culmination of both a radical commitment to a plebeian mentality and the promotion of individuality. Wang and his group, known as the Taizhou school and later labeled the "left-wing" of Wang Yangming's followers, emancipated themselves from the scholastic constraints of the tradition of endless commentaries on the Confucian canon and drew lessons instead from a daily involvement with the world. A rejuvenated orientation on the untainted, child-like origins of individuality had emerged in singularity.[45]

Most scholars of the seventeenth century believed that heavenly principles—then considered the foundation of moral actions and social order—could only be upheld by keeping selfish and egoistic impulses (*si*) in check. This ensured the perpetuation of patriarchal hierarchy, gender segregation, political loyalty, filial piety, and other characteristics of Chinese society. A gradual but fundamental change regarding the discourse of self (*ji* or *si*) took place in the sixteenth century: egoistic impulses (*si yu*) were recast as social, economic, or biological needs. This significant change in the Confucian moral discourse did not result in the elimination of all talk of heavenly principles. On the contrary, thinkers like Li Zhi, Lü Kun (1536–1618), Chen Que (1604–77), Huang Zongxi (1610–95), and Gu Yanwu (1613–82) spoke of heavenly principles as an aggregate of individual social, economic, or biological needs. For them, these principles had ceased to be transcendental or impersonal. They reflected instead the true needs of an authentic self. A good society based upon heavenly principles would be, above all, a community of authentic individuals. Huang and Gu, in particular, described as ideal that society in which all individual needs

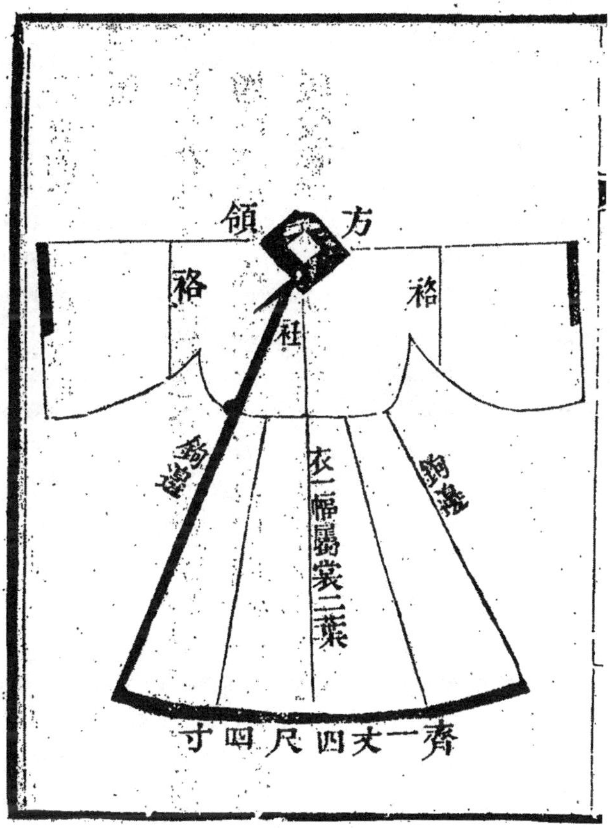

FIGURE 8.1. The scholar's robe in the Ming dynasty. Features include a collar decorated with a band of fabric that extends from the neck down both sides of the robe, wide sleeves, and decorative stitching extending down the lower half of the robe. Huang Zongxi pointed out that his predecessors had identified the scholarly fashion of the Ming dynasty with the original style of the "long garment" in antiquity, a situation arising from contradictory depictions in classical texts. Huang upheld an ethnocentric view of Chinese superiority and political distinction between Chineseness and foreignness by stressing the garment's *ren* 衽 (lapel; at the upper center of this illustration). From Huang Zongxi, *Shenyi kao*.

and desires (*si*), whether social, economic, or biological, would be not only fulfilled but guaranteed. Dai Zhen was beyond doubt an extension of this long conversation on the idea of "self" in the seventeenth century.[46]

The hat and dress regulation, recorded in detail in *The Collected Statutes of the Great Ming Dynasty* (Da Ming huidian) and later set down in *The Ming History* (Ming shi), established the symbolic hierarchy of the ruling elites while announcing the ethnicity of the ruling dynasty. This was one of the reasons the Qing dynasty (1644–1911) revamped the regulation and instituted a Manchu one. The Manchu conquerors had arrived in the Ming capital looking drastically different from the newly deposed rulers. Still, it was not until 1759 that the Qing imperial house published *The Illustrated Compendium of Qing Rituals* (Huangchao liqi tushi), the formal inauguration of the hat and dress regulation.[47] All exhibited a distinctive and exotic character and served as striking reminders of Manchu domination.

The official clothing of both the Ming and Qing was only relevant to bureaucrats and state rituals. As Shen Congwen suggested, the vast majority of China's people would have barely noticed the symbolic changes mandated by the state.[48] While the Manchu conquest of China continued, the partnership between the Chinese political elites and the Qing state was consolidated. Still, most of the literati destined never to hold office expressed subtle disagreements with the Qing state. This was the group that led the extensive commercialization that marked several wealthy regions. That process had been blurring the gap between literati and professional merchants since the beginning of the sixteenth century. Consequently, the members of these barely distinguishable groups became acutely aware of commercial trends as part of the endless struggle over status. Both literati and professional merchants learned to position themselves in the fluid dynamics of status formation. Not only official status markers but also symbols derived from the classical scriptures were creatively deployed to signify wealth, culture, and education, similar to conspicuous consumption.[49]

From the sixteenth to the seventeenth century, a rising interest in fashion had also brought changes to men's clothes as commercial expansion gradually reached an unprecedented scale. Suzhou acquired a reputation for producing fine and novel garments, as designers and tailors competed to produce clothes and accessories for men and women. A sense of style became de rigueur among literati and wealthy merchants. Even as the imperial authority prescribed a set of wearable markers to signal one's place in the symbolic hierarchy, the wealthy succumbed to the lure of conspicuous consumption, undermining the hierarchies the imperium insisted on.[50] Among the various fashions of the day, antiquarianism marked one as sophisticated. Wealthy merchants, in particular, bought up rare books and antiques in order to claim some of the luster of literati-officials. When even the clothing of faraway times became commercialized, how did a scholar with a taste for antiquity set himself apart from the philistines?

This question assumed a different configuration when the Ming dynasty imploded in May 1644. Later that year, a large Manchu army penetrated Beijing, driving the peasant army who had overthrown the Ming from the capital. Loyalists in South China resisted the Manchu conquest for the next four decades. Huang Zongxi, a polymath and epigone of the Ming's famous—and infamously martyred—Donglin partisans, identified as a Ming loyalist. In his view, Confucian civilization faced an imminent existential threat, and he cast himself as its champion. His diagnosis and treatment plan appear in *Waiting for the Dawn: A Plan for the Prince* (Ming yi dai fang lu) and in a collection of critical biographies, *The Records of Ming Scholars*.[51] Why did the author of these impressive political writings also devote a treatise to the style and function of the long garment?

To be sure, the long garment was an important topic throughout the Ming dynasty, and it was mentioned in the titles of eleven monographs listed in *The Bibliography of Vast Hall* (Qian qing tang shumu) by Huang Yuji (1629–91), the most important catalogue of books compiled in Ming China.[52] The catalogue was a record of Huang's private library, a collection inherited from his father, Huang Juzhong, then considerably expanded. The bibliography lists 14,907 works, many of which have since been lost. It circulated in manuscript for 250 years and was printed only after the fall of the Qing dynasty in 1912.[53] Huang's catalogue has eleven titles that contained the term *shenyi* (table 8.1); none of them can be found today.

This short bibliography indicates, however, that the style and function of the long garment were not obscure topics in Ming China. In *A Study of the Long Garment* (Shenyi kao), Huang Zongxi engaged many of the works cited—in addition to others—offering a remarkable interpretation of what the long garment meant to his contemporaries. As Huang

TABLE 8.1. Treatises on the scholar's robe (*shenyi*) in the Ming dynasty

Author	Life Dates	Degree Date	Treatise
Zhu You 朱右	1314–1376	n/a	深衣考 *Shenyi kao*
Huang Rungyu 黄潤玉	1389–1477	1420 *juren*	考定深衣古制 *Kaoding shenyi gu zhi*
Yue Zheng 岳正	1418–1472	1448 *jinshi*	深衣纂疏 *Shenyi zuan shu*
Xia Shizheng 夏時正	1412–1499	1445 *jinshi*	深衣考 *Shenyi kao*
Yang Lian 楊廉	1452–1525	1487 *jinshi*	深衣纂要 *Shenyi zuan yao*
Gao Jun 高均	n/a	n/a	深衣考 *Shenyi kao*
Zuo Zan 左贊	n/a	1457 *jinshi*	深衣考正 *Shenyi kaozheng*
Wang Tingxiang 王廷相	1474–1544	1508 *jinshi*	深衣圖論 *Shenyi tu lun*
Xu Pan 許泮	n/a	n/a	古深衣訂 *Gu shenyi ding*
Huang Daozhou 黄道周	1585–1646	1622 *jinshi*	緇衣集傳 *Ziyi ji zhuan*
Zheng Guan 鄭瓘	n/a	1490 *jinshi*	深衣圖說 *Shenyi tu shuo*

had spent a year living under the roof of Huang Juzhong and making daily use of his library, it seems likely he began studying the long garment then.[54] After the fall of the Ming dynasty, Huang refused to serve the Qing dynasty, a position that changed his life and his thinking. His scholarship was marked by deep reflections on the twilight of the Ming. Haunted by this heartbreaking failure, Huang was determined to study something concrete and useful—a decision shared by many Ming loyalists. If, as he hoped against all odds, the remnant Southern Ming state (1644–62) managed to drive the Manchus out of China, he would apply his practical knowledge to serving the rejuvenated Ming. And so he worked and waited.

Huang often claimed that he conducted his scholarly labors in melancholy and solitude. Here is what he wrote in 1647 about the conditions under which he completed his treatise on mathematical astronomy: "I used to reside alone near a deep valley with a pair of waterfalls outside of my window. At midnight, when outside my house apes wailed and the ghosts of those devoured by tigers shrieked wildly, I studied trigonometry with counting rods and told myself that I was truly a silly man. After I completed my study of trigonometry, I discovered that my mathematical knowledge—like the extraordinary skills of dragon slayers—was entirely useless and I could discuss it with no one."[55]

In addition to his search for something concrete, useful, perhaps even therapeutic, Huang scrutinized the work of his Ming predecessors. Refusing to speak of the dynastic failure in terms of conscience, principles, meanings, or human nature, he turned to academic disciplines like historical geography, hydraulic engineering, and mathematical astronomy. The study of the long garment belongs to this impressive foray, part of what was called "substantive learning" (*shixue*), an intellectual trend of the seventeenth and eighteenth centuries that Huang largely initiated. The hope of one day passing his useful knowledge along to a revived Ming turned out to be vain. But he never gave up, and one cannot help thinking that he hoped his enterprise might promote resistance to the Manchu domination.

After the collapse of armed resistance in the south, Huang projected a new despair onto the long garment. He concentrated on what it meant for literati values. It is likely he began his research in 1641 and continued to study it until 1677. His conclusions differed from those of both Huang Rungyu (1389–1477), a leading intellectual of eastern Zhejiang who lectured at the prestigious Nanshan Academy, and Wang Tingxiang (1474–1544), a political philosopher who had engaged in sharp debates with Wang Yangming. For Huang Zongxi, the long garment symbolized the transmission of literati political values, something distinct from dynastic politics. The style and function of the long garment directly corresponded, he said, to the "grand implication" (*da yi*) of literati values.[56]

In 1677 Huang Zongxi contributed a preface to a new book entitled *Queries from Studies of the Ritual Classics* (Xue li zhi yi) by his friend Wan Sitong (1638–1702). There he argued for including ritual studies in the newly fashionable category of practical or substantive learning. Huang's apology was based upon two premises: maintenance of the ritual order benefited everybody, not just a few sages, and the ritual order aligned with the grand pattern of the cosmos and was therefore natural, humane, and applicable to all human societies. As he explained, "In ancient times, every ritual had a corresponding performance (*yi*). All rituals were passed down by officials and universally practiced. Everyone could learn the rituals; they were not monopolized by the sages. The grand rituals, such as the imperial sacrifices and travel, serve as substantive politics. The minor rituals, such as properly greeting a superior or a host, serve as substantive actions."[57] The ritual order, as Huang envisioned it, was a ritualistic utopia with a strong antimercantile streak. By addressing the vanishing principles of a Confucian civilization embodied in the long garment, he offered a brilliant rationale for limiting tyrannical power, simultaneously rejecting the coercive hat and garment regime and the wave of commercial expansion, represented by the fashions that threatened to disrupt the symbolic order and class hierarchy.

The classical scholars in the eighteenth century rejected Huang's view all together. The editors of *The Annotated Catalogue of "The Complete Works of the Four Treasuries"* explained why they included Huang Zongxi's *A Study of the Long Garment* in the *Four Treasuries* but viciously attacked Huang's description of the long garment. *The Annotated Catalogue* has been required scholarly reading since the 1790s; even today it influences the assessment of many works.[58] Huang's work, it was explained, was included only because the mistakes of a famous erudite ("sloppiness" and "speculation" were his main offenses) had to be exposed and identified.[59] The editors argued that Huang's readings were farfetched. They seem to be imputing to him an ulterior motive, but they say nothing about his political convictions. Instead they simply caution their readers: "We were concerned that Huang Zongxi's reputation as a classical scholar might mislead later scholars. We decided to identify his mistakes and include it [i.e., *A Study of the Long Garment*] in the imperial library, so that future readers would be able to judge for themselves."[60] For the editors of *The Annotated Catalogue*, "identifying his mistakes" meant little but opening Jiang Yong's *Correcting Errors in the Tailoring of the Long Garment* (Shenyi kaowu), as their comments are taken straight from that work.

Negation of Huang's interpretation of scholastic terms was satisfactory to neither Jiang Yong nor Dai Zhen, both of whom had discussed the long garment extensively. The most significant feature of Huang's vision was the political and cultural distinction between Chinese civilization and inferior counterparts. This was highlighted in his explanation that the scholar's robe should be donned by folding the *ren* toward the right (i.e., Chinese) direction. Jiang Yong's *Correcting Errors in the Tailoring of the Long Garment* points out that Huang mistakenly identified the *ren* part as the crucial marker for the Chineseness and shows that, instead, both the right *ren* and the left *ren* were symmetrical parts of the lower part of scholar's robe. Since both were present, neither should be identified as Chinese or foreign. Behind this seemingly arcane and tedious analysis lies Jiang Yong's political message that the quintessential ritual garment in ancient China—the long garment—does not embody the distinction between Chinese and foreign cultures or customs. Huang's view, as powerful and pervasive as it had been, was merely an incorrect understanding of the ancient ritual system. The classical scholarship that thrived and cumulated in Qing China (mostly in the eighteenth century) had, in Jiang Yong's view, revealed Huang's dated and sloppy research, which in turn led to an ideological distortion of the classical world.

Dai Zhen also endorsed Jiang's view but took the issue in a different and more cosmopolitan direction. In his short unpublished manuscript dubbed "An Explanation of the Scholar's Robe" (Shenyi jie), Dai adopted a more radical and systematic explanation of the *ren* part than had Jiang. Dai argued that the *ren* part systematically appeared in many occasions across all Five Classics and should be considered as an integral aspect of the state rituals in ancient China. The scholar's robe, or the long garment, was simply one component in the entire ritual system.[61] Dai's political philosophy, revealed in this long trail of discussion on the scholar's robe, was to bypass or ignore the idea of Chineseness entirely. Instead, the ritual apparel system in ancient China could best approximate or mimic the physical pattern of cosmological order (like the Palace of Light) and should be universally applicable to states in East and Southeast Asia. Similar to the mobility of the Sanskrit *cosmopolis*, Dai Zhen's Confucian *cosmopolis* should be universally shared, recognized, respected, and applied in all regions that operated in the classical tradition of the Confucian canon.[62]

Dai's comments on the scholars' robe not only expressed a different political vision but also engaged an age-long dialogue on the scholar's vocation. Dai inherited Jiang's politics and methodology and expanded on them in several important ways, while departing from him with a coherent formulation of transcendent sources of mathematical cosmology and historical geography. Dai's classical vision, evident not just in his philosophical discussion of individuality but also in his studies of astronomy, cosmology, geography, the Palace of Light, and the scholar's robe, decisively departed from the long commentary dialogue of classical knowledge and arrived at a cohesive depiction of cosmological and political order. Given his decisive break from the Neo-Confucian orthodoxy, however, Dai still operated within the framework and crucial assumptions of the Confucian canon. The formation and reformation of Dai's classical vision as well as how it differed from the imperial state's tunnel vision constitute the bulk of this book.

In actuality, some components of Dai's classical vision became legible and resourceful for the Qing imperial ideology. It was precisely in this process that Dai's classical vision became ideological. It was also for this reason that Dai's vision went beyond the horizon of intellectual history and entered the realm of the politics. In the hand of his followers and critics, Dai's vision became a viable political alternative to the Neo-Confucian orthodoxy for the largest and longest-lived polity in the early modern era. Beginning with the Opium Wars and climaxing with the Sino-Japanese War of 1895, China fell before the modern armories of overseas foes. As late as 1899, the classical scholar Sun Yirang (1848–1908) nonetheless wrote in the prologue to his magnum opus, *The Correct Interpretation of the Rituals of Zhou* (Zhouli zhengyi): "Were we to bypass politics and culture and discuss only how to pursue wealth and power, it would be like destroying our ports while making plans to sail out onto the ocean! When today's political discussions cause us despair, there is no better guidance than [*The Rituals of Zhou*]."[63] In 1899, Sun certainly belonged to the intellectual minority to make such an assertion, but the Confucian text corpora he adamantly held in esteem remained powerful and respectable, still read by millions of literati. The core of this attitude should not be attributed to the intractable character of "Chinese conservatism" but to the malleability and creativity of the blueprint of ancient cosmopolis.

Before China was reduced from a vibrant and cosmopolitan civilization to a weak nation, Dai Zhen had already transcended the dichotomy between technical know-how and classical value and provided a forward-looking Confucian cosmopolis with considerable technical detail. After China entered the maelstrom of the modern world, the scientific and technological horizon dramatically shifted to the industrial mode and Dai was rapidly transformed into someone entirely different: an intellectual prophet of modern humanism and individuality.

NOTES

1. THE MAN AND HIS TIMES

1 Yasuda Jirō's analysis of Dai's book was one of the first to point to its fusion of individualism and objective methodology; see Yasuda Jirō and Kondō Mitsuo's *Tai Shin shū*, 3-26. Yasuda also translated the entire *Mengzi ziyi shuzheng* into Japanese and provided detailed commentary. Two dissertations explore Dai's moral philosophy based on Dai's *Mengzi ziyi shuzheng*: John Ewell, "Re-inventing the Way: Dai Zhen's Evidential Commentary on the Meanings of Terms in Mencius (1777)," and Justin Tiwald, "Acquiring 'Feelings That Do Not Err': Moral Deliberation and the Sympathetic Point of View."

2 In his reconstruction of a group of ancient mathematical texts known as *The Nine Chapters of Mathematics* (Jiuzhang suanshu), Dai Zhen occasionally speculated on how the original mathematical texts should look like in order to bring them into line with his overall conception of the classical world; see Qian Baozong's critique of Dai Zhen in Li Yan and Qian Baozong, *Li Yan Qian Baozong kexueshi quanji*, 9:145–46.

3 Mizoguchi, "Mōshi jigi soshō no rekishi teki kōsatsu," 144–65, and *Chūgoku zen kindai shisō no kussetsu to tenkai*, 283–315.

4 Hu Shi argued that none of Dai's followers and critics could truly understand and appreciate the critical spirit of his philosophy; see *Dai Dongyuan de zhexue*, 80–197. Both Chen Pingyuan and Qiu Weijun portrayed Hu, along with Zhang Taiyan, as the most influential interpreters of Dai in the twentieth century; see Chen Pingyuan, *Zhongguo xiandai xueshu zhi jianli*, and Qiu Weijun, *Dai Zhen xue de xingcheng*.

5 Mizoguchi, *Chūgoku zen kindai shisō no kussetsu to tenkai*, 42. On the idea of public principles in Yan Fu, Kang Youwei, Liang Qichao, and Zhang Taiyan, see Wang Hui, *Xiandai Zhongguo sixiang de xingqi*, 753–1103.

6 Louis Dumont provides a useful discussion of this subject—without any reference to China—in his *Essays on Individualism*, 60–103. Dumont's discussion of natural law and the genesis of individualism can be compared with Mizoguchi's discussion in *Chūgoku zenkindai shisō no kussetsu to tenkai*, 332–65.

7 Mizoguchi, *Chūgoku zenkindai shisō no kussetsu to tenkai*, 311–13. See also Yamanoi, *Min-Shin shisōshi no kenkyū*, 361–77.

8 Nishi, *Chūgoku shisō ronshū*, 356–71.

9 I used the *Dai Dongyuan xiansheng quanji* edition for these calculations.

10 China historians in Japan, including Kondō Mitsuo, Yabuuchi Kiyoshi, and Kawahara Hideki, pioneered this integration.

11 Hu Shi, *Dai Dongyuan de zhexue*, 121.

12 This thesis—hermeneutics of linguistic purism—is advanced in Kai-wing Chow, *The Rise of Confucian Ritualism*, 161–86.

13 For detailed elaboration of the purist hermeneutics of classical scholarship, see Kai-wing Chow's *The Rise of Confucian Ritualism in Late Imperial China*. I differ from Chow in that I only use it to characterize one trend—Hui Dong and his followers—among others in the overall evidential scholarship movement.

14 Hu Shi, *Dai Dongyuan de zhexue*, 119.

15 Chow, *The Rise of Confucian Ritualism*, 161-86.

16 Cited in Hu Shi, *Dai Dongyuan de zhexue*, 119–20.

17 Genyou Wu, "A Preliminary Discussion of Dai Zhen's Philosophy of Language."

18 Ian Hacking, *Representing and Interveneing*, 21-31. For an intriguing and opposing view, see Richard Rorty, *Philosophy and the Mirror of Nature*, 17-69.

19 Stephen Toulmin's *Cosmopolis: The Hidden Agenda of Modernity* is certainly an inspiration. This rather obscured and ancient name designates two distinct and yet closely related kinds of "order:" the order of Heaven and Earth and the order of human realm. In *The Language of the Gods in the World of Men*, Sheldon Pollock speaks of Sanskrit cosmopolis to depict a similar totality of language, culture and power in premodern India.

20 Zhang Xuecheng, *Wenshi tongyi*, 523–24. Susan Mann has written on women writers in both Zhedong and Zhexi; see her *Precious Records*, 201–5. Wen-hsin Yeh provides a detailed historical and cultural geography of Zhedong and Zhexi, see her *Provincial Passages*, 11-50.

21 Zhang Shunhui, *Qingdai Yangzhou xueji*, 2-3.

22 Liang Qichao, *Jindai xuefeng zhi dili de fenbu*, 3.

23 Yamamoto Eishi's *Dentō Chūgoku no chiikizō* gives many good examples on local history. See also Kojima Tsuyoshi, *Chūgoku kinsei ni okeru rei no gensetsu*, 14–15. Yang Nianqun's *Rujia diyuhua de jindai xingtai* (Modern patterns of regional Con fucianism) also takes a regional approach.

24 Elman, *Classicism, Politics, and Kinship*, 145–85. For a comprehensive survey of Donglin politics, see Frederic Wakeman's "The Price of Autonomy," in *Telling Chinese History*, 146-63.

25 Duara, *Sovereignty and Authenticity*, 209–37; in Duara's study the locality was Manchukuo.

26 On Jesuits and science, see Dear, *Discipline and Experience*, 32–62; Feingold, *Jesuit Science and the Republic of Letters*, 1–45; Hsia, *Sojourners in a Strange Land*, 51–128. On Jesuits in general, see O'Malley, *The First Jesuits*. Standaert's *Handbook of Christianity in China: 635–1800* is a good bibliography.

27 Rowbotham, *Missionary and Mandarin*, 132.
28 See, for example, Minamiki, *The Chinese Rites Controversy*, 65–82; Huang Yi-long, *Liangtoushe*, 387–435.
29 Qiong Zhang has challenged this triad in her "About God, Demons, and Miracles," 1–36.
30 For recent treatments of the narratives constructed around the Jesuit sciences, see Elman, *On Their Own Terms*; and Jami, *The Emperor's New Mathematics*.
31 Elman, *On Their Own Terms*, 190–221; Jensen, *Manufacturing Confucianism*, 31–133; Standaert, *The Interweaving of Rituals*, 81–206; Brockey, *Journey to the East*, 57–203; Menegon, *Ancestors, Virgins, and Friars*, 260–356.
32 Xi Zezong, "Lun Kangxi kexue zhengce de shiwu," 18. I met Xi Zezhong by chance at Beijing's Institute for the History of Natural Sciences (IHNS) in 2004. I had been speaking with his student Zhang Baichun, now the head of IHNS, when Xi happened by. He invited me to dinner and after a lively conversation about the history of science he presented me with his newest publication. China's most senior historian of science at the age of seventy-eight, Xi treated me, a newly minted PhD without a job, like a colleague. I never had the privilege of speaking with him again before he died in 2008.
33 In January 2012, *The Economist* launched a new section devoted to China's growing economic might, its internal transformations, and its impact on the world. Many of the ensuing reports have examined science, technology, and medicine.
34 Xi Zezong, "Lun Kangxi kexue zhengce de shiwu," 18–20.
35 Ibid., 27.
36 Lattis, *Between Copernicus and Galileo*, 86–144.
37 Elman, *On Their Own Terms*, 24–60. On how the Jesuits cast their ideas and beliefs in terms of heavenly learning, see Peterson, "Learning from Heaven," 789–839.
38 Kang-I Sun Chang and Stephen Owen, *The Cambridge History of Chinese Literature*, 1:140–49.
39 Yutang Lin, *The Gay Genius*.
40 Wei Liaozi, 3b.
41 I had access to two electronic databases: the Chinese Full-Text Electronic Database at the Institute of History and Philology at Taiwan's Academia Sinica and the electronic database for *Siku quanshu* (Wenyuange edition).
42 Liu Shipei, *Liu Shipei xinhai qian wenxuan*, 153.
43 Ibid.
44 Lin Tingkan, *Jiaolitang wenji*, 39.
45 Ibid.
46 Ruan Yuan, "Xu (Prologue)," *Chouren zhuan*, 2a.
47 Ibid.
48 Ibid., 1a–1b.
49 Tan Tai, "Chouren jie," 1a–5b.
50 Ruan Yuan, "Xu (Prologue)," *Chouren zhuan*, 2a.
51 Ibid.
52 For details, see Peterson, "Squares and Circles," 47–60.
53 Ruan Yuan, *Chouren zhuan*, 42:22a–22b.
54 Elman, *Classicism, Politics, and Kinship*, 10–11.
55 Wang Shiduo, "Xu," 1b. For an analysis of the radical views Wang Shiduo set out in his private diary, see Wang, *Zhongguo jindai sixiang yu xueshu de xipu*, 61–94.

56 Qiu Weijun, *Dai Zhen xue de xingcheng*, 97–105. For criticism of Qiu's account, see Zheng Jixiong, "Ping Qiu Weijun *Dai Zhen xue de xingchen*," 197–213.

57 Needham, "Author's note," *Science and Civilisation in China*, vol. 5, pt. 3: xxxi.

58 Jami's *The Emperor's New Mathematics* is a case in point.

59 Liu Shu-hsien, regarded in Taiwan and Hong Kong as a third-generation Neo-Confucian philosopher, published *Understanding Confucian Philosophy* in 1998 and its sequel, *Essentials of Contemporary Neo-Confucian Philosophy*, in 2003 to advance this view.

60 This comes from Wolfgang Schluchter's title *The Rise of Western Rationalism: Max Weber's Developmental History*.

2. HOW JESUIT SCIENCE CONQUERED THE KANGXI COURT

1 A few examples of works that treat the Jesuit purge are: Gernet, *China and the Christian Impact*; Zhu Weizheng, *Coming Out of the Middle Ages*, 81–112; Elman, *On Their Own Terms*, 133–49; Huang Yi-long, *Shehui tianwenshi shijiang*, 195–237; Zurndorfer, "'One Adam,'" 141–68; Jami, *The Emperor's New Mathematics*, 57–81.

2 Zurndorfer, "'One Adam,'" 141–68.

3 Yang Guangxian, *Bu de yi*, 195.

4 Ibid.

5 Ruan Yuan, *Chouren zhuan*, 36:2a–4b.

6 Lloyd, *Adversaries and Authorities*, 165–89; see also Lloyd, *The Ambitions of Curiosity*, 21–43.

7 Kawahara Hideki identifies what was called heavenly learning in ancient China as a system of philosophy, mathematical astronomy, cosmology, and cosmography. See his *Chūgoku no kagaku shisō*, 114–95, 298–91.

8 Yamada Keiji, *Jujireki no michi*, 76–119; Elman, *On Their Own Terms*, 63–106; Sivin, *Granting the Seasons*, 56–60.

9 Mingshi, *Siku quanshu*, 31:3a–3b; Chen Jiujin, *Huihui tianwenxue shi yanjiu*, 248-52.

10 Florence Hsia calls into question the Jesuit persona as mathematical astronomers in China. See her *Sojourners in a Strange Land*, 13–50.

11 Crossley, *The Manchus*, 75–80.

12 Naquin, *Peking*, 289–95.

13 Elliott, *The Manchu Way*, 89–132.

14 *Xiyang xinfa lishu*, "Zoushu," 1:3.

15 Martini's *De bello tartarico* was the most popular account of the Manchu conquest in seventeenth-century Europe. See Mungello, *Curious Land*, 106–16; Lach and Van Kley, *Asia in the Making of Europe*, 3:1664–68.

16 For the Southern Ming's relations with Catholicism, see Huang Yi-long, *Liangtoushe*, 311–85.

17 *Xiyang xinfa lishu*, "Zoushu," 1:3.

18 Huang Yi-long, "Cong Tang Ruowang suobian minli shixi Qingchu Zhong-Ou wenhua de chongtu yu tuoxie"; Huang Yi-long, "Qingchu Qintianjian zhong ge minzu tianwenxuejia de quanli qifu," 75–118.

19 I follow Nathan Sivin by dividing what the Chinese literati simply referred to as *li* into four categories: mathematical astronomy, which computed both past and

future celestial phenomena; systematic astronomy, step-by-step computations that generated forecasts of celestial phenomena and assembled them into a complete ephemeris; a computational treatise, meaning the textual embodiment of the astronomical system; and published ephemerides and almanacs, which were the product of computational treatises. See Sivin, *Granting the Seasons*, 38–42.

20 *Xiyang xinfa lishu*, "Zoushu," 1:2.

21 The Collected Statutes of the Qing Dynasty, first compiled in 1687 (the project was initiated by the Kangxi emperor), was revised in 1734 (in the Yongzheng emperor's reign), 1757 (in the Qianlong emperor's reign), and 1801 (in the Jiaqing emperor's reign). This compendium shows that the state rituals depended heavily on imperial ephemerides and astronomy more generally throughout the first half of the Qing. Much of the section on the Astrocalendrical Bureau from the Yongzheng era is concerned with mathematical astronomy, the astronomical system, and how to combine them with ritual procedures. See also Hucker, *A Dictionary of Official Titles in Imperial China*, 86–88; Elman, "Imperial Politics and Confucian Societies in Late Imperial China," 379–418.

22 For an extensive discussion of the symbolic content of ritual, see Bourdieu, *Language and Symbolic Power*, 107–36. David Kertzer stresses the legitimacy, mystification, and ambiguity of rituals; see his *Ritual, Politics, and Power*, 35–76. While none of these addresses Chinese rituals, each sheds light on different aspects of Qing practices. For a comparative analysis of funeral ritual practices, see Standaert, *The Interweaving of Rituals*, 10–36.

23 Rawski, *The Last Emperors*, 197–294.

24 Guy, "Who Were Manchus: A Review Essay," 154-57; Hevia, *Cherishing Men from Afar*, 116–33.

25 Rawski, *The Last Emperors*, 197–230. See also Zito, *Of Body and Brush*, 121–84. When Kai-wing Chow discussed "Confucian ritualism" in *The Rise of Confucian Ritualism in Late Imperial China*, he focused almost exclusively on what Qing literati called *jia li* (family rites), particularly those rites performed in honor of ancestors. His discussion of ancient state rituals, on 167–70, is somewhat misleading.

26 *Da Qing huidian Kangxi chao*, 71:1a–1b.

27 Ibid., 1b–3a.

28 *Da Qing huidian Yongzheng chao*, 246:1a.

29 Ibid., 3b.

30 Wu Zhenyu, *Yangjizhai conglu*, 77.

31 Paul Ricoeur discusses narrative temporality in *Time and Narrative*, 1:5–90; 2:100–152.

32 Elman, *On Their Own Terms*, 63–106.

33 Jiang Xiaoyuan and Niu Weixing, *Tianwen Xixue Dongjian ji*, 317–29.

34 Dardess, *Blood and History*, 31–169.

35 Hashimoto, *Hsü Kuang-ch'i and Astronomical Reform*, 84–103.

36 Ibid., 104–42.

37 I came to this conclusion by examining many incomplete parts of *Xiyang xinfa lishu*, such as Giacomo Rho, Borgia Cinese 318 (1), *Bili guijie*. The Italian Jesuit Claudio Filippo Grimaldi (1639–1712) modified one of the astronomical tables found in the "tabulation" (*fashu bu*) section. See Claudio Grimaldi, BN Chinois 5007, *Biaoyuan tushuo*; Chu, "Archiving Knowledge," 157–84.

38 For an extensive discussion of the Tychonic cosmos in China, see Jiang Xiaoyuan

and Niu Weixing, *Tianwen Xixue Dongjian ji*, 269–316. Peng Yoke Ho, who often collaborated with Joseph Needham (1900–1995), honored the visit of the queen of Denmark to Hong Kong by delivering a lecture in 1981 on Tycho Brahe's cosmography centered on illustrations 2.1 and 2.2. See *Hong Kong–Denmark Lectures on Science and Humanities*, 53–60.

39 Huang Yi-long, "Qingchu Qintianjian zhong ge mingzu tianwenxuejia de quanli qifu," 77–81.

40 *Yuzhi Lixiang kaocheng shangbian*, 1:3a–4a.

41 Joseph Needham made no distinction between the seventeenth-century Latin West and the emerging Protestant West. See Needham, *Science and Civilisation in China*, 3:150–68. Toby Huff reframed the Needham question in Weberian terms. See his *The Rise of Early Modern Science*, 8–46. Unlike Needham and Huff, I do not ask whether China had developed or absorbed the "spirit" of modern science.

42 Needham, *The Grand Titration*, 15.

43 Huff championed this historiography. See his *The Rise of Early Modern Science*, 8–46. For a critical reflection on this historiography, see Bala, *The Dialogue of Civilizations in the Birth of Modern Science*, 7–51.

44 State rituals were intimately linked to imperial politics. See McDermott, ed., *State and Court Ritual in China*, 1–19. For several concrete case studies, see Gan Huaizhen, *Huangquan, liyi, yu jingdian quanshi*, 35–206.

45 Hu Tiezhu, "Dayanli jiaoshi jisuan jingdu," 316. Shigeru Nakayama, *A History of Japanese Astronomy*, 65–69. During the Tang, there were at least two mathematical formulas for computing eclipses. See Sivin, *Granting the Seasons*, 116–18.

46 Sivin, *Granting the Seasons*, 227–37.

47 Huang Yi-long, *Shehui tianwenshi shijiang*, 28–48.

48 Sivin, *Granting the Seasons*, 151–70.

49 Dianjiaoben ershisishi, *Mingshi*, 3:518.

50 Ibid., 519–20; Tian, *Zhongguo shuxue de xihua licheng*, 12–13.

51 *Da Qing huidian Kangxi chao*, 71:3b–5a.

52 *Da Qing huidian Yongzheng chao*, 246:5a–5b; Deane, "The Chinese Imperial Astronomical Bureau," 272–304.

53 Among historians of Western science, Thomas Kuhn stressed the problems associated with planets. See his classic account in *The Copernican Revolution*, 45–99. For a different picture that stresses the administrative problems encountered by the Roman Catholic Church, see Heilbron, *The Sun in the Church*, 3–46.

54 In contrast, early Greek astronomers measured the tropical year from one vernal equinox to the next. The alternative measurement, the sidereal year, is measured according to the Sun's motion relative to the background stars. That motion takes place along the ecliptic, an imaginary line running across the sky from west to east that early Chinese astronomers called the "celestial perimeter" (*zhoutian*). Along the ecliptic, all twenty-eight Chinese constellations spread out. As recent archaeoastronomic studies show, most astronomers in ancient Greece, Mexico, and China charted the Sun's position against the constellations at sunrise and sunset. Atmospheric refraction in the morning and evening made it difficult to establish accurate charts of the Sun's position. For details, see Aveni, *Skywatchers*, 49–126.

55 Dagmar Schäfer reinterpreted Song Yingxing's *The Works of Heaven and the Inception of Things* (Tiangong kaiwu) in light of this Neo-Confucian premise. See her

The Crafting of the 10,000 Things, 50–90. For an interesting discussion of the Jesuit understanding of *qi*, see Elman, *On Their Own Terms*, 116–22.

56 Sivin, *Granting the Seasons*, 408–12.

57 For centuries the Roman Catholic Church was determined that Easter, which should be celebrated on the first Sunday after the first full moon following the vernal equinox, would not overlap with Passover. Jesus was arrested after arriving in Jerusalem for the annual observance of Passover. He was crucified on a Friday and rose from the dead on a Sunday. If the date of Easter overlapped with that of Passover, then the entire sequence of the passion would have made no sense. See Heilbron, *The Sun in the Church*, 12–23; Elman, *On Their Own Terms*, 80–106; Mosshammer, *The Easter Computus and the Origins of the Christian Era*, 40–55.

58 John Heilbron provided a detailed analysis of how equinoxes of precession were used by Catholic Church. See *The Sun in the Church*, 144–75.

59 For a contemporary description of precession, see Smart, *Textbook on Spherical Astronomy*, 226–48.

60 Huang Yi-long, "Yang Guangxian jiashi yu shengping kao"; Zurndorfer, "'One Adam,'" 141–68. Zhu Weizheng identified Yang as a litigation master. See his *Zouchu zhongshiji*, 173–74. For a detailed discussion of litigation masters in Qing China, see Macauley, *Social Power and Legal Culture*, 100–194.

61 For details, see Oxnam, *Ruling from Horseback*, 166–203.

62 The Manchu way represented an alternative to compromise with Han culture and customs; it was marked by a chauvinistic adherence to preconquest traditions, which in the political realm amounted to autocracy. Mark Elliott argued that the institutional base of the Manchu way, the Banner system, persisted to the end of the Qing dynasty. See *The Manchu Way*, 345–61.

63 Jami, *The Emperor's New Mathematics*, 57.

64 This *Qing Code* (1646) was revised in 1670 and again more extensively in 1723–27. It was issued in its final form in 1740. Only in 1905 was it discarded in favor of a modern legal system. See Bodde and Morris, *Law in Imperial China*, 57.

65 Yang Guanxian, *Bu de yi*, 5.

66 Ibid., 5–6. See also Hu, "Provenance in Contest," 17–29; Elman, *On Their Own Terms*, 139–41.

67 Yang Guanxian, *Budeyi*, 6, 11–14.

68 Ibid.

69 Jones, *The Great Qing Code*, 238–39. For the Chinese text, see Shen Zhiqi, *Da Qinglü jizhu*, 548–49. I use William Jones's translation in his erudite work, *The Great Qing Code*, in the main text.

70 Jones, *The Great Qing Code*, 239–40. For the Chinese text, see Shen Zhiqi, *Da Qinglü jizhu*, 550–52.

71 Sai was a set of temple-based liturgical rituals that evolved from the existing models provided by the Buddhist, Daoist, and Confucian ritual regimes. It provides detailed information of both spectacle and sacrifice—the processions, the operas, and the food offerings of the great temple festivals of southern Shanxi. See Johnson, *Spectacle and Sacrifice*, 303–19.

72 See Elman, *On Their Own Terms*, 142–44, for some of the other political implications of Yang's complaint.

73 The charges brought against Schall and his fellow Jesuits differed from those leveled against the "soulstealers" studied by Philip Kuhn. In the latter case, the

"sorcery scare of 1768," the defendants were accused of having violated sacrificial norms as defined in statute 162 from *The Qing Code*'s section on ritual (*lilü*). See Kuhn, *Soulstealers*, 84–93.

74 Ibid.

75 Shen Zhiqi, *Da Qinglü jizhu*, 396–97; Jones, *The Great Qing Code*, 176.

76 Yang Guangxian, *Bu de yi*, 43–47. Cf. Huang Yi-long, *Shehui tianwenshi shijiang*, 94–120.

77 Liu Fengyun, *Qingdai sanfan yanjiu*, 128–75.

78 For details of the contest, see Chu, "Scientific Dispute in the Imperial Court."

79 Väth, *Johann Adam Schall von Bell SJ*, 241–66. For the disarray of the missionaries after 1664, see Brockey, *Journey to the East*, 125–42.

80 Verbiest, *Lifa budeyi bian*, 139–42, 144–45.

81 Ibid., 139–42.

82 Ibid., 144–45.

83 Ibid., 139. This line of thinking was already present in the policy questions set in the civil service examination of the late Ming. See Elman, *A Cultural History of Civil Examinations in Late Imperial China*, 468–71.

84 Verbiest, *Lifa budeyi bian*, 189–93.

85 Golvers, *The Astronomia Europaea of Ferdinand Verbiest, S. J.*, 62–73; Martzloff, "Space and Time in Chinese Texts," 81–83.

86 Verbiest, *Lifa budeyi bian*, 179–89.

87 Ibid., 139.

88 Needham, *Science and Civilisation in China*, 3:228.

89 George Saliba, *Islamic Science and the Making of the European Renaissance*, 193-232.

3. SEARCHING FOR TRUTH IN THE ORIGINS OF CIVILIZATIONS

1 Richard Smith provides an overall survey on *Xieji bianfang shu*, see his *Fortune-tellers and Philosophers*, 49-91.

2 Ibid., 98-108.

3 Most historians with Jesuit affiliations avoid this judgment.

4 Li Zubai, *Tianxue chuangai*, 1055–62. See also Mungello, *The Forgotten Christians of Hangzhou*, 85–88; Wang Yangzong, "'Xixue zhongyuan' shuo zai Ming Qing zhiji de youlai ji qi yanbian," 42–43.

5 Buglio, *Bu de yi bian*, 134–35.

6 Lu Jiuyuan, *Xiangshan ji*, 22:10a. Lu Jiuyuan's original formulation included a temporal dimension: he claimed that a wise man of ten thousand years ago would have arrived at the same set of abstract principles as a wise man would today.

7 Joseph Levenson's classic account, in *Confucian China and its Modern Fate*, captures this narrative succinctly but fails to account for the struggles that broke out along ethnic lines.

8 Liu Ning, *Jue si lu*, 3b; Lundbaek, *Joseph de Prémare*, 142–45.

9 Liu Ning, *Jue si lu*, 3b.

10 Zhang Xingyao, *Tianru tongyi kao*, 433–35, 439–41; Mungello, *The Forgotten Christians of Hangzhou*, 95–167.

11 Zhang Xingyao, *Tianru tongyi kao*, 444–45.

12 Von Collani, "Introduction: The Author of the 'Journal,'" 3–23; Landry-Deron, "Les mathématiciens envoyés en Chine par Louis XIV en 1685," 423–63.
13 Jami, *The Emperor's New Mathematics*, 139-59.
14 This is widely known due to Bouvet's hagiography of the Kangxi emperor. See Mungello, *Curious Land*, 303.
15 Walker, *The Ancient Theology*, 1. Some specialists insist on using the term "Figurist" to describe Bouvet and his colleagues. See Von Collani, *P. Joachim Bouvet S. J. Sein Leben und sein Werk*, 1–9.
16 Hermes Trismegistus and Orpheus were legendary figures, and their names were associated with these texts only to lend them an antique aura. For the intellectual milieu out of which Bouvet emerged, see Von Collani, *P. Joachim Bouvet S. J. Sein Leben und sein Werk*, 1–9; Rule, "Moses or China," 306–22; Pocock, *Barbarism and Religion*, 4:104–9.
17 Mungello, *Curious Land*, 329–53; Van Kley, "Europe's 'Discovery' of China and the Writing of World History," 358–85.
18 Rule, *K'ung-tzu or Confucius?*, 134–37; Walker, *The Ancient Theology*, 194–230.
19 Von Collani, *P. Joachim Bouvet S. J. Sein Leben und sein Werk*, 145–203. Tripartite historical models were common in early modern Europe and China. For example, see Elman, *Classicism, Politics, and Kinship*, 231–37.
20 Von Collani, "Introduction," 23–55.
21 Bouvet to an unknown correspondent, French Jesuit Archives, Vanves, Paris, APPSI, ms. Vivier 52.
22 Walker, *The Ancient Theology*, 199. Leibniz's fascination with China was clearly driven by his interest in binary logic, universal language, and natural theology. See Perkins, *Leibniz and China*, 108–57.
23 Mungello, *Leibniz and Confucianism*, 46–62.
24 Bouvet to an unknown correspondent, French Jesuit Archives, Vanves, Paris, APPSI, ms. Vivier 52.
25 The letter is listed as entry no. 78 in "Bibliographie des Ouvrages et des lettres du P. Bouvet," compiled by Janette Gatty, in *Voyage de Siam du Père Bouvet*.
26 Rule, *K'ung-tzu or Confucius?*, 154–67; Witek, *Controversial Ideas in China and in Europe*, 143–53.
27 Von Collani, "Claudio Filippo Grimaldi S. J. zur Ankenft des Päpstilichen Legaten Charles-Thomas Maillard de Tournon in China."
28 I found two editions of *The Essentials of the Heavenly Learning* (Tianxue benyi) in the Biblioteca Apostolica Vaticana: Borg.Cinese.357(9) and Borg.Cinese.317(15). Two other editions are in the Bibliothèque Nationale: BN Chinois 7160 and BN Chinois 7162. I shall rely on Borg.Cinese.317(15), dated 1707, as it is the most complete and includes a preface by Han Tan (1637–1704), minister of rites and a Hanlin scholar. For more on the various editions of Bouvet's work, see Mungello, "Unearthing the Manuscripts of Bouvet's 'Gujin' after Nearly Three Centuries."
29 Von Collani, *P. Joachim Bouvet S. J. Sein Leben und sein Werk*, 52-54.
30 Rule, *K'ung-tzu or Confucius?*, 162 (my emphasis).
31 Bouvet, *Gujin jingtianjian tianxue benyi*, 5a–5b, 86b–99a.
32 Kessler, *K'ang-hsi and the Consolidation of Ch'ing Rule*, 137–46; Struve, "The Hsü Brothers and Semiofficial Patronage of Scholars in the K'ang-hsi Period," 249.
33 Bouvet, *Gujin jingtianjian tianxue benyi*, 83b, 84a–b, 85a–86b.
34 Von Collani, "Charles Maigrot's Role in the Chinese Rites Controversy," 161–83.

35 *Kangxi chao Manwen zhupi zouzhe quanyi*, docs. 872, 875, 882.
36 Rule, *K'ung-tzu or Confucius?*, 156–58.
37 Bouvet submitted the first installment of his research on *The Book of Changes* to the emperor in the fifth month of 1711 and continued until the second month of 1712. See *Kangxi chao Manwen zhupi zouzhe quanyi*, docs. 1711, 1719, 1724, 1725, 1738, 1741, 1752, 1755, 1759, 1760, 1764, 1776.
38 Bouvet, *Yiyin yuangao*, 1–3.
39 Walker, *The Ancient Theology*, 227.
40 Shelford, *Transforming the Republic of Letters*, 174.
41 Walker, *The Ancient Theology*, 227–28.
42 Lackner, "A Figurist at Work," 23–56; Von Collani, *P. Joachim Bouvet S. J. Sein Leben und sein Werk*, 145–203.
43 Mungello, *Curious Land*, 312–28.
44 Kangxi chao Manwen zhupi zouzhe quanyi, docs. 1719, 1724, 1738; cf. Luo Lida, "Baijin yanjiu Yi jing shishi jikao," 173–85.
45 Wu Boya, "Yesu huishi Baijin dui 'Yi jing' de yanjiu," 57–58.
46 Li Guangdi, "Yuzhi xu" (Imperial preface) and "Fanli" (User's guide), *Yuzuan Zhou yi zhezhong*; cf. Zhu Xi, *Zhouyi benyi.*
47 On Foucquet, see *Kangxi chao Manwen zhupi zouzhe quanyi*, doc. 1755; *Kangxi chao Hanwen zhupi zouzhe huibian*, doc. 856. On Prémare, see Lundbaek, *Joseph de Prémare*, 114.
48 Lundbaek, *Joseph de Prémare*, 115.
49 Ibid.
50 Wang Xichan, *Xiao'an yishu zhisi: Zazhu*, 1a–1b.
51 Qian Baozong, "Mei Wu'an xiansheng nianpu," 608–38; Hang Shijun, "Mei Wending zhuan," 1a–16a.
52 Qian Baozong, "Mei Wu'an xiansheng nianpu," 610–17.
53 Hang Shijun, "Mei Wending zhuan," 1a–16a; *Guangxu Xuancheng xianzhi*, 10.
54 Mei, *Jixuetang shiwenchao*, 3–8.
55 Lu Longqi, *Sanyutang riji*, 28–29, 33, 138–39.
56 Wei Xi (1624–81), for instance, wrote a preface for Mei's *Questions in Mathematical Astronomy* in which he argued that only in mathematical astronomy could the moderns supersede the ancients. Wang Yuan (1648–1710), who also wrote a preface, stressed that Mei illuminated technical details and exposed the immorality of charlatans, a laudable Confucian act. For his part, Mei recalled how Wan Sitong had encouraged him to finish this work. Wei, Wang, and Wan, all of them from the lower Yangzi, applauded Mei's mathematical astronomy as part of the "practical and concrete learning" (shixue) they were promoting. All these comments were made prior to Li Guangdi's patronage.
57 Hashimoto, "Bai Buntei no rekisangaku"; Hashimoto, "Bai Buntei no sūgaku kenkyu."
58 Elman, *From Philosophy to Philology*, 100–112.
59 Han Qi, "Patronage scientifique et carrière politique," 7–37.
60 Li Guangdi, "Xu," Lixue yiwen, 3a.
61 Li Guangdi, *Rongcun yulu; Rongcun xuyulu*, 431, 485, 775–76.
62 Arai Shinji, "Astronomical Studies by Zhao Youqin."
63 Yamada Keiji, *Shushi no shizengaku*, 413–22.
64 Mei, *Lixue yiwen*, 3:1a–19b.

65 Ibid., 11a–12b; Jami, "History of Mathematics in Mei Wending's (1633–1721) Work."
66 Mei, *Lixue yiwen*, 10a–10b.
67 According to Catherine Jami, this was a new argument for Mei Wending: "It is not their mediocrity in mathematics that he reproaches them for, but their aspiration to do better than the ancients. Denouncing the vanity of this pretension, he associated it with their objective—absurd and dishonorable, it goes without saying—to transform China by their (religious) doctrine. It is a moral condemnation, not an intellectual criticism that he formulates against them." See Jami, "Légitimité dynastique et reconstruction des sciences," 708.
68 Hui Dong, *Zhou yi shu*, 224.
69 Lynn, *The Classic of Changes*, 445.
70 Elman, *A Cultural History of Civil Examinations*, 470.
71 Ibid., 468–73.
72 *Lixue yiwen bu*, Mei's supplement to *Lixue yiwen*, was written at the emperor's request. See Jami, "Légitimité dynastique et reconstruction des sciences," 715, 724–25.
73 Mei, *Lixue yiwen*, 10a–10b. See also Jami, "Légitimité dynastique et reconstruction des sciences," 716.
74 Mei, *Lixue yiwen*, 13a–15b.
75 Ibid., 11a–12b.
76 Ibid., 18b–20a.
77 Ibid., 20a.
78 *Shengzu renhuangdi yuzhi wenji*, 19:6a–13b.
79 Ibid.
80 Ibid.
81 Chemla, "Similarities between Chinese and Arabic Mathematical Writings," 207–66.
82 Sivin, *Granting the Seasons*, 218–25.
83 Chen Jiujin, *Huihui tianwenxue shi yanjiu*, 231–47, 309–75.
84 Saliba, *Islamic Science and the Making of the European Renaissance*, 131–232. For more on Islamic spherical trigonometry, see Van Brummelen, *The Mathematics of the Heavens and the Earth*, 166–215.
85 Huang Yi-long, "Qingchu Qintianjian zhong ge minzu tianwenxuejia de quanli qifu," 75–108.
86 Ben-Dor Benite, *The Dao of Muhammad*, 115–62; Murata, Chittick, and Tu, *The Sage Learning of Liu Zhi*, 3–19; Frankel, "Uncontrived Concord," 46–54.
87 Ben-Dor Benite, *The Dao of Muhammad*, 165–68.
88 For a detailed exposition of Liu Zhi's work, see Frankel, *Rectifying God's Name*.
89 Murata, Chittick, and Tu, *The Sage Learning of Liu Zhi*, 49–80, 581–617.
90 Xu Yuanzheng, "Xu," 1a–2a.
91 Ibid.
92 Ibid, 4a–4b.
93 For details, see Naquin, *Peking*, 500–554.
94 Cf. Pan Xinghui, "Dai Zhen lingyiguan fawei."
95 Ji Yun, *Yuewei caotang biji*, 10:11b.
96 Ibid., 7:2b.
97 Ibid.
98 Ibid., 5:7a.

99 Ibid., 6:4a.

100 Ibid., 23:1a–1b.

4. HOW TO BUILD A COALITION AROUND SCIENCE

1 The term *jiaohua wang*, applied to the pope, appears frequently in *Qing zhong qian qi xiyang tianzhujiao zai Hua huodong dangan shiliao* and *Kangxi chao Manwen zhupi zouzhe quanyi*.

2 *Qing shilu*, 5.139:30a–33b.

3 Crossley, *A Translucent Mirror*, 55–128.

4 Joseph Needham identified the Old Man Star as Canopus (Carina), the second brightest star in the sky; see Needham, *Science and Civilisation in China*, 3:238.

5 *Shengzu renhuangdi yuzhi wenji*, 29:3a–3b; Sima, *Shiji suoyin*, 4:8a.

6 Jami, "Imperial Control and Western Learning." The script for the performance at the Nanjing observatory was similar to that for the astronomy contest held in 1670, described in chapter 2.

7 Elman, *A Cultural History of Civil Examinations in Late Imperial China*, 1–61.

8 Xie, "Qingchu Shunzhi-Kangxi jian zhi dangzheng"; Struve, "The Hsü Brothers and Semiofficial Patronage of Scholars in the K'ang-hsi Period"; Wang Jingqi, "Gao Wenque yishi," 36a–37a.

9 Li Guangdi, *Rongcun yulu; Rongcun xuyulu*, 741–43.

10 Prospev Intorcetta (1625–96), an enthusiastic Italian Jesuit, introduced Fontenay and Gabiani to Kangxi when the emperor was visiting Intorcetta's church in Hangzhou, at the southern end of Kangxi's tour; see Mungello, *The Forgotten Christians of Hangzhou*, 41–67.

11 Hoang, *Zhengjiao fengbao*, 98–101.

12 Li Guangdi, *Rongcun yulu; Rongcun xuyulu*, 736–43. According to Li's account, Xu and Gao had taken advantage of his absence from the court (he had been dismissed from his post as grand secretary) to carry out a surreptitious power grab. When they attempted to persuade Xiong Cilu (1635–1709), Kangxi's previous teacher and Li's trusted ally, to join their camp, he hesitated but eventually yielded, proving his loyalty to the Xu-Gao faction by declaring to the emperor that Li was ignorant. According to Han Qi, Li's disgrace at Nanjing was a set-up by the emperor. Having spoken with Jean de Fontenay prior to his meeting with Li, "Kangxi had carefully prepared for his conversation with Li Guangdi. Not only was he playing some kind of trick on Li, but he took advantage of the situation to display his knowledge, as he often liked to do." After his return to Beijing, Kangxi relieved Li of his post at the Hanlin Academy because of his poor performance; see Han Qi, "Patronage scientifique et carrière politique: Li Guangdi entre Kangxi et Mei Wending," 10–12.

13 *Qing shilu*, 5.139:32a.

14 Huang Yi-long, "Qingqianqi dui Zui-Can liangsu xianhou cixu de zhengzhi," 71–93.

15 Kangxi also argued that the retrograde movement of Mars described in *Celestial Officials in the Historical Record* lacked empirical support. He argued that Mars did not display any retrograde motion. He was wrong, but on this occasion he nevertheless demonstrated that calendrical constants and methods presented in

historical records cannot be taken for granted; they need to be verified empirically; see *Qing shilu*, 5.132:32b.

16 Because the Earth's axis of rotation wobbles, tracing a circle over a period of 26,000 years, the location of the Sun at biannual equinoxes changes relative to the background of distant stars. This apparent motion is known as the "precession of the equinoxes."

17 Yang Guanxian, *Bu de yi*, 39–40.

18 *Qing shilu*, 5.132:33a.

19 Wu, *Communication and Imperial Control in China*, 34. The positive outcome of this ideology was a system of national granaries for emergencies; see Will, *Bureaucracy and Famine in Eighteenth-Century China*, 127–48.

20 Elvin, *The Retreat of the Elephants*, 414–20.

21 Ibid., 420.

22 Ibid., 420–36.

23 Qu Dajun, *Guangdong xinyu*, 7.

24 Elvin, *The Retreat of the Elephants*, 73.

25 Ibid., 72–78.

26 Guy, *Qing Governors and Their Provinces*, 79–110.

27 Feng, *Yongzheng zhuan*, 9–49; Spence, *Emperor of China*, 123–34.

28 Feng, *Yongzheng zhuan*, 9–65.

29 Spence, *Emperor of China*, 134. I have made small changes to the translation for the sake of consistency.

30 Spence, *Chinese Roundabout*, 98.

31 *Kangxi chao Hanwen zhupi zouzhe*, doc. 930.

32 *Shengzu renhuangdi yuzhi wenji* 3.3:5a. Even after their setbacks in the Rites Controversy, the Jesuits held important posts in the Astrocalendrical Bureau, the Academy of Mathematics, and a number of imperial workshops that benefited from Western technology and know-how.

33 Elman, *On Their Own Terms*, 177–80.

34 *Qinding guozijianzhi*, 2a; *Qinding huangchao wenxian tongkao*, 66:25a–25b.

35 Jami, "Learning Mathematical Sciences during the Early and Mid-Ch'ing," 238–39.

36 Jami, *The Emperor's New Mathematics*, 284–311.

37 Han Qi, "Patronage scientifique et carrière politique: Li Guangdi entre Kangxi et Mei Wending," 24–29.

38 Li Guangdi, *Rongcun yulu; Rongcun xuyulu*, 643.

39 *Kangxi chao Manwen zhupi zouzhe quanyi*, doc. 1779.

40 Little has been written about the encounter between Mezzabarba and the Kangxi emperor. For context, see Minamiki, *The Chinese Rites Controversy*, 62–67; Brockey, *Journey to the East*, 169–98.

41 *Qing zhong qian qi xiyang Tianzhujiao zai Hua huodong dang'an shiliao*, 1:31-36.

42 Ibid., 1:36–37.

43 Ibid., 1:37.

44 Ibid.

45 Ibid., 1:37–38.

46 Ibid., 1:38–39.

47 Ibid., 1:39.

48 Ibid., 1:40–41.

49 Ibid., 1:36–37.

50 Ibid., 1:47.

51 Menegon, *Ancestors, Virgins, and Friars*, 116–53. For a remarkable collection of primary documents regarding local persecution of missionaries, see Wu Min and Han Qi, *Ouzhou suo zang Yongzheng Qianlong chao Tianzhujiao wenxian huibian*, 19–197.

52 Jami provides the best and most detailed analysis of *Shuli jingyun* to date; see her *The Emperor's New Mathematics*, 315–63.

53 Hashimoto, "Bai Buntei no rekisangaku," 491–518; "Bai Buntei no sūgaku kenkyu," 233–79; "Rekisho kōsei no seiritsu," 49–92.

54 Yinti was commanding the military campaigns against both the Zunghars and Tibet when his father died on December 21, 1722. Peter Perdue argues that Yinti's absence made it easier for Yinzhen to claim the throne; see his *China Marches West*, 227–40.

55 Bartlett, *Monarchs and Ministers*, 169–99.

56 *Yuzhi lixiang kaocheng houbian*, "Zouyi," 1a–1b. Joseph Needham noted many observations of eclipses made by Jacques-Philippe Simonelli and published jointly by Kögler and Melchior della Briga between 1744 and 1747; see Needham, *Science and Civilisation in China*, 3:454.

57 Elman, *On Their Own Terms*, 190–221.

58 For a detailed analysis of the *Compendium* and the *Supplement*, see Hashimoto Keizō, "Rekisho kōsei no seiritsu," 49–92; "Daen hō no tenkai," 245–72.

59 *Zhongguo kexue jishu dianji tonghui: Shuxue juan*, 5:1148–1150; Jami, *The Emperor's New Mathematics*, 315-19.

60 *Yuzhi lixiang kaocheng houbian*, "Zouyi," 2b–3b.

61 *Yuzhi lixiang kaocheng houbian*, "Zouyi," 4b–7a. For details on this episode, see Guo Shirong and Li Di, "Putaoya chuanjiaoshi Xu Maode zai qintianjian de tianwen gongzuo," 20–23.

62 Huters, "From Writing to Literature," 67–76.

63 Durand, *Dai Mingshi nianpu*, 563.

64 Ōtani, *Shindai seiji shisōshi kenkyū*, 125–28.

65 Pierre-Henri Durand offers a documentation of this episode in *Dai Mingshi nianpu*.

66 Fang Bao, *Fang Bao ji*, 869. During Fang's visit to Li's mansion, Fang also met Li's disciple, Yang Mingshi (1660–1736), who had helped his master satisfy the Kangxi emperor's demand for an orthodox interpretation of *The Book of Changes* at a time when Bouvet was offering his own radical synthesis of that ancient classic and modern astronomy. Yang became a leading voice of the science faction, and he later weathered the Yongzheng emperor's purges. He was to recall members of the science faction to the court in the first year of the Qianlong emperor's reign.

67 Ōtani, *Shindai seiji shisōshi kenkyū*, 140–41.

68 Fang Bao, *Fang Bao ji*, 335.

69 When it came to recognition, Mei and his lineage stood out, a truly rare case of a renowned lineage possessing specialized and technical knowledge. In 1712, when the Kangxi emperor opened the Studio for the Cultivation of Youth, he summoned Mei Wending's grandson, Mei Juecheng (1680–1763), to teach there. On being released from prison, Fang became his colleague. Rescued by Li and sheltered by the glory of the Mei lineage, Fang had many reasons to associate with the group

whose promotion of Western astronomy linked his two saviors. And their fame had everything to do with it. Fang Bao, *Fang Bao ji*, 336.

70 Zhao Erxun, *Qing shi gao*, juan 290, liezhuan 77, 2642.

71 Fang Bao, *Fang Bao ji*, 263–64; Lu Wenzhao, *Baojingtang wenji*, 352–53.

72 Yang Mingshi's conflict with the Yongzheng emperor was the subject of many rumors and disputes; see Guo Chengkang, "Zhengzhi chongtu yu wenhua gehe."

73 Fang Bao, *Fang Bao ji*, 262; Yuan Mei, *Yuan Mei quanji*, 2:23–25.

74 *Qing shilu*, 9.14:13b–14a. The seven literati were: Zhuang Henyang (*jinshi* degree), Pan Yongji, Cai Dejun, Qin Huitian (1702–64), Wu Nai (*juren* degree), Guan Xianyao, and Xia Zonglan (*shengyuan*).

75 *Wenxian congbian quanbian*, 2:498–99.

76 Ibid., 2:499–503.

77 The young cohort included Cai Dejin (*juren* 1726) and the brothers Wu Nai (*jinshi* 1736) and Wu Ding (*jinshi* 1751).

78 Team members included Fang Guancheng, Lu Jianzeng, and Song Zhongyuan.

79 Qian Xuan, *Sanli tonglun*, 21–33.

80 Chang So-an, *Shiba shiji lixue kaozheng de sixiang huoli.*

81 *Siku quanshu zongmu*, 22:16b–17a.

5. AN OUTSIDER ENTERS THE MAINSTREAM

1 Duan, "Dongyuan nianpu dingbu," 650; Hong Bang, "Dai xiansheng xingzhuan," 1b–2a; Elman, *On Their Own Terms*, 258; cf. Chin and Freeman, *Tai Chen on Mencius*, 1–2.

2 Wang Yangming challenged Zhu Xi's emendations to the text of *The Great Learning*, triggering widespread debates over the book's exact meaning; see Huang Jinxing, *You ru sheng yu*, 351–91; Rusk, *The Rogue Classicist*, 282–61.

3 Wei Jiangong dismissed this anecdote as the sort of hagiography common to most biographical annals (*nianpu*) in late imperial China; see *Wei Jiangong wenji*, 5:72. The eminent reformer Liang Qichao, credited as the founder of modern Dai Zhen scholarship, considered this anecdote an indication that as a child Dai already possessed "scientific spirit," a combination of healthy skepticism and positivist methodology; see Liang, *Qingdai xueshu gailun*, 56–57. Benjamin Elman, in contrast, reads the anecdote as illustrating the way Dai joined the chorus of the evidential scholarship movement, using philology to defy Zhu Xi's classical authority; see Elman, *On Their Own Terms*, 258–59.

4 The historical idea of the child genius has been explored in a number of recent biographies; see, for example, DeNora, *Beethoven and the Construction of Genius*, 5–9; Shields, *Logic and Sin in the Writings of Ludwig Wittgenstein*, 1–7; Michael Fisher, "Autobiographical Voices (1, 2, 3) and Mosaic Memory."

5 Zurndorfer, *Change and Continuity in Chinese Local History*, 167–69.

6 Ibid., 126–35.

7 Classic accounts of Huizhou merchants are found in Ping-ti Ho, "The Salt Merchants of Yang-chou"; Fujii, "Shinan shōnin no kenkyū"; Fu Yiling, *Ming Qing shidai shangren ji shangye ziben*, 3–88. After 1980, studies of Huizhou quickly multiplied as more and more commercial and legal documents became available for researchers; see Wang Zhenzhong, *Huizhou shehui wenhuashi tanwei*, 312–445.

The emergence of a distinct scholarship in eighteenth-century China, encouraged and sponsored by the merchants of Huizhou, was also significant. On the expansion of Huizhou merchant culture, see Qitao Guo, *Ritual Opera and Mercantile Lineage*, 50–86.

8 A number of different terms were used in late imperial China to refer to examination essays. Some of the more common were *zhiyi* (implementation arts), *shiwen* (contemporary prose), *baguwen* (eight-legged prose), and *zhijuwen* (implementation of examination prose). I will use only the term "examination essays."

9 Miyazaki, *China's Examination Hell*, 18–38; Rawski, *Education and Popular Literacy in Ch'ing China*, 24–53; Elman, *A Cultural History of Civil Examinations in Late Imperial China*, 260–76; Mann, "The Education of Daughters in the Mid-Ch'ing Period," 19–26.

10 Duan, "Dongyuan nianpu dingbu," 650.

11 Ibid., 651.

12 Xu Chengyao, *Shexian zhi*, 7:7a.

13 Leung, "Elementary Education in the Lower Yangtze Region," 384–87. For further details on community school curricula in Ming China, see Schneewind, *Community Schools and the State in Ming China*, 94–111.

14 Woodside and Elman, "The Expansion of Education in Ch'ing China," 525–28.

15 Ibid., 393–96; Elman, *A Cultural History of Civil Examinations in Late Imperial China*, 260–76.

16 Leung, "Elementary Education in the Lower Yangtze Region," 389–91. For a lively description of a clan school, see Cao Xueqin, *The Story of the Stone*, 1:202–28.

17 Philip A. Kuhn, *Origins of the Modern Chinese State*, 2–8; Elman, *A Cultural History of Civil Examinations in Late Imperial China*, 125–72.

18 Esherick and Rankin, "Introduction"; Yamamoto, *Dentō Chūgoku no chiikizō*, 1–9.

19 Miyazaki, *China's Examination Hell*, 18–38; Elman, *A Cultural History of Civil Examinations in Late Imperial China*, 143.

20 Dai, *Dai Zhen quanshu*, 6:541. Dai claimed that he determined to pursue the way at the age of seventeen *sui*, which would be his sixteenth year. Therefore, I give the date as 1740.

21 Ibid., 6:440. As Dai wrote: "Although most people in Huizhou are merchants, they all share the literati ethos." For an elaborate discussion of how merchants adopted the literati ethos, see Yü Ying-shih, *Zhongguo jinshi zongjiao lunli yu shangren jingshen*, 121–36.

22 Ainajushi, *Doupeng xianhua*, 67. This evidence is also cited in Kinoshita, "Tai Shin to Wanpai no gakujutsu," 542n24.

23 Duan, "Dongyuan nianpu dingbu," 653.

24 Ibid.

25 Wang Chang, "Dai Dongyuan xiansheng muzhi ming," 1b; Hong Bang, *Dai xiansheng xingzhuan*, 1a.

26 The weakness of Dai Zhen's branch of the family may be inferred from the result of a property dispute in which Dai was embroiled in 1754; see Lu Wenzhao, *Baojingtang wenji*, 259.

27 Qian Mu wrote that Dai "speculated and searched in the dark" to initiate his research agenda; see Qian, *Zhongguo jinsanbainian lai xueshushi*, 1:314.

28 Duan, *Jingyunlou ji*, 201.

29 Hong Bang, *Dai xiansheng xingzhuan*, 2a. During the first half of the seventeenth

century, Mao Jin (1599–1659) sponsored the publication of the Song edition of the paleographical dictionary *Shuowen jiezi* as well as all of the Thirteen Classics; see Ye Dehui, *Shulin qinghua*, 7:16b–18b. Ye complained of the poor quality of the Mao editions. In the eighteenth century Zhu Yun (1729–81) and Sun Xingyan (1753–1818) each sponsored a more refined version of *Shuowen jiezi*; see Sun Xingyan, *Sunshi chongkan songben shuowen xu*, 3a.

30 It took not only time but also specific training to master the elegant style of a good examination essay; see Elman, *A Cultural History of Civil Examinations in Late Imperial China*, 380–403.

31 Hong Bang, *Dai xiansheng xingzhuan*, 4b–5a.

32 Yü Ying-shih's note on the twentieth-century discovery of the *Jing kao* manuscript remains the best discussion available; see Yü, *Lun Dai Zhen yu Zhang Xuecheng*, 192–96.

33 For more on this study, see chapter 6.

34 Liang, *Qingdai xuezhe zhengli jiuxue zhi zongchengji*, 228.

35 Dai, *Dai Zhen quanshu*, 5:275–76. On the date of this piece, see Duan, "Dongyuan nianpu dingbu," 659.

36 Cheng Yaotian, "Wuyou ji," 5b.

37 Elman, *A Cultural History of Civil Examinations in Late Imperial China*, 287.

38 Duan Yucai mistakenly dated the encounter to 1757. Qian Mu noted this mistake; while he believed that the meeting must have taken place prior to 1753, he was uncertain as to whether it occurred in 1749 or 1750; see Qian, *Zhongguo jinsanbainian lai xueshushi*, 312–13. Yang Yingjin confirmed Qian Mu's correction and suggested that the encounter took place in the fifth month of 1750; see *Shunshan Shi Zhongming xiansheng nianpu*, 45b.

39 Dai, *Dai Zhen quanshu*, 6:370.

40 Almost all of the books on Dai Zhen or Qing intellectual history use this text to show a new scholarly orientation. But I have not seen a serious attempt to historicize this text.

41 *Shunshan Shi Zhongming xiansheng nianpu*, 1a–1b.

42 Shi Jing was involved in three associations: the Association for Overcoming and Restoring (Kefu) in Chengjiang, the Association for the Restoration of Seven (Fuqi) in Xishan, and the Association for Moral Action (Gongxing) in Jijiang; see *Shunshan Shi Zhongming xiansheng nianpu*, 2a.

43 *Shunshan Shi Zhongming xiansheng nianpu*, 3b–4a. It was his mother who first took Shi Jing to a meeting of Gu Pei's Association for the Restoration of Seven. Gu was devoted to the principles of Donglin activism; see Qian Mu, *Zhongguo jinsanbainian lai xueshushi*, 20.

44 Qian Mu, *Zhongguo jinsanbainian lai xueshushi*, 8–21.

45 A proliferation of private Confucian academies during the late Ming, the increasing gap between the number of examination graduates and the number of available administrative openings, and significant inflation affecting urban areas combined to incite political dissent among the gentry.

46 The Donglin faction has long been a focus for China historians. Studies include Ono, *Minki Tōsha kō*; Dardess, *Blood and History in China*; Elman, "Imperial Politics and Confucian Societies in Late Imperial China"; and Mizoguchi, "Iwayuru Tōrinha jinshi no shisō."

47 A brilliant and fascinating exposition of this position is Huang Zongxi's *Min-*

gyi daifang lu. An English translation was published in De Bary, "A Plan for the Prince." For a contextual analysis, see Struve, "Huang Zongxi in Context," 475–79.

48 Liu Shipei, *Liu Shipei shixue lunzhu xuanji*, 396–97.

49 Wang Fansen, "Mingmo Qingchu de renpu yu shengguohui," 679–712; Lu Miaw-fen, *Xiaozhi tianxia*, 155–84.

50 Huang Jinxing, *You ru sheng yu*, 217–311; Lu Miaw-fen, *Yangmingxue shiren shequn*, 327–68.

51 *Shunshan Shi Zhongming xiansheng nianpu*, 24b–25a.

52 Ibid., 29a–31b.

53 For details on mourning rituals, see Kutcher, *Mourning in Late Imperial China*, 120–89; Lu Miaw-fen, *Xiaozhi tianxia*, 187–213.

54 Gaozong chunhuangdi shilu, *Qing shilu*, 11:208:12b.

55 *Shunshan Shi Zhongming xiansheng nianpu*, 32a.

56 Ibid., 33a.

57 Wang Zhonghan, *Qing shi liezhuan*, 16:2b–3a.

58 *Shunshan Shi Zhongming xiansheng nianpu*, 32a–43b.

59 Ibid., 44b–45a.

60 Ruan Kuisheng, *Chayu kehua*, 60.

61 Dong Chao, *Donggao zachao*, 22.

62 Jiang Han, *Shiwengshanfang zhaji* 9:8b, For a detailed discussion, see He Zehan, *Rulin waishi renwu benshi kaolue*, 1–14.

63 Gao Xiang, *Jindai de chushu*, 94–161.

64 Yuan Mei, *Yuan Mei quanji*, 6:319–20.

65 Ibid., 2:252.

66 Ibid., 2:253.

67 For this assessment of Dai's methodology, see Wang Hui, *Xiandai Zhongguo sixiang de xingqi*, 417–18; Elman, *From Philosophy to Philology*, 29–34.

68 Yuan Mei, *Yuan Mei quanji*, 2:253. Cf. Kinoshita, "Tai Shin to Wanpai no gakujutsu," 528; Elman, "From Value to Fact," 498–500.

69 Dai, *Dai Zhen quanshu*, 6:370.

70 Ibid., 6:371.

71 Dai, *Dai Zhen quanshu*, 6:371.

72 Ibid.

73 Ibid. On the "proto-trigonometry" in use before 1680, see Sivin, *Granting the Seasons*, 66–67.

74 Pioneering Japanese studies of Dai's synthesis of classical research and mathematical astronomy are: Yabuuchi, "Taishin no rekisan gaku," 27–34; Kawahara, "Taishin to seiyō lekisan gaku."

75 Dai, *Dai Zhen quanshu*, 6:371.

76 Ibid.

77 Ibid.

78 Yü Ying-shih, *Lun Dai Zhen yu Zhang Xuecheng*, 19–36.

79 Cf. Yamanoi, *Minshin shisōshi no kenkyū*, 361–77.

80 Dai was clearly not an outsider to the civil service examination system. I am speaking strictly of his methodology and his classical knowledge.

81 Yü Ying-shih, *Lun Dai Zhen yu Zhang Xuecheng*, 95–150.

82 Ibid., 191–210. Yü selected many examples from Dai's *Evidential Analysis of the Classics* to illustrate his point. He rarely used examples from Dai's many more

technical writings on such topics as mathematical astronomy, analytical phenology, and historical geography.

83 Yü Ying-shih, *Lun Dai Zhen yu Zhang Xuecheng*, 199.

84 Ibid., 108–21.

6. HOW TO DETHRONE JESUIT SCIENCE

1 The centrality of Dai Zhen's intervention has been acknowledged by many. Some of the most important assessments are: Yasuda and Kondō, *Tai Shin shū*; Mizoguchi, "Mōshi jigi soshō no rekishi teki kōsatsu"; Yamanoi, *Min-Shin shisōshi no kenkyū*; Yü Ying-shih, *Lun Dai Zhen yu Zhang Xuecheng*.

2 A recent dissertation is the first major English-language study of Qian Daxin: Sela, "Qian Daxin (1728–1804)."

3 The most eminent members of the technically proficient group that passed the palace examination in 1754 were Ji Yun (1724–1805), Qian Daxin, Wang Mingsheng (1722–98), Wang Chang (1725–1806), and Zhu Yun (1729–81).

4 Ruan Yuan, *Chouren zhuan*, 42:22a–22b. For a brief discussion of practical learning, see Rowe, *Saving the World*, 133–37.

5 Dai Zhen's relationship with Jiang Yong has long been a controversial one. Most historians, following Qian Mu and Yü Ying-shih, continue to focus on the moral dimension of their relationship; see Yü Ying-shih, *Lun Dai Zhen yu Zhang Xuecheng*, 210–29; Cai Jinfang, *Dai Zhen shengping yu zoupin kaolun*, 3–27. Too many have worried over the question of Dai's loyalty to his mentor by asking whether the younger man defied the Confucian hierarchy between master and disciple when he scorned his old master. This seems to me a rather counterproductive dispute. Instead I focus on their intellectual relationship.

6 Several historians have argued that Dai and Jiang met after 1742. For a detailed rehearsal (and rebuttal) of these arguments, see Cai Jinfang, *Dai Zhen shengping yu zuopin kaolun*, 3–15. I follow Cai in dating the meeting to 1742.

7 Jiang Yong, *Shuxue*, 1. Chu Pingyi explains Jiang Yong's ambivalent relationship with Mei Wending; see Chu, "Ch'eng-Chu Orthodoxy, Evidential Studies and Correlative Cosmology," 75–80.

8 Jiang Yong, *Shuxue*, 1.

9 Ibid.

10 Dai Zhen later changed the title to *Studies of Numbers* (Shuxue; this word has become the modern Chinese term for mathematics) and included it in *Siku quanshu*; see *Siku quanshu zongmu*, "Tianwen lifa lei," 83–84.

11 Jiang Yong, *Shuxue*, 3–4.

12 Ibid.

13 Dai's explanation of the use of this device, first named *Arithmetic by Counting Rods* (Chou suan), was later renamed *Napier's Bones Arithmetic* (Ce suan) and published as an addition to his compilation of *The Nine Chapters of Mathematics* (Jiuzhang suanshu).

14 Enrolled at Ziyang Academy were Cheng Yaotian, Zheng Mu (1714–92), Fang Ju (1729–89), Jin Bang (1735–1801), and Wang Zhaolong (1721–80).

15 Dai chose to label this collection of ancient texts "Taifu" in honor of Dai De, the Han dynasty scholar who compiled them; Dai De's official title was *taifu*, or "grand

mentor," and he served as the tutor to the heir apparent. This collection had barely survived to Dai's times and was widely considered obscure and unreadable.

16 Jiang Yong, *Shanyutang wenji*, cited in Qi Yongxiang, "Xinfaxian Dai Zhen 'Jiang Shenxiu xiansheng qishi shouxu' yiwen yipian," 122–23.

17 All of Dai's manuscripts were rediscovered and authenticated in the twentieth century. Yü Ying-shih described him as breathing new life into Confucianism; see Yü, *Lun Dai Zhen yu Zhang Xuecheng*, 191–210.

18 Dai, *Dai Zhen quanshu*, 5:313.

19 Wang Zhong, *Wang Zhong ji*, 262.

20 Hacking, *Representing and Intervening*, 21. Joseph Levenson, following Alfred North Whitehead, characterized the early Qing empiricists as nominalists. I think Jiang Yong fits Levenson's description quite well; see Levenson, *Confucian China and Its Modern Fate*, part 1, 6–9.

21 On Dai Zhen's misfortune: Lu Wenzhao, *Baojingtang wenji*, 259; Duan, "Dongyuan nianpu dingbu," 666. Dai Qinquan, a descendent of the Dai clan, recounted this event in the early twentieth century; see Cai Jinfang, *Dai Zhen shengping yu zuopin kaolun*, 109. I do not consider this information reliable.

22 The high cost of travel was usually a significant hurdle for many scholars taking the provincial or metropolitan examination; see Yang Liansheng, *Yang Liansheng lunwenji*, 93.

23 The first lodge of this sort was built in Beijing in 1560. Propelled by commercializing forces and networking among members of the national and local elites, more and more lodges sprang up in the capital, catering to sojourners who came from the same place, spoke the same dialect, and liked the same foods. Naquin, *Peking*, 598–612.

24 Qian Daxin, *Jiading Qian Daxin quanqi*, 9:672–73.

25 Kondō, *Shinchō kōkyogaku no kenkyū*, 278.

26 Cf. Yü Ying-shih, "Tai Chen's Choice between Philosophy and Philology," 80–81.

27 During the Manchu conquest of the Lower Yangzi Delta, the Manchu army slaughtered many in the cities of Yangzhou, Nanjing, and Jiading. Most of the loyalist forces in Jiading were organized and led by local licentiates, like Qian Daxin's father and grandfather. For almost a century after that, keeping or circulating any records of the slaughter was punishable by execution of the entire clan. Yet *A Concise Record of the Jiading Slaughter* (Jiading tucheng jilue), an essential document of the horrible event, was quietly passed on from generation to generation. Qian Daxin had heard of the cruelty of the Manchu conquest from his father and grandfather and probably read *A Concise Record* as a young man. This experience seeded an ambivalent attitude in young Qian's heart toward the Manchu conquest elite, which grew alongside his scholarly ambition; see Ōtani, *Shindai seiji shisōshi kenkyū*, 305–10; Kinoshita, *"Shinchō kōshogaku" to sono jidai*, 74–76. On *A Concise Record of the Jiading Slaughter*, see Struve, *The Ming-Qing Conflict*, 242–43.

28 Qian Daxin, *Qian Xinmei xiansheng nianpu*, 3–7. For more details, see Sela, *Qian Daxin (1728–1804)*, 34–52.

29 Uxorilocal marriage was a good tactic for social mobility—a talented young man who married into a wealthy family could access more resources for examination success and build a familial alliance. When Qian Daxin agreed to marry into the Wang family, the Wangs abruptly had two young men, not one, on the rise. The marriage also provided Qian the financial support he needed for the repeated costs of travel and lodging incurred while taking examinations, as well as improved

access to the network of literati who informed him about the critical academic issues of the times.

30 Qian Daxin, *Jiading Qian Daxin quanji*, 829–30.

31 Chai Degeng, *Shixue congkao*, 255–56.

32 Qian Daxin, *Qian Xinmei xiansheng nianpu*, 3–7.

33 Among these examiners, Gao Bin (d. 1755), a Manchu Bannerman who sat on the Grand Council, was especially fond of this new group and invited them to stay at his mansion in Qingjiangpu, the southern port of the Grand Canal, along with other prominent scholars. From there, a boat hired by Gao took these young men north along the Grand Canal to Beijing; see Qian Daxin, *Qian Xinmei xiansheng nianpu*, 11.

34 Ibid., 11–12.

35 Ibid., 12.

36 Ibid., 12-13.

37 Qian Daxin, *Jiading Qian Daxin quanqi*, 9:672–73.

38 Innovative use of illustrations has recently become a focal point of historical investigation. See the essays collected in Bray, Dorofeeva-Lichtmann, and Métailié, *Graphics and Text in the Production of Technical Knowledge in China.*

39 Dai, *Dai Dongyuan xiansheng quanji*, 201–2.

40 Ibid., 1098–99.

41 Yao, *Xibaoxuan shiwen ji*, 76–77. For details regarding the intellectual tension between Dai Zhen and Yao Nai, see Wang Damin, *Yao Nai yu Qian-Jia xuepai*, 11–30.

42 Dai, *Dai Zhen quanshu*, 6:277–79. For a detailed analysis of this dispute, see Kondō, *Shinchō kōkyogaku no kenkyū*, 328–40. Chen Zhifeng provides an excellent contextual analysis of this dispute in "Lun Wang Mingsheng, Dai Zhen jia 'Yao dian' 'guangbeisibiao' ji xiangguang wenti," 6–18.

43 Dai, *Dai Zhen quanshu*, 6:278.

44 Wang Mingsheng, *Eshu pian*, 4:8a–8b.

45 Wu Sixiao, "Xu," *Dai Dongyuan xiansheng quanji*, 832.

46 This is the main theme running through Hong Wansheng and Liu Dun's edited volume *Tan tian san you*.

47 Qian Daxin, *Qianyantang wenji*, 33: 2b. Ori Sela translated this letter into English; see Sela, *Qian Daxin (1728–1804)*, 207–11.

48 Qian Daxin, *Qianyantang wenji*, 33:2b.

49 Sivin, "Copernicus in China," 10–11.

50 Qian Daxin, *Qianyantang wenji*, 33:4b.

51 Nakayama, *A History of Japanese Astronomy*, 126–27.

52 Jiang Yong, *Shuxue*, 55.

53 Ibid., 56–58.

54 Sivin, "Copernicus in China," 44.

55 Shi Yunli, "Reforming Astronomy and Compiling Imperial Science in the Post-Kangxi Era," 51–69.

56 Jiang Yong, *Shuxue*, 62–68.

57 Ibid., 125–27.

58 Mao Yuesheng, "Buxu," 2b.

59 Qian Daxin, *Qianyantang wenji*, 33:3a.

60 Ibid., 33:4a–4b.

61 Ibid., 33:3b.

62 For historians in twentieth-century China, the difference between Dai Zhen and Jiang Yong was usually framed in moral terms: did Dai betray his mentor? This naturally made Dai look bad. I think it counterproductive and wrongheaded to understand their difference in moral terms. For a survey of the debates, see Cai Jinfang, *Dai Zhen shengping yu zoupin kaolun*, 16–27.

7. BRINGING IT HOME TO THE PALACE OF LIGHT

1 Ye Shuxian, *Zhongguo shenhua zhexue*, 145–76; Lewis, *The Construction of Space in Early China*, 271; Yang Hongxun, "Mingtang fanlun—Mingtang de kaoguxue yanjiu"; Wechsler, *Offerings of Jade and Silk*, 195–211.

2 Creel, *The Origins of Statecraft in China*, 1:81–100, 493–506. The date of the Zhou conquest of Shang has been a topic of long-lasting debate among scholars. I follow Shaughnessy and Nivison in dating it to 1045 BCE; see Loewe and Shaughnessy, *The Cambridge History of Ancient China*, 22–24; Xia Shang Zhou duandai gongcheng zhuanjiazu, *Xiashangzhou duandai gongcheng 1996–2000 nian jieduan chengguo baogao*, 48–49; Jiang Xiaoyuan and Niu Weixing, *Hui tian: Wuwang fa Zhou yu tianwen lishi niandaixue*, 144–46.

3 Shiji, "Fengshanshu," 28:3b. For this event and the making of the Duke of Zhou as a canonical figure, see Nylan, "The Many Dukes of Zhou in Early Sources."

4 Guo Maoqian, *Yuefu shiji*, 25:14b.

5 Yang Hongxun, *Yang Hongxun jianzhu kaoguxue lunwenji*, 494–506.

6 Lundbaek, *The Traditional History of the Chinese Script*, 27.

7 Wagner, "The 'Zhouli' as the Late Qing Path to the Future," 359–87.

8 Zhang Yibing, *Mingtang zhidu yanjiu*, 33–36.

9 Soothill, *The Hall of Light*, 78–83. Michael Nylan calls it "Spirit Hall" in Nylan, "Many Dukes of Zhou in Early Sources," 95. Wu Hung calls it "Bright Hall" and construes the meanings of *Ming tang* in terms of its monumentality; see Wu, *Monumentality in Early Chinese Art and Architecture*, 176–87.

10 On Fang Wuru, see *Qing shilu*, 71:31a–31b. He Zhuo was a disciple of Li Guangdi and a friend of Yan Ruoju; see *Qing shilu*, 71:27b–28b.

11 *Rijiang liji jieyi*, 35:4a–4b.

12 For details on the Institute of the Three Ritual Classics, see Lin Cunyang, *San Li Guan*, 63–72; Chang So-an, *Shiba shiji lixue kaozheng de sixiang huoli*, 55–61.

13 Ōtani, *Shindai seiji shisōshi kenkyū*, 140.

14 Fang Bao, *Fang Bao ji*, 26–27.

15 The archaeological site of Wang Mang's Palace of Light was discovered in 1956; see Yang Hongxun, "Ming tang fanlun," 55–63; Wu Hung, *Monumentality in Early Chinese Art and Architecture*, 176–87.

16 Kawahara, *Chūgoku no kagaku shisō*, 148–95.

17 Ibid., 288–91.

18 Fang Bao, *Fang Bao ji*, 47.

19 Ibid., 47-48.

20 Ibid.

21 Bol, "Government, Society, and the State"; Bol, "Wang Anshi and the *Zhouli*," 229–51.

22 Elman, "The Story of a Chapter," 330–55.

23 The first one was Wang Mang's Palace of Light in 5 CE; the second, Liu Xiu's Palace of Light in 56 CE (rebuilt in 526); the third, Wu Zetian's Palace of Light in 688. Each time rebuilding was proposed, controversies broke out that lasted for years; see Yang Hongxun, "Ming tang fanlun," 55–88.

24 Anthony Barbieri-Low combines archaeological data with details from "Records of Ancient Technology" to reconstruct the ancient world of artisans, see his *Artisans in Early Imperial China.*

25 The editors of *Sikuquanshu* later identified Li Rugui (1479-1547) as its author.

26 Jiang Yong, *Yili shigong zengzhu*, 26a.

27 Rawski, *The Last Emperors*, 197–215.

28 Elliott, *Emperor Qianlong*, 107–24; Kahn, *Monarchy in the Emperor's Eyes*, 159–60.

29 Cai Yong, *Cai Zhonglang ji*, 3:6a–12b.

30 Ibid.

31 Yuan Mei, *Yuan Mei quanji*, 5:200.

32 Ibid.

33 Yuan Mei was acquainted with many scholars in the Suzhou area; see Wang Biao, *Chengshi zhishifenzi de shehuixingtai*, 122-54.

34 Carney Fisher, *The Chosen One*, 49–83.

35 Chang So-an, *Shiba shiji lixue kaozheng de sixiang huoli*, 224.

36 Ibid., 224–25.

37 Zhao Erxun, *Qing shi gao*, juan 83, zhi 58, li 2, 2507.

38 Ibid., 2510.

39 Zhu Xi, *Sishu zhangju jizhu*, 1:20b–21a.

40 Ibid.

41 Mao Qiling, *Sishu gaicuo*, 4:11b; Soothill, *The Hall of Light*, 84–88.

42 *Qing shilu*, 68:14a–15b.

43 Mao, *Ming tang wen*, 1a–4b.

44 Lu Longqi, *Sishu jiangyi kunmianlu*, 25:13b–14b.

45 Qi Yongxiang, *Dong Wu san Hui shiwenji*, 302–3.

46 Hui Dong, *Ming tang da dao lu*, 2:1a–2a.

47 Ibid., 1:1a–2a.

48 Ibid, 3: 1a.

49 Ibid., 1:13b.

50 *Qing shilu*, 71:40b–41a. For details, see Cao Jianghong, "Dai Zhen yu Lu Jianzeng mufu yanjiu."

51 Intellectual historians believe that Dai was more or less a follower of Zhu Xi before his encounter with Hui and that the master of Han learning induced a sort of gestalt shift, after which Dai forever acted as an implacable foe of Zhu Xi and Neo-Confucianism more generally. See Qian Mu, *Zhongguo jinsanbainian lai xueshushi*, 322–24; Yü Ying-shih, *Lun Dai Zhen yu Zhang Xuecheng*, 127–29. But this is a misleading interpretation. Dai was always an outsider to the imperial orthodoxy and would never be a supporter of Han learning, as we can tell from his candid and critical assessment of *The Grand Way of the Palace of Light.* To corroborate, we should also look at Dai's other provocative claim, that "among contemporary scholars, I consider Qian Daxin as number two." Historians usually take this as an expression of Dai's arrogance, because it implies that he thought of himself as number one. Trying not to be dismissive, historian Yü Ying-shih explained that Dai made this judgment based upon his intellectual agenda rather than what the

textual evidence might dictate. Instead of condemning him for his arrogance, we should note that he placed Qian Daxin above everyone else, including Hui Dong. If we compare this assessment of Qian with Dai's letter to Shi Jing, in which he critically dismissed almost all classical scholars as unable to meet his standards of (1) erudition and profundity, (2) cognition and judgment, and (3) veracity and accuracy, then it is remarkable that Dai considered Qian to be superior to everyone else.

52 Historians who see Dai in terms of the labels "Han learning" or "Song learning," or who argue that Dai shifted from a pro-Song to an anti-Song stance, miss the point entirely.

53 Wu Xiu, *Zhaodai mingren chidu xiaozhuan*, 31–511.

54 Qi Yongxiang, *Dong Wu san Hui shiwenji*, 345.

55 Wechsler, *Offerings of Jade and Silk*, 44–49.

56 Hui Dong, *Ming tang da dao lu*.

57 Ibid., 2:4a–7a.

58 Dai, *Dai Zhen quanshu*, 6:251–52.

59 Ibid.

60 Ibid., 6:252.

61 Ibid., 6:251.

62 Ibid., 6:252.

63 Arthur Conan Doyle, *A Study in Scarlet*, searchable on www.literature.org.

64 Dai, *Dai Zhen quanshu*, 6:370.

65 Dai, *Dai Zhen quanshu*, 6:318–21.

66 For a list of authors and their conjectures about the floor plan of the Palace of Light, see Zhang Yibing, *Mingtang zhidu yanjiu*, 347–54.

67 Jiang Fan, *Jiang Fan ji*, 2–7.

68 Wang Zhong, *Wang Zhong ji*, 22.

69 Ruan Yuan, *Yan jing shi ji*, 981–99.

70 *Qing shi gao*, juan 83, zhi 58, li 2, 2508.

71 Wang Guowei, *Wang Guowei yi shu*, *Guangtang jilin*, 3:2a.

72 Wang Fansen, *Zhongguo jindaisixiang yu xueshudexipu*, 307.

73 Wang Guowei, *Wang Guowei yi shu*, *Guangtang jilin*, 12:31a–34b.

74 Zhang Yibing, *Ming tang zhidu yanjiu*, 19–20.

75 Yang Hongxun has argued that the Palace of Light, according to archaeological evidence, was built on an elevated platform; see Yang, *Gongdian kaogu tonglun*, 104–19. Neither Dai Zhen nor Wang Guowei had the faintest suspicion of this configuration. Yang argued that Wang Mang's Palace of Light was indeed very close to the description in "Records of Ancient Technology"; see *Yang Hongxun jianzhu kaoguxue lunwenji*, 257–79.

76 Elman, *From Philosophy to Philology*, 93–122; Yamanoi, *Min-Shin shisōshi no kenkyū*, 361–77.

77 Wang Hui, *Xiandai Zhongguo sixiang de xingqi*, 382–410.

8. LEGIBILITY OF VISIONARY SCHOLARS

1 For example, see R. Kent Guy, *Qing Governors and Their Provinces*, 183-351.

2 The list is taken from Joseph Fletcher's pioneering essay, "Integrative History: Parallels and Interconnections in the Early Modern Period, 1500-1800," 37-57. A body

of excellent scholarship on the malleability of the Qing empire has emerged. See Guy, *Qing Governors and Their Provinces*; Rowe, *Saving the World*; Hirano, *Shin Teikoku to Chibetto mondai*; Kuroda, *Chūka teikoku no kōzō to sekai keizai*; Elliott, *The Manchu Way*; and Crossley, *A Translucent Mirror.*

3 For a wide range of imperial publications during High Qing, see Wu Zhenyu, *Yangjizhai conglu*, 259-273.

4 James Scott, *Seeing Like a State*, 11.

5 My analysis here is built on the intellectual and cultural history pioneered by Benjamin Elman, R. Kent Guy, Mizoguchi Yūzō, and Yü Ying-shih. See, in particular, Elman's *From Philosophy to Philology* and *On Their Own Terms*, Guy's *The Emperor's Four Treasuries*, Mizoguchi Yūzō's *Chūgoku zen kindai shisō no kussetsu to tenkai*, and Yü Ying-shih's *Lun Dai Zhen yu Zhang Xuecheng.*

6 Chow, *The Rise of Confucian Ritualism*, 3.

7 Chang So-an, *Shiba shiji lixue kaozheng de sixiang huoli*, iv. See also Kojima Tsuyoshi's analysis of the religiosity and political nature of the ritual practice in his *Chūgoku kinsei ni okeru rei no gensetsu.*

8 For a parallel discussion, see Itō, *Shisō toshite no Chūgoku kinsei*, 79–135.

9 In a sense, I am speaking to Onogawa Hidemi's classic *Shinmatsu seiji shisō kenkyū*, esp. chapters 1-3, in which she meticulously documented the formation of the new dichotomy between China and the West, essence and value (*dao*) versus tool and application (*qi*), as well as how these intellectual and political movements eventually led to the reform movement in 1898, epitomized by Kang Youwei.

10 Michael Mann, *The Sources of Social Power*, 1:519.

11 Guo Shuchun, *Huijiao jiuzhang suanshu*, 647–49. Guo appreciates Dai Zhen's pioneering efforts in rebuilding the complete text of *Zhoubi suanjing* but also dismisses his agenda as distortion rather quickly.

12 Guo, *Huijiao jiuzhang suanshu*, 649–54.

13 Qian Qianyi (1582–1664) was Mao Jin's mentor. Both were from Changshou in the Lower Yangzi Delta (in the southeast of today's Jiangsu province).

14 Ye Dehui, *Shulin qinghua*, 7:18b.

15 Kai-wing Chow, *Publishing, Culture, and Power in Early Modern China*, 40–42. Inoue Susumu extensively contextualizes the pricing strategies of book sales in the sixteenth century. See his *Chūgoku shūppan bunkashi*, 262-80.

16 Hu Zhenheng, "Xu."

17 *Qing shi gao*, juan 485, liezhuan 272, wenyuan 2, 13404–5.

18 Cover page of *Zhoubi suanjing*. Rare book collection in the Palace Museum, Taipei, Taiwan.

19 The fear was cogently formulated in Fang Dongshu's *Hanxue shangdui*, especially in 2:1a–17a. See also Wang Fansen, *Zhongguo jindaisixiang yu xueshudexipu*, 19.

20 Cheng Yaotian, *Cheng Yaotian quanji*, 2:408.

21 Peterson, "The Life of Ku Yen-wu, part 2," 210–13; Wilkinson, *Chinese History: A New Manual*, 209.

22 Qian Yong, *Luyuan conghua*, 89.

23 Ibid., 88-89.

24 Ibid., 94.

25 Dai Zhen, *Dai Zhen quanji*, 4:482–84.

26 Cheng Yaotian, *Cheng Yaotian quanji*, 2:441–57.

27 Ibid., 2:415–18.

28 Ibid., 2:460–67.
29 Ruan Yuan, *Yan jing shi ji*, 1:265–77.
30 *Siku quanshu zongmu*, 21:29a.
31 Huang Zongxi, *Shenyi kao*, 17a.
32 Yü Ying-shih, *Zhu Xi de lishi shijie*, 1:287-387 ; Peter Bol, *Neo-Confucianism in History*, 138-52.
33 Ma Duanlin, *Wenxian tongkao*, 110:26a–b.
34 Ibid., 110:27a.
35 Ibid., 110:27a–b.
36 Ibid.
37 Ibid.
38 Ibid.
39 Gu Jiegang, *Qin Han de fangshi yu rusheng*, 1–16. For a recent elaboration and revision of the ancient cosmological theory, see Li Ling, *Zhongguo fangshukao*, 174–76.
40 Wang Yingdian, *Zhouli tushuo*, 2:50a-52a.
41 Lin Li-yueh, "Yishang yu fengjiao," 114–19; Wu Jen-shu, "Mingdai pingmin fushi de liuxing fengshang yu shidafu de fanying," 57–65.
42 A sizable literature already touches upon this issue. See Craig Clunas, *Superfluous Things*, 141–65; Timothy Brook, *The Confusions of Pleasure*, 219–33; Kishimoto, "'Laoye' to 'xianggong'—Koshō kara mita chihō shakai no kaisō kankakuku," 54–76; and Kishimoto, *Chūgoku no mibunsei to shakai chitsujyo*, 1-33. Cf. Pierre Bourdieu, *Distinction*, 260–317.
43 Confucius, *The Analects of Confucius*, 185.
44 Huang Zongxi, *Ming Ru xue'an*, 2:709. The translation is based on Julia Ching's version, *The Records of Ming Scholars*, 174.
45 Shimada Kenji, *Chūgoku ni okeru kindai shii no zasetsu*, 134–37; Kishimoto, *Minshin kōtai to Kōnan shakai*, 59–100.
46 This is one of several main arguments in Mizoguchi Yūzō's *Chūgoku zen kindai shisō no kussetsu to tenkai*.
47 *Huangchao liqi tushi*, 14–94.
48 Shen Congwen, *Zhongguo gudai fushi yanjiu*, 414–61.
49 Yü Ying-shih, *Zhongguo jinshi zongjiao lunli yu shangren jingshen*, 97-166; Kishimoto, "Meishi no kōyō: Minshin jidai ni okeru shitaifu no kōsai," 243–76; Kishimoto, "Min-Shin jidai no shinbun kankaku", 403–27.
50 For details, see Lin Li-yueh, "Yishang yu fengjiao," 119-27.
51 Lynn Struve, "Huang Zongxi in Context," 475–84.
52 Huang Yuji, *Qian qing tang shumu*, 41–42.
53 Wilkinson, *Chinese History*, 804.
54 Huang Binghou, *Huang Zongxi nianpu*, 19.
55 Huang Zongxi, *Huang Zongxi quanji*, 10:36. This statement is also cited in Zhao Yuan, *Ming Qing zhi ji shidaifu yanjiu*, 406. Zhao provides an interesting characterization of "Ming loyalist" scholarship on pages 402–67. For dating this event, see Huang Binghou, *Huang Zongxi nianpu*, 26.
56 Huang Zongxi, *Shenyi kao*, 17b.
57 Huang Zongxi, *Huang Zongxi quanji*, 10:24.
58 For details and complexity on scholarly partisanship, see R. Kent Guy, *The Emperor's Four Treasuries*, 121–56.
59 *Siku quanshu zongmu tiyao*, 21:19a–20a.

60 Ibid., 21:20b.

61 For Dai's discussion on the long garment and other ritual clothes, see *Dai Zhen quanshu*, 6:256–73. For Jiang's discussion, see Jiang Yong's *Shenyi kaowu*.

62 For the Sanskrit *cosmopolis*, see Pollock's *The Language of the Gods in the World of Men*.

63 Sun, *Zhouli zhengyi*, 1:5.

BIBLIOGRAPHY

Ainajushi 艾衲居士. *Doupeng xianhua* 豆棚閒話 (Idle talk under the bean canopy). Guben xiaoshuo jicheng 331. Shanghai: Shanghai Guji Chubanshe, 1990.

Arai Shinji. "Astronomical Studies by Zhao Youqin." *Taiwanese Journal for Philosophy and History of Science* 8, no. 1 (1996): 59–102.

Aveni, Anthony F. *Skywatchers.* Austin: University of Texas Press, 2001.

Baik Yong-seo 白永瑞. *Sixiang Dongya: Hanbandao de lishi yu shijian* 思想東亞：韓半島的歷史與實踐 (East Asia as an idea: History and practice from the perspective of the Korean Penisula). Taipei: Taiwan Shehui Yanjiu Zazhishe, 2009.

Bala, Arun. *The Dialogue of Civilizations in the Birth of Modern Science.* New York: Palgrave Macmillan, 2006.

Barbieri-Low, Anthony. *Artisans in Early Imperial China.* Seattle: University of Washington Press, 2007.

Bartlett, Beatrice S. *Monarchs and Ministers: The Grand Council in Mid-Ch'ing China, 1723–1820.* Berkeley: University of California Press, 1991.

Ben-Dor Benite, Zvi. *The Dao of Muhammad: A Cultural History of Muslims in Late Imperial China.* Cambridge, MA: Havard University Asia Center, 2005.

Bodde, Derk, and Clarence Morris, eds. *Law in Imperial China: Exemplified by 190 Ch'ing Dynasty Cases (Translated from the "Hsing-an hui-lan") with Historical, Social, and Juridical Commentaries.* Cambridge, MA: Harvard University Press, 1967.

Bol, Peter K. "Government, Society, and State: On the Political Visions of Ssu-ma Kuang (1019–1086) and Wang An-shih (1021–1086)." In *Ordering the World: Approaches to State and Society in Sung Dynasty China*, edited by Robert P. Hymes and Conrad Schirokauer, 128–92. Berkeley: University of California Press, 1993.

———. *Neo-Confucianism in History.* Cambridge, MA: Harvard University Asia Center, 2008.

———. "The Rise of Local History: History, Geography, and Culture in Southern Song and Yuan Wuzhou." *Harvard Journal of Asiatic Studies* 61, no. 1 (2001): 37–76.

———. "Wang Anshi and the *Zhouli*." In *Statecraft and Classical Learning: The Rituals of Zhou in East Asian History*, edited by Benjamin A. Elman and Martin Kern, 229–51. Leiden: Brill, 2010.

Bourdieu, Pierre. *Distinction: A Social Critique of the Judgment of Taste*. Translated by Richard Nice. Cambridge, MA: Harvard University Press, 1984.

———. *Language and Symbolic Power*. Translated by Gino Raymond and Matthew Adamson. Cambridge, MA: Harvard University Press, 1991.

Bouvet, Joachim. *Gujin jingtianjian tianxue benyi* 古今敬天鑒天學本義 (Mirrors of ancient and modern worship of Heaven: The essentials of Heavenly learning). Borg. Cinese.317 (15), Vatican Library.

———. *Journal des voyages*. Edited by Claudia von Collani. Taipei: Taipei Ricci Institute, 2005.

———. *Voyage de Siam du Père Bouvet*. Leiden: Brill, 1963.

———. *Yi yin yuangao* 易引原稿 (Original draft of extending "The Book of Changes"). Borg.Cinese.317 (6), Vatican Library.

Bray, Francesca, Vera Dorofeeva-Lichtmann, and Georges Métailié, eds. *Graphics and Text in the Production of Technical Knowledge in China: The Warp and the Weft*. Leiden: Brill, 2007.

Brockey, Liam Matthew. *Journey to the East: The Jesuit Mission to China, 1579–1724*. Cambridge, MA: Harvard University Press, 2007.

Brook, Timothy. *The Confusions of Pleasure: Commerce and Culture in Ming China*. Berkeley: University of California Press, 1998.

Buglio, Louis. *Bu de yi bian* 不得已辯 (A rebuttal to "I Cannot Do Otherwise"). In *Bu de yi: Fu erzhong* 不得已： 附二種 (I cannot do otherwise, with two comments), by Yang Guangxian 楊光先 et al., compiled by Chen Zhanshan 陳占山, 93–135. Hefei: Huangshan, 2000.

Cai Jinfang 蔡錦芳. *Dai Zhen shengping yu zuopin kaolun* 戴震生平與作品考論 (An investigation and discussion of Dai Zhen's life and work). Guilin: Guangxi Shifan Daxue Chubanshe, 2006.

Cai Yong 蔡邕. *Cai Zhonglang ji* 蔡中郎集 (A collection by Cai Yong). In *Siku quanshu*. Hong Kong: Zhongwen Daxue Chubanshe, 1999.

Cao Jianghong 曹江紅. "Dai Zhen yu Lu Jianzeng mufu yanjiu" 戴震與盧見曾幕府研究 (A study of Dai Zhen and Lu Jianzeng's private secretariat). *Qingshi luncong* (2010): 238–56.

Cao Xueqin. *The Story of the Stone*. Translated by David Hawkes and John Minford. 5 vols. Penguin: Harmondsworth, 1973–1986.

Cen Yicheng 岑溢成. *Shi buzhuan yu Dai Zhen jiejing fangfu* 詩補傳與戴震解經方法 (A supplemental commentary on "The Book of Songs" and Dai Zhen's classical exegeses). Taipei: Wenjin Chubanshe, 1992.

Chai Degeng 柴德賡. *Shixue congkao* 史學叢考 (Collected historical studies). Beijing: Zhonghua, 1982.

Chang, Kang-i Sun, and Stephen Owen. *The Cambridge History of Chinese Literature*. 2 vols. Cambridge: Cambridge University Press, 2010.

Chang So-an 張壽安. "Dai Zhen yili sixiang de jichu jiqi tuizhan" 戴震義理思想的基礎及其推展 (The foundation of Dai Zhen's philosophy and its application). *Hanxue yanjiu* 15, no. 1 (1997): 173–85.

———. *Shiba shiji lixue kaozheng de sixiang huoli: Lijiao lunzheng yu lizhi chongxing* 十八世紀禮學考證的思想活力： 禮教論爭與禮秩重省 (The vibrant thought of eighteenth-century evidential research on rituals: The debates over ritual orthodoxy and the reconstruction of the ritual order). Taipei: Zhongyang Yanjiuyuan Jindaishi Yanjiusuo, 2001.

Chemla, Karine. "Similarities between Chinese and Arabic Mathematical Writings: (I) Root Extraction." *Arabic Sciences and Philosophy* 2, no. 4 (1994): 207–66.

Chen, Jiang-ping. "The Evolution of Transformation Media in Spherical Trigonometry in Seventeenth- and Eighteenth-Century China, and Its Relation to 'Western Learning.'" *Historia Mathematica* 37, no. 1 (2010): 62–109.

Chen Fangzheng 陳方正. *Jicheng yu panni: Xiandai kexue weihe chuxian yu xifang* 繼承與叛逆：現代科學為何出現於西方 (Heritage and betrayal: A treatise on the emergence of modern science in Western civilization). Beijing: Shenghuo, Dushu, Xinzhi Sanlian Shudian, 2009.

Chen Jiujin 陳久金. *Huihui tianwenxue shi yanjiu* 回回天文學史研究 (Studies of the history of Islamic astronomy in China). Nanning: Guangxi Kexue Jishu Chubanshe, 1996.

Chen Pingyuan 陳平原. *Zhongguo xiandai xueshu zhi jianli: Yi Zhang Taiyan, Hu Shizhi wei zhongxin* 中國現代學術之建立：以章太炎，胡適之為中心 (The establishment of a modern academic system in China: Centered on Zhang Taiyan and Hu Shizhi). Taipei: Maitian, 2000.

Chen Zhifeng 陳志鋒. "Lun Wang Mingsheng, Dai Zhen jie *Yao dian* 'guangbeisibiao' ji xiangguang wenti" 論王鳴盛，戴震解《堯典》「光被四表」及相關問題 (On Wang Mingsheng's and Dai Zhen's exegeses of "Guangbeisibiao" in *Yao dian* and some related issues). *Zhongguo wenxue yanjiu* 30 (2010): 184–214.

Cheng Yaotian 程瑤田. *Cheng Yaotian quanji* 程瑤田全集 (The complete works of Cheng Yaotian), collated by Chen Guanming 陳冠明. Hefei: Huangshan, 2008.

———. "Wuyou ji" 五友記 (Story of five friends). In *Cheng Yaotian quanji* 程瑤田全集 (The complete works of Cheng Yaotian), collated by Chen Guanming 陳冠明, 310–16. Hefei: Huangshan, 2008.

Chin, Ann-ping, and Mansfield Freeman, trans. *Tai Chen on Mencius: Explorations in Words and Meaning.* New Haven, CT: Yale University Press, 1990.

Ching, Julia (trans). *The Records of Ming Scholars.* Translation of Huang Zongxi's *Ming Ru xue'an.* Honolulu: University of Hawai'i Press, 1987.

Chow, Kai-wing. *Publishing, Culture, and Power in Early Modern China.* Stanford, CA: Stanford University Press, 2004.

———. *The Rise of Confucian Ritualism in Late Imperial China: Ethics, Classics and Lineage Discourse.* Stanford, CA: Stanford University Press, 1994.

Chu Pingyi. "Archiving Knowledge: A Life History of the *Calendrical Treatises of the Chongzhen Reign* (*Chongzhen Lishu*)." *Extrême-Orient, Extrême-Occident* 2007 (hors série): 159–84.

———. "Ch'eng-Chu Orthodoxy, Evidential Studies and Correlative Cosmology: Chiang Yung and Western Astronomy." *Philosophy and the History of Science: A Taiwanese Journal* 4, no. 2 (1995): 71–108.

———. "Scientific Dispute in the Imperial Court: The 1664 Calendar Case." *Chinese Science* 14 (1997): 7–34.

Clunas, Craig. *Superfluous Things: Material Culture and Social Status in Early Modern China.* Honolulu: University of Hawai'i Press, 1991.

Confucius. *The Analects of Confucius.* New York, N.Y.: Columbia University Press, 2007.

Creel, Herrlee G. *The Origins of Statecraft in China*. Vol. 1, *The Western Chou Empire*. Chicago: University of Chicago Press, 1970.

Crossley, Pamela Kyle. *The Manchus*. Cambridge, MA: Blackwell, 1997.

———. *A Translucent Mirror: History and Identity in Qing Imperial Ideology*. Berkeley: University of California Press, 1999.

Da Qing huidian Kangxi chao 大清會典康熙朝 (The collected statutes of the Qing dynasty: Kangxi reign). Taipei: Wenhai, 1992.

Da Qing huidian Yongzheng chao 大清會典雍正朝 (The collected statutes of the Qing dynasty: Yongzheng reign). Taipei: Wenhai, 1992.

Dai Zhen 戴震. *Dai Dongyuan xiansheng quanji* 戴東原先生全集 (The complete works of Dai Dongyuan [aka Dai Zhen]). Taipei: Dahua, 1936.

———. *Dai Zhen quanshu* 戴震全書 (The complete works of Dai Zhen). Edited by Zhang Dainian 張岱年. 7 vols. Hefei: Huangshang, 1994.

Dardess, John W. *Blood and History in China: The Donglin Faction and Its Repression, 1620–1627*. Honolulu: University of Hawai'i Press, 2002.

De Bary, William Theodore. "Individualism and Humanitarianism in Late Ming Thought." In *Self and Society in Ming Thought*, edited by Wm. Theodore de Bary and the Conference on Ming Thought, 145–247. New York: Columbia University Press, 1970.

———. *The Liberal Tradition in China*. Hong Kong: Chinese University of Hong Kong, 1983.

Deane, Thatcher Elliott. "The Chinese Imperial Astronomical Bureau: Form and Function of the Ming Dynasty Qintianjian from 1365 to 1627." PhD diss., University of Washington, 1989.

Dear, Peter. *Discipline and Experience: The Mathematical Way in the Scientific Revolution*. Chicago: University of Chicago Press, 1995.

DeNora, Tia. *Beethoven and the Construction of Genius: Musical Politics in Vienna, 1792–1803*. Berkeley: University of California Press, 1995.

Dianjiaoben ershisishi 點校本二十四史 (The twenty-four standard histories, a punctuated edition). Beijing: Zhonghua, 1974.

Dong Chao 董潮. *Donggao zachao* 東皋雜鈔 (Miscellaneous jottings from East Bank). In *Congshu jicheng chubian* 叢書集成初編 (First collection of collactanea), vol. 2963. Shanghai: Shangwu, 1935–1940.

Duan Yucai 段玉裁. "Dongyuan nianpu dingbu" 東原年譜訂補 (The chronological biography of Dai Zhen, with additions and corrections). In *Dai Zhen quanshu* 戴震全書 (The complete works of Dai Zhen), edited by Zhang Dainian 張岱年, 6:649–718. Hefei: Huangshang, 1994.

———. *Jingyunlou ji: Fu bubian nianpu* 經韻樓集: 附補編年譜 (A collection from the Hall of Classics and Rhymes: An annotated chronological biography). Shanghai: Shanghai Guji Chubanshe, 2008.

Duara, Prasenjit. *Sovereignty and Authenticity: Manchkuo and the East Asian Modern*. Lanham, MD: Rowman and Littlefield, 2003.

Dumont, Louis. *Essays on Individualism: Modern Ideology in Anthropological Perspective*. Chicago: University of Chicago Press, 1986.

Durand, Pierre-Henri. *Dai Mingshi nianpu* 戴名世年譜 (A chronological biography of Dai Mingshi). Beijing: Zhonghua, 2004.

Elliott, Mark C. *Emperor Qianlong: Son of Heaven, Man of the World*. New York: Pearson Longman, 2009.

———. *The Manchu Way: The Eight Banners and Ethnic Identity in Late Imperial China*. Stanford, CA: Stanford University Press, 2001.

Elman, Benjamin A. *Classicism, Politics, and Kinship: The Ch'ang-chou School of New Text Confucianism in Late Imperial China*. Berkeley: University of California Press, 1990.

———. *A Cultural History of Civil Examinations in Late Imperial China*. Berkeley: University of California Press, 2000.

———. *From Philosophy to Philology: Intellectual and Social Aspects of Change in Late Imperial China*. 2nd rev. ed. Los Angeles: UCLA Asian Pacific Monograph Series, 2001.

———. "From Value to Fact: The Emergence of Phonology as a Precise Discipline in Late Imperial China." *Journal of the American Oriental Society* 102, no. 3 (1982): 493–500.

———. "Imperial Politics and Confucian Societies in Late Imperial China: The Hanlin and Donglin Academies." *Modern China* 15, no. 4 (1989): 379–418.

———. *On Their Own Terms: Science in China, 1550–1900*. Cambridge, MA: Harvard University Press, 2005.

———. "Political, Social, and Cultural Reproduction via Civil Service Examinations in Late Imperial China." *Journal of Asian Studies* 50, no. 1 (February 1991): 7–28.

———. "The Story of a Chapter: Changing Views of the 'Artificer's Record' and the *Zhouli*." In *Statecraft and Classical Learning: The Rituals of Zhou in East Asian History*, edited by Benjamin Elman and Martin Kern, 330–55. Leiden: Brill, 2010.

Elvin, Mark. *The Retreat of the Elephants: An Environmental History of China*. Stanford, CA: Stanford University Press, 2004.

Esherick, Joseph W., and Mary B. Rankin. "Introduction." In *Chinese Local Elites and Patterns of Dominance*, edited by Joseph W. Esherick and Mary B. Rankin, 1–24. Berkeley: University of California Press, 1990.

Ewell, John, Jr. "Re-inventing the Way: Dai Zhen's *Evidential Commentary on the Meanings of Terms in Mencius* (1777)." PhD diss., University of California, Berkeley, 1990.

Fang Bao 方苞. *Fang Bao ji* 方苞集 (A collection by Fang Bao). Edited by Liu Jigao 劉季高. 2 vols. Shanghai: Shanghai Guji Chubanshe, 1983.

Feingold, Mordechai, ed. *Jesuit Science and the Republic of Letters*. Cambridge, MA: MIT Press, 2003.

Feng Erkang 馮爾康. *Yongzheng zhuan* 雍正傳 (Biography of the Yongzheng emperor). Beijing: Renmin Chubanshe, 1985.

Fisher, Carney T. *The Chosen One: Succession and Adoption in the Court of Ming Shizong*. Sydney: Allen and Unwin, 1990.

Fisher, Michael. "Autobiographical Voices (1, 2, 3) and Mosaic Memory: Experimental Bondages in the (Post)modern World." In *Autobiography and Postmodernism*, edited by Kathleen Ashley, Leigh Gilmore, and Gerald Peters, 79–129. Boston: University of Massachusetts Press, 1994.

Fletcher, Joseph F. "Integrative History: Parallels and Interconnections in the Early Modern Period, 1500-1800." *Journal of Turkish Studies* 9 (1985): 37-57..

Frankel, James D. *Rectifiying God's Name: Liu Zhi's Confucian Translation of Monotheism and Islamic Law*. Honolulu: University of Hawai'i Press, 2011.

———. "Uncontrived Concord: The Eclectic Sources and Syncretic Theories of Liu Zhi, A Chinese Muslim Scholar," *Journal of Islamic Studies* 20, no. 1 (2009): 46–54.

Freedman, Maurice, ed. *Family and Kinship in Chinese Society*. Stanford, CA: Stanford University Press, 1970.

Fu Daiwie 傅大為. *Yishikong li de zhishi zhuizhu* 異時空裡的知識追逐 (The pursuit of knowledge in alternative time and space). Taipei: Dongda, 1992.

Fu Yiling 傅衣凌. *Ming Qing shidai shangren ji shangye ziben* 明清時代商人及商業資本 (Merchants and merchant capital in the Ming and Qing dynasties). Beijing: Renmin Chubanshe, 1956.

Fujii Hiroshi 藤井宏. "Shinan shōnin no kenkyū" 新安商人の研究 (Studies of the merchants of Huizhou). *Tōhō gakuhō* 36 (1953): 1–4.

Fujitsuka Akinao 藤塚明直 and Fujitsuka Chikashi 藤塚鄰. *Shinchō bunka tōden no kenkyū: Kakei Dōkō gakudan to Richō no Kin Gendō* 清朝文化東傳の研究： 嘉慶道光學壇と李朝の金阮堂 (Studies of cultural transmission from the Qing Dynasty to Korea: Scholarship in the Jiaqing and Daoguang reigns and Kim Gendo in the Yi dynasty). Tokyo: Kankōkai, 1975.

Funakoshi Akio 船越昭生. *Sakoku Nihon ni kita "Kōki-zu"* no chirigakushiteki kenkyū 鎖国日本にきた「康熙図」の地理学史的研究 (Studies from the history of geography on the "Kangxi Atlas" and its impact on Japan during the closed-door era). Tokyo: Hōsei Daigaku shuppankyoku, 1986.

Fung Yu-lan. *A History of Chinese Philosophy*. 2 volumes, translated by Derk Bodde with introduction, notes, bibliography and index. Princeton, NJ: Princeton University Press, 1952.

Furth, Charlotte. *A Flourishing Yin: Gender in China's Medical History, 960–1665*. Berkeley: University of California Press, 1999.

Gan Huaizhen 甘懷真. *Huangquan, liyi, yu jingdian quanshi: Zhongguo gudai zhengzhishi yanjiu* 皇權、禮儀與經典詮釋: 中國古代政治史研究 (Imperial power, rituals, and classical interpretation: A political history of ancient China). Taipei: Taiwan Daxue Chuban Zhongxin, 2004.

Gernet, Jacques. *China and the Christian Impact: A Conflict of Cultures*. Translated by Janet Lloyd. Cambridge: Cambridge University Press, 1995.

Gao Xiang 高翔. *Jindai de chushu: Shiba shiji Zhongguo guannian bianqian yu shehui fazhan* 近代的初曙：十八世紀中國觀念變遷與社會發展 (The dawn of modernity: Intellectual change and social development in eighteenth-century China). Beijing: Shehui Kexue Wenxian Chubanshe, 2000.

Ge Zhaoguang 葛兆光. *Zhongguo sixiangshi* 中國思想史 (Chinese intellectual history). 2 vols. Shanghai: Fudan University Press, 2000.

Golvers, Noel, comp. *The Astronomia Europaea of Ferdinand Verbiest, S. J.: (Dillingen, 1687); Text, Translation, Notes, and Commentaries*. Nettetal: Steyler Verlag, 1993.

Gong Zizhen 龔自珍. *Gong Zizhen quanji* 龔自珍全集 (The complete works of Gong Zizhen). Shanghai: Shanghai Guji Chubanshe, 1999.

Goodman, Howard, and Anthony Grafton. "Ricci, the Chinese, and the Toolkits of Textualists." *Asia Major* 3, no. 2 (1990): 95–148.

Grimaldi, Claudio Filippo. *Biaoyuan tushuo* 表圓圖說 (Explanation and Illustration of the Foundation of Astronomical Tables). Chinois 5007, Bibliothèque Nationale de France.

Gu Jiegang 顧頡剛. *Qin Han de fangshi yu rusheng* 秦漢的方士與儒生 (Magicians and Classicists in Qin and Han dynasties). Shanghai: Shanghai Guji Chubanshe, 2005.

Gu Yanwu 顧炎武. *Rizhilu* 日知錄 (Records of daily reckoning). *Siku quanshu*.

Guangxu Xuancheng xianzhi 光緒宣城縣誌 (Xuancheng gazetteer, Guangxu reign). 1888; reprint, Nanjing: Jiangsu Guji Chubanshe, 1998.

Gujin tushu jicheng 古今圖書集成 (Complete collection of illustrations and writings from the earliest to current times). Chen Menglei and Jiang Tingxi. 1628 volumes. Shanghai: Tushu Jichen Ju, 1884. Fu Sinian Library collection.

Guo, Qitao. *Ritual Opera and Mercantile Lineage: The Confucian Transformation of Popular Culture in Late Imperial Huizhou*. Stanford, CA: Stanford University Press, 2005.

Guo Chengkang 郭成康. "Zhengzhi chongtu yu wenhua gehe: Yang Mingshi an toushi" 政治衝突與文化隔閡：楊名時案透視 (Political conflicts and cultural gaps: Perspectives on the case of Yang Mingshi). *Qingshi yanjiu* 4 (November 2002): 33–48.

Guo Maoqian 郭茂倩. *Yuefu shiji* 樂府詩集 (Collection of yuefu poems). *Siku quanshu*.

Guo Shirong 郭世榮 and Li Di 李迪. "Putaoya chuanjiaoshi Xu Maode zai qintianjian de tianwen gongzuo" 葡萄牙傳教士徐懋德在欽天監的天文工作 (The contributions of the Portuguese Jesuit Tomé Pereira to the Astrocalendrical Bureau). *Shandong keji daxue xuebao* 13, no. 2 (2011): 18–24.

Guo Shuchun 郭書春. *Huijiao jiuzhang suanshu* 匯校九章算術 (Collective collation of the nine chapters of mathematics). Taipei: Jiuzhang, 2004.

Guo Shuchun and Liu Dun, eds. *Suanjing shishu* 算經十書 (Ten volumes of mathematical classics). Taipei: Jiuzhang, 2001.

Guochao qixian leizheng chubian 國朝耆獻類徵初編 (Venerable documents and similar testimonies from the Qing dynasty. First series). Compiled by Li Huan 李桓. Reprint. Taipei: Wenyou, 1966.

Guy, R. Kent. *The Emperor's Four Treasuries: Scholars and the State in the Late Ch'ien-lung Era*. Cambridge, MA: Council on East Asian Studies, distributed by Harvard University Press, 1987.

———. "Fang Pao and the Ch'in-ting Ssu-shu-wen." In *Education and Society in Late Imperial China, 1600–1900*, edited by Benjamin Elman and Alexander Woodside, 150–82. Berkeley: University of California Press, 1994.

———. *Qing Governors and Their Provinces: The Evolution of Territorial Administration in China, 1644–1796*. Seattle: University of Washington Press, 2010.

———. "Who Were the Manchus: A Review Essay." *Journal of Asian Studies* 61, no. 1 (February 2002): 151–64.

———. "Zhang Tingyu and Reconciliation: The Scholar and the State in the Early Qianlong Reign." *Late Imperial China* 7, no. 1 (June 1986): 50–62.

Hacking, Ian. *Representing and Intervening: Introductory Topics in the Philosophy of Natural Science*. Cambridge: Cambridge University Press, 1983.

Hamaguchi Fujiō 浜口富士雄. *Shindai Kōkyogaku no Shisōshi teki Kenkyū* 清代考拠学の思想史的研究 (Research on the intellectual history of Qing dynasty evidential studies). Tokyo: Kokusho Kankokai, 1994.

Han Qi 韓琦. "Bai Jing de *Yi jing* yanjiu he Kangxi shidai de 'Xixue Zhongyuan' shuo" 白晉的《易經》研究和康熙時代的「西學中源」說 (Bouvet's research on *The Book of Changes* and the doctrine of the "Chinese origins of Western learning" in Kangxi's reign). *Hanxue yanjiu* 16, no. 1 (1998): 185–201.

———. "Cong *Ming shi* lizhi de zuanxiu kan Xixue zai Zhongguo de chuanbo" 從《明史》曆志的纂修士看西學在中國的傳播 (Investigating the dissemination of Western learning in China by examining the editing process of the calendar section of the *Ming History*). In *Keshi xinzhuan—Qingzhu Du Shiran xiansheng congshi kexueshi yanjiu sishi zhounian xueshu lunwenji* 科史薪傳—慶祝杜石然先生從事科學史研究四十周年學術論文集 (Passing the torch in history of science: A Festschrift for Mr. Du Shiran in honor of his forty-year devotion to history of science), edited by Liu Dun and Han Qi, 61–70. Shenyang: Liaoning Jiaoyu Chubanshe, 1997.

———. "Patronage scientifique et carrière politique: Li Guangdi entre Kangxi et Mei Wending" (Scientific patronage and political careerism: Li Guangdi between Kangxi and Mei Wending). Études chinoises 16, no. 2 (1997): 7–37.

———. "'Shuli gezhi de faxian'—Jianlun shiba shiji Niu Dun xiangguan zhuzuo zai Zhongguo de chuanbo" 《數理格致》的發現—兼論十八世紀牛頓相關著作在中國的傳播 (The discovery of "Sciences of Numbers and Principles": A discussion of the dissemination of Issac Newton's work in China). *Zhongguo keji shiliao* 19, no. 2 (1998): 78–85.

———. "Zhang Xingyao and the Collected Discussions on the Imperial Decrees concerning Missionaries 張星曜與欽命傳教約述." *Sino-Western Cultural Relations Journal* 22 (2000): 1–10.

———. "'Zili' jingshen yu lisuan huodong" 「自立」精神與歷算活動 (The spirit of "independence" and mathematical astronomy). *Ziran kexue shi yanjiu* 21, no. 3 (2002): 210–21.

Hang Shijun 杭世駿. *Daogutang wenji* 道古堂文集 (Selected works from the Hall of Dao and Antiquity). *Xuxiu Siku* quanshu, vol. 1426.

———. "Mei Wending zhuan" 梅文鼎傳 (Biography of Mei Wending). In *Daogutang wenji* 道古堂文集, 30:1a–16a.

Hartwell, Robert. "Demographic, Political, and Social Transformations of China, 750–1550," *Harvard Journal of Asiatic Studies* 42, no. 2 (1982): 365–442.

Hashimoto Keizō 橋本敬造. "Bai Buntei no rekisangaku" 梅文鼎の暦算学 (Mei Wending's mathematical astronomy). *Tōhō gakuhō* 41 (1970): 491–518.

———. "Bai Buntei no sūgaku kenkyu" 梅文鼎の数学研究 (Mei Wending's mathematical research). *Tōhō gakuhō* 44 (1973): 233–79.

———. "Daen hō no tenkai: 'Rekisho kosei gohen' no naiyō ni tsuite" 楕円法の展開—「暦象考成後編」の内容について (The unfolding of elliptical method-concerning the content of the "Sequel of the Critical Compendium of Calendrical Phenomena"). *Tōhō gakuhō* 42 (1971): 245–72.

———. *Hsü Kuang-ch'i and Astronomical Reform: The Process of Chinese Acceptance of Western Astronomy 1629–1635*. Osaka: Kansai University Press, 1988.

———. "Rekisho kōsei no seiritsu" 暦象考成の成立 (The establishment of the "Critical Compendium of Calendrical Phenomena"). In *Min-Shin jidai no kagaku gijutsu shi* 明清時代の科学技術史 (A history of science and technology in the Ming and Qing periods), edited by Yabuuchi Kiyoshi and Yoshida Mitsukuni, 49–92. Kyotō: Kyotō daigaku jinbun kagaku kenkyūjō, 1970.

———. "Telescope and Observation in Late Ming China." *Kansai daigaku shakai gakubu kiyō* 19, no. 2 (1988): 91–101.

Hazama Naoki 狭間直樹, ed. *Ryōkeichō: Seiyō kindai shisō jyuyō to Meiji Nihon* 梁啓超—西洋近代思想受容と明治日本 (Liang Qichao: The reception of modern Western thought and Meiji Japan). Tokyo: Misuzu, 1999.

He Yikai 何奕愷. Qingdai xuezhe xiangzhuan yanjiu 清代學者象傳研究 (A study of the portraits of Qing scholars). Shanghai: Shanghai guji chubanshe, 2010.

He Zehan 何澤翰. *Ru lin waishi renwu benshi kaolue* 儒林外史人物本事考略 (A study of the inspirations for the characters in "An Unofficial History of Scholars"). Shanghai: Gudian Wenxue Chubanshe, 1957.

Heilbron, John. *The Sun in the Church: Cathedrals as Solar Observatories*. Cambridge, MA: Harvard University Press, 1999.

Henderson, John. *The Development and Decline of Chinese Cosmology*. New York: Columbia University Press, 1984.

———. *Scripture, Canon, and Commentary: A Comparison of Confucian and Western Exegesis*. New York: Princeton University Press, 1991.

———. "Strategies in Neo-Confucian Heresiography." In *Imaging Boundaries: Changing Confucian Doctrines, Texts, and Hermeneutics*, edited by Kai-wing Chow, On-cho Ng, and John Henderson, 107–20. Albany: State University of New York Press, 1999.

Henrickson, George. "Ch'üan Tsu-wang (1705-1755) and Hang Shih-chün (1696–1773): The Controversial Relationship." *Paper on Far Eastern History* 23 (1981): 63–82.

Hevia, James. *Cherishing Men from Afar: Qing Guest Ritual and the Macartney Embassy of 1793*. Durham, NC: Duke University Press, 2005.

Hirano Satoshi 平野聡. *Shin Teikoku to Chibetto mondai: Taminzou tōgō no seiritsu to gakai* 清帝国とチベット問題—多民族統合の成立と瓦解 (Qing empire and Tibetan issue: Establishment and collapse of the multiethnic coalition). Nagoya-shi: Nagoya daigaku shuppankai, 2004.

Hiraoka Takeo 平岡武夫. *Keisho no seiritsu* 經書の成立 (The establishment of the Confucian canon). Tokyo: Sōbunsha, 1983.

Ho, Ping-ti. *The Ladder of Success in Imperial China: Aspects of Social Mobility, 1368–1911*. New York: Columbia University Press, 1962.

———. "The Salt Merchants of Yang-chou: A Study of Commercial Capitalism in Eighteenth-Century China." *Harvard Journal of Asiatic Studies* 17, no. 1/2 (June 1954): 130–68.

Ho Peng Yoke. "Tycho Brahé (1546–1601) and China." *Hong Kong–Denmark Lectures on Science and Humanities*, 53–60. Hong Kong: Hong Kong University Press, 1983.

Hoang, Pierre [Huang Feimo黃斐默], ed. *Zhengjiao fengbao* 正教奉褒 (Imperial praise of the orthodox religion). Shanghai: Cimutang, 1894.

Hong Bang 洪榜. "Dai xiansheng xingzhuang" 戴先生行狀 (Description of Dai Zhen). In *Dai Dongyuan xiansheng quanji* 戴東原先生全集 (The complete works of Dai Zhen), 53–59. Reprint. Taipei: Dahua, 1978.

Hong Wansheng 洪萬生, ed. "Chinese Mathematics at the Turn of the Nineteenth Century: Jiao Xun, Wang Lai and Li Rui." In *Philosophy and Conceptual History of Science in Taiwan*, edited by Cheng-hung Lin and Daiwie Fu, 167–208. Boston Studies in the Philosophy of Science 141. Boston: Kluwer Academic Publishers.

———. "Qingchu xifang daishu zhi shuru 清初西方代數之輸入" (Importing Western algebra to the early Qing dynasty). In *Kongzi yu shuxue: Yige renwen de huaixiang* 孔子與數學: 一個人文的懷想 (Confucius and mathematics: A humanistic rethinking), 2nd edition, 179–204. Taipei: Mingwen, 1999.

———. *Tan tian san you* 談天三友 (Three friends who study heaven). Taipei: Mingwen, 1993.

Hou Wailu 侯外廬. *Jindai Zhongguo sixiang xueshuo shi* 近代中國思想學說史 (History of modern Chinese thought and theory). Shanghai: Shenghuo, 1947.

Hsia, Florence. *Sojourners in a Strange Land: Jesuits and Their Scientific Missions in Late Imperial China*. Chicago: University of Chicago Press, 2009.

Hu, Minghui. "Impossible Choices." *International Journal of Asian Studies* 4, no. 2 (2007): 259–73.

———. "Provenance in Contest: Searching for the Origins of Jesuit Astronomy in Late Imperial China." *International History Review* 24, no. 1 (2002): 1–36.

Hu Shi 胡適. *Dai Dongyuan de zhexue* 戴東原的哲學 (The philosophy of Dai Dongyuan). Taipei: Taiwan shangwu yinshuguan, 1963.

Hu Tiezhu 胡鐵珠. "Dayanli jiaoshi jisuan jingdu" 大衍曆交食計算精度 (The precision of the calculation of eclipses in the Great Extension Calendar). *Ziran kexueshi yanjiu* 自然科學史研究 20, no. 4 (2001): 312–19.

Hu Zhenheng 胡震亨. "Xu" (Prologue) of *Zhoubi suanjing* 周髀算經, Jiguge version. Included in *Jindai mishu* 津逮秘書 (Private collection of Sailing to the Other Side), vol. 4, no. 59–60. Sold from Shanghai Bogu rare book resaler in 1922. Collection of Fusinian Library, Academia Sinica.

Huang, Yi-long, and Chih-ch'eng Chang. "The Evolution and Decline of the Ancient Chinese Practice of Watching for the Ethers." *Chinese Science* 13 (1996): 82–106.

Huang Binghou 黄炳垕. *Huang Zongxi nianpu* 黄宗羲年譜 (Chronological biography of Huang Zongxi). Punctuated and edited by Wang Zhengyao. Beijing: Zhonghua, 1993.

Huang Jinxing 黄進興. *You ru sheng yu: Quanli, xingyang yu zhengdangxing* 優入聖域：權力，信仰與正當性 (Entering the master's sanctuary: Power, belief and legitimacy in traditonal China). Taipei: Yunchen Wenhua, 1994.

Huang Yi-long 黄一農. "Cong Tang Ruowang suobian mingli shixi qingchu Zhong Ou wenhua de chongtu yu tuoxie" 從湯若望所編民曆試析清初中歐文化的衝突與妥協 (An analysis of conflicts and compromises between Chinese and European cultures in early Qing: A case study of the popular calendar made by Adam Schall). *Qinghua xuebao* 26, no. 2 (1996): 189–220.

———. "Kangxi chao Hanren shidafu dui liyu de taidu jiqi suoyansheng de chuanshuo" 康熙朝漢人士大夫對曆獄的態度及其所衍生的傳說 (The attitude of Chinese scholars toward the persecution of Jesuit mathematical astronomy in the Kangxi reign and the stories that ensued). *Hanxue yanjiu* 11, no. 2 (1993): 137–61.

———. *Liangtoushe: Mingmo Qingchu de diyidai Tianzhujiao tu* 兩頭蛇：明末清初的第一代天主教徒 (Two-headed snakes: The first generation of Catholic converts during the Ming-Qing transition). Xinzhushi: Guoli Qinghua Daxue Chubanshe, 2005.

———. "Qingchu Qintianjian zhong ge minzu tianwenxuejia de quanli qifu" 清初欽天監中各民族天文學家的權力起伏 (The power structure of the Directorate of Celestial Surveillance in the early Qing, considered according to ethnicity). *Xin shixue* 5, no. 1 (1991): 75–108.

———. "Qingqianqi dui Zui-Can liangsu xianhou cixu de zhengzhi" 清前期對觜參兩宿先後次序的爭執 (The controversy on the sequence of the Zui and Can constellations in the early Qing). In *Jindai Zhongguo kejishi lunji* 近代中國科技史論集 (Collected studies of the history of science and technology in modern China), edited by Yang Cuihua and Huang Yi-long, 71-93. Taipei: Zhongyang Yanjiu Yuan Jindaishi Yanjiusuo, 1991.

———. *Shehui tianwenshi shijiang* 社會天文史十講 (Ten lectures on the social history of astronomy). Shanghai: Fudan Daxue Chubanshe, 2004.

———. "Yang Guangxian jiashi yu shengping kao" 楊光先家世與生平考 (An inquiry into Yang Guangxian's family and life). *Guoli bianyiguan guankan* 19, no. 2 (1900): 15–28.

Huang Yuji 黄虞稷. *Qian qing tang shumu* 千頃堂書目 (A bibliography of Vast Hall). Shanghai: Shanghai Guji Chubanshe, 2001.

Huang Zongxi 黄宗羲. *Huang Zongxi quanji* 黄宗羲全集 (The complete works of Huang Zongxi). Hangzhou: Zhejiang Guji Chubanshe, 1985.

———. *Ming Ru xue'an* 明儒學案 (The records of Ming scholars). 2 vols. Beijing: Zhonghua Shuju, 1985.

———. *Mingyi daifang lu* 明夷待訪錄 (Record of a Ming barbarian waiting for a visitor). In *Huang zongxi quanji* 黄宗羲全集 (The complete works of Huang Zongxi), 1:1–47. Hangzhou: Zhejiang Guji Chubanshe, 1985.

———. *Shenyi kao* 深衣考 (A study of the long garment). Reprint. Taipei: Yiwen Yinshuguan, 1967.

Huangchao liqi tushi 皇朝禮器圖式 (Illustrated compendium of Qing rituals). Edited by Yinlu 允祿 and Wan Sitong 萬斯同. In *Siku quanshu zhenben liuji* 四庫全書珍本六集, edited by Wang Yunwu, vols. 122–29. Taipei: Taipei Shangwu Yinshuguan, 1976.

Qinding huangchao wenxian tongkao 欽定皇朝文獻通考 (Imperially sanctioned studies of documents in the Qing dynasty). *Siku quanshu*. This body of administrative records was compiled from 1644 to the 1730s and published in the name of the Qianlong emperor.

Hucker, Charles. *A Dictionary of Official Titles in Imperial China*. Stanford, CA: Stanford University Press, 1985.

Huff, Toby. *The Rise of Early Modern Science: Islam, China, and the West*. 2nd ed. Cambridge: Cambridge University Press, 2003.

Hui Dong 惠棟. *Di shuo* 禘說 (Explanation of Di). In *Xuxiu Siku quanshu* 續修四庫全書 (The sequel to the complete collection of the four treasuries), vol. 108. Reprint. Shanghai: Shanghai Guji Chubanshe, 2002.

———. *Ming tang da dao lu* 明堂大道錄 (Records of the Great Dao and the Palace of Light). *Xuxiu Siku quanshu* 續修四庫全書 (The sequel to the complete collection of the four treasuries), vol. 108. Reprint. Shanghai: Shanghai Guji Chubanshe, 2002.

———. *Zhou yi shu* 周易述 (The transmission of "The Book of Changes"). Reprint. Tainan: Dafu, 1994.

Hulsewé, A. F. P. "Notes on the Historiography of the Han Period." In *Historians of China and Japan*, edited by W. G. Beasley and E. G. Pulleyblank, 31–43. London: Oxford University Press, 1961.

Hummel, Arthur W., ed. *Eminent Chinese of the Ch'ing Period*. Taipei: SMC: 1991.

Huters, Theodore. "From Writing to Literature: The Development of Late Qing Theories of Prose." *Harvard Journal of Asiatic Studies* 47, no. 1 (1987): 51–96.

Hymes, Robert, and Conrad Schirokauer, eds. *Ordering the World: Approaches to State and Society in Sung Dynasty China*. Berkeley: University of California Press, 1993.

Inoue Susumu 井上進. *Chūgoku shūppan bunkashi: Shomotsu sekai to chi no fūkei* 中国出版文化史： 書物世界と知の風景 (A cultural history of Chinese publishing: The world of books and the landscape of knowledge). Nagoya: Nagoya Daigaku shuppankai, 2002.

———. "Kangaku no seiritsu" 漢学の成立 (The establishment of Han learning). *Tōhō gakuhō* 東方学報 61 (1989): 223–319.

Itō Takayuki 伊東貴之. *Shisō toshite no Chūgoku kinsei* 思想としての中国近世 (Considering early modernity in Chinese thought). Tokyo: Tokyo Daigaku Shubankai, 2005.

Ivanhoe, Philip, trans. and ed. *Confucian Moral Self Cultivation*. Indianapolis: Hackett, 2000.

———. *On Ethics and History: Essays and Letters of Zhang Xuecheng.* Stanford, CA: Stanford University Press, 2010.

Ivanhoe, Philip, and David Nivison. *Chinese Language, Thought, and Culture: Nivison and His Critics.* Chicago: Open Court, 1996.

Jacob, Margaret C. *Strangers Nowhere in the World: The Rise of Cosmopolitanism in Early Modern Europe.* Philadelphia: University of Pennsylvania Press, 2006.

Jami, Catherine. *The Emperor's New Mathematics: Western Learning and Imperial Authority during the Kangxi Reign (1662–1722).* Oxford: Oxford University Press, 2012.

———. "History of Mathematics in Mei Wending's (1633–1721) Work." *Historia Scientiarum* 4, no. 2 (1994): 159–74.

———. "Learning Mathematical Sciences during the Early and Mid-Ch'ing." In *Education and Society in Late Imperial China, 1600–1900*, edited by Benjamin Elman and Alexander Woodside, 223–56. Berkeley: University of California Press, 1994.

———. "Légitimité dynastique et reconstruction des sciences: Mei Wending (1633–1721)." *Annales: Histoire, Sciences Sociales* 59 (2004): 701–27.

Javary, Geneviève. "Hou ji, prince Millet, l'agriculteur divin: Interprétation du mythe chinois par le père Joachim Bouvet, S. J." In *Actes du IIIe Colloque international de sinologie.* Paris: Les Belles Lettres, 1983.

Jensen, Lionel. *Manufacturing Confucianism: Chinese Traditions and Universal Civilization.* Durham, NC: Duke University Press, 1997.

Ji Yun 紀昀. *Ji Xiaolan wenji* 紀曉嵐文集 (The collected writings of Ji Yun). Shijiazhuang: Hebei Jiaoyu Chubanshe, 1991.

———. *Yuewei caotang biji* 閱微草堂筆記 (Jottings from the straw hut for reading subtleties). Reprint. Taipei: Guangwen, 1991.

Jiang Fan 江藩. *Hanxue shichen ji qianshi* 漢學師承記箋釋 (Interpretative comments on the records of the genealogy of Han learning). Comments by Qi Yongxiang. Shanghai: Shanghai Guji Chubanshe, 2006.

———. *Jiang Fan ji* 江藩集 (Collected works of Jiang Fan). Compiled by Qi Yongxiang. Shanghai: Shanghai Guji Chubanshe, 2006.

Jiang Guanghui 姜廣輝. *Yili yu kaoju: Sixiangshi yanjiuzhong de jiazhi guanhuai yu shizheng fangfa* 義理與考據： 思想史研究中的價值關懷與實證方法 (Philosophy and philology: Value orientation and empirical method in history of ideas). Beijing: Zhonghua, 2010.

Jiang Han 江瀚. *Shiwongshanfang zhaji* 石翁山房札記 (Sketches from the mountain retreat of Stone Old Man). In *Changting Jiangxiansheng zhushu wuzhong* 長汀江先生著書五種 (Five books by Jiang of Changting), possibly put together and published by Jiang Han himself, vol. 7. Taiyuan: 1864, n.p.

Jiang Xiaoyuan 江曉原. *Tianxue zhenyuan* 天學真原 (The truth of the theory of Heaven). Shenyang: Liaoning Jiaoyu Chubanshe, 2004.

Jiang Xiaoyuan 江曉原 and Niu Weixing 鈕衛星. *Hui tian: Wuwang fa Zhou yu tianwen lishi niandaixue* 回天： 武王伐紂與天文歷史年代學 (Returning to Heaven: King Wu's conquest of Emperor Zhou and [its relation to] astronomical and historical chronology). Shanghai: Shanghai Renmin Chubanshe, 2000.

———. *Tianwen Xixue Dongjian ji* 天文西學東漸集 (The Eastern transmission of Western astronomy). Shanghai: Shanghai shudian, 2001.

Jiang Yong 江永. *Shenyi kaowu* 深衣考誤 (Correcting errors in [the tailoring of] the long garment). Taipei: Yiwen, 1968.

———. *Shuxue* 數學 (Studies of numbers). Shanghai: Shangwu, 1936.

———. *Yili shigong zengzhu* 儀禮釋宮增注 (Commentary on the chapter of explanation of buildings in "The Ceremonial Rituals"). *Siku quanshu.*

Jiao Xun 焦循. *Diaoguji* 雕菰集 (Collected writings from [the Studio of] Engraved Bamboo). In *Wenxuanlou congshu* 文選樓叢書, compiled by Ruan Yuan 阮元, case 9. Reprint. Taipei: Yiwen, 1967.

———. *Litang daoting lu* 里堂道聽錄 (A record of what Litang heard). In *Beijing tushuguan guji zhenben congkan* 北京圖書館古籍珍本叢刊 (Collectenea of rare books at the Bejing Library), vol. 69. Beijing: Shumu Wenxian Chubanshe, 1987.

Jin Guantao 金觀濤. *Guannianshi yanjiu: Zhongguo xiandai zhongyao zhengzhi shuyu de xingcheng* 觀念史研究： 中國現代重要政治術語的形成 (Studies in the history of ideas: The formation of important political terms in modern China). Hong Kong: Chinese University of Hong Kong, 2008.

Jin Guantao 金觀濤 and Liu Qingfeng 劉青峰. *Zhongguo xiandai sixiang de qiyuan: Chaowending jiegou yu Zhongguo zhengzhi wenhua de yanbian* 中國現代思想的起源： 超穩定結構與中國政治文化的演變 (The origins of modern Chinese thought: Ultra-stable structure and the evolution of political culture in China). Hong Kong: Chinese University Press, 2000.

Johnson, David. *Spectacle and Sacrifice: The Ritual Foundations of Village Life in North China.* Cambridge, MA: Harvard East Asian Monographs, 2009.

Jones, William. *The Great Qing Code.* Oxford: Clarendon Press, 1994.

Kahn, Harold L. *Monarchy in the Emperor's Eyes: Image and Reality in the Ch'ien-lung Reign.* Cambridge, MA: Harvard University Press, 1971.

Kangxi chao Hanwen zhupi zouzhe huibian 康熙朝漢文朱批奏折彙編 (Collection of Chinese-language memorials with vermilion comments from the Kangxi emperor's reign). Beijing: Dang'an, 1985.

Kangxi chao Manwen zhupi zouzhe quanyi 康熙朝滿文朱批奏折全譯 (Complete translation of Manchu-language memorials with vermilion comments from the Kangxi emperor's reign). Beijing: Zhongguo Shehui Kexue Chubanshe, 1996.

Kawahara Hideki 川原秀城. *Chūgoku no kagaku shisō: Ryō Kan tengaku kō* 中国の科学思想—両漢天学考 (Scientific thought in China: An inquiry into celestial learning in the Western and Eastern Han dynasties). Tokyo: Sōbunsha, 1996.

———. "Chūgoku seiritsu shōshi" 中国聲律小史 (A short history of Chinese musicology). In *Shin hatsugen Chūgoku kagakushi shiryō no kenkyū* 新発現中国科学史資料の研究 (An investigation of the newly discovered materials concerning the history of science in China), ed. Yamada Keiji 山田慶児. Kyoto: Kyoto daigaku jinbun kagaku kenkyūjo, 1985.

———. "Taishin to Seiyō rekisan gaku" 戴震と西洋暦算学 (Dai Zhen and Western calendrical mathematics). *Chūgoku shisōshi kenkyū* 12 (1989): 1–35.

Kertzer, David. *Ritual, Politics, and Power.* New Haven, CT: Yale University Press, 1988.

Kessler, Lawrence. *K'ang-hsi and the Consolidation of Ch'ing Rule, 1661–1684.* Chicago: University of Chicago Press, 1976.

Kim Taejun 金泰俊. *Kogaku kara jitsugaku he:* Jūhachi seiki chōsen chishikijin Hong-Tae-yong no pekin ryokō 虚学から実学へ—十八世紀朝鮮知識人洪大容の北京旅行 (From empty to concrete studies: The travel of a Korean intellectual to eighteenth-century Beijing). Tokyo: Tokyo Daigaku Shuppankai, 1988.

Kinoshita Tetsuya 木下鉄矢. *"Shinchō kōshogaku" to sono jidai* 「清朝考証学」とその時代 (The so-called "Qing dynasty evidential scholarship" and its time). Tokyo: Sōbunsha, 1996.

———. "Tai Shin no ongaku" 戴震の音学 (Dai Zhen's phonology). Tōhōgaku 東方学 58 (July 1979): 128-42.

———. "Tai Shin to Wanpai no gakujutsu" 戴震と皖派の学術 (Dai Zhen and the scholarship of Wan). *Tōyōshi kenkyū* 45, no. 3 (December 1986): 519–42.

Kishimoto Miō 岸本美緒. "Chinese History and the Concept of 'Early Modernities.'" Paper presented in the Early Modern World Conference, University of Chicago, 2004.

———. *Chūgoku no mibunsei to shakai chitsujyo* 中国の身分制と社会秩序 (Status system and social order in China). Tokyo: Tokyo Daigaku Daigakuin Jinbun Shakaikei Kenkyūkai, 2007.

———. "'Laoye' to 'xianggong'—Koshō kara mita chihō shakai no kaisō kankaku 「老爺」と「相公」—呼称からみた地方社会の階層感覚 ("Laoye" and "xianggong": The perception of class in local society viewed from the use of address)." In *Dentō Chūgoku no chiikizō* 伝統中国の地域像 (Images of regions in traditional China) edited by Yamamoto Eishi, 51-84. Tokyo: Keiō Gijuku Daigaku Shuppankai, 2000.

———. "Meishi no kōyō: Minshin jidai ni okeru shitaifu no kōsai" 名刺の効用—明清時代における士大夫の交際 (The use of name cards: Social interactions of literati in Ming and Qing periods). In *Hito to hito no chiikishi* 人と人の地域史 (A local history of human interaction) edited by Kimura Seiji 木村靖二 and Ueda Makoto 上田信, 243-76. Tokyo: Yamakawa Shuppansha, 1997.

———. "Min-Shin jidai no shinbun kankaku" 明清時代の身分感覚 (The perception of status in the Ming and Qing periods). In *Min-shin jidai no kihon mondai shiri-zu 4* (The series of basic issues in Ming and Qing periods, no. 4), edited by Mori Masao, Noguchi Tetsuro, Hamashima Atsutoshi, Kishimoto Mio, and Satake Yasuhiko, 403–27. Tokyo: Kyūko shoin, 1997.

———. *Min-Shin kōtai to Kōnan shakai: 17-seiki Chūgoku no chitsujo mondai* 明清交替と江南社会—17世紀中国の秩序問題 (The Ming-Qing transition in Jiangnan: The problem of social order in seventeenth-century China). Tokyo: Tokyo Daigaku Shuppankai, 1999.

———. *Shindai Chūgoku no bukka to keizai hendō* 清代中国の物価と経済変動 (Price and economic change in Qing China). Tokyo: Kenbun Shuppan, 1997.

Kojima Tsuyoshi 小島毅. *Chūgoku kinsei ni okeru rei no gensetsu* 中国近世における礼の言説 (Ritual discourse in twelfth- to sixteenth-century China). Tokyo: Tokyo Daigaku Shuppankai, 1996.

Kondō Mitsuo 近藤光男. *Shinchō kōkyogaku no kenkyū* 清朝考證学の研究 (Studies of Qing evidential scholarship). Tokyo: Kenbun Shuppan, 1987.

Kuhn, Philip A. *Origins of the Modern Chinese State*. Stanford, CA: Stanford University Press, 2002.

———. *Soulstealers: The Chinese Sorcery Scare of 1768*. Cambridge, MA: Harvard University Press, 1990.

Kuhn, Thomas. *The Copernican Revolution: Planetary Astronomy in the Development of Western Thought*. Cambridge, MA: Harvard University Press, 1957.

Kuroda Akinobu 黒田明伸. *Chūka teikoku no kōzō to sekai keizai* 中華帝国の構造と世界経済 (Structure of the Chinese empire and the world economy). Nagoya-shi: Nagoya Daigaku Shuppankai, 1994.

Kutcher, Norman. *Mourning in Late Imperial China: Filial Piety and the State*. New York: Cambridge University Press, 1999.

Lach, Donald F., comp. *The Preface of Leibniz's Novissima Sinica: Commentary, Translation, Text*. Honolulu: University of Hawai'i Press, 1957.

Lach, Donald, and Edwin Van Kley. *Asia in the Making of Europe*. Vol. 3, *A Century of Advance*. Chicago: University of Chicago Press, 1965.

Lackner, Michael. "A Figurist at Work: The Vestigia of Joseph de Prémare S. J." In *L'Europe en Chine: Interactions scientifiques, religieuses et culturelles aux XVIIe et XVIIIe siècles*, 23–56. Paris: Collège de France, Institut des Hautes Études Chinoises.

Landry-Deron, Isabelle. *La preuve par la Chine: La "Description" de J.-B. Du Halde, Jesuite, 1735*. Paris: Editions de l'Ecole des Hautes Études en Sciences Sociales, 2002.

———. *Les leçons de sciences occidentales de l'empereur de Chine Kangxi (1662–1722) par les pères Bouvet et Gerbillon*. Paris: Mémoire de diplôme de l'EHESS, 1995.

———. "Les mathématiciens envoyés en Chine par Louis XIV en 1685." *Archive for History of Exact Sciences* 55 (2001): 423–63.

Lattis, James. *Between Copernicus and Galileo: Christoph Clavius and the Collapse of Ptolemaic Cosmology*. Chicago: University of Chicago Press, 1994.

Legge, James. *The Chinese Classics: With a translation, critical and exegetical notes, prolegomena, and copious indexes*. 7 volumes. Hong Kong: Print Office of London Missionary Society, 1861.

Leibniz, Gottfried Wilhelm. *Discourse on the Natural Theology of the Chinese*. Honolulu: University of Hawai'i Press, 1977.

Leung, Angela Ki Che. "Elementary Education in the Lower Yangtze Region." In *Education and Society in Late Imperial China, 1600–1900*, edited by Benjamin Elman and Alexander Woodside, 381–416. Berkeley: University of California Press, 1994.

Levenson, Joseph. *Confucian China and Its Modern Fate: A Trilogy*. Berkeley: University of California Press, 1968.

Lewis, Mark Edward. *The Construction of Space in Early China*. Albany: State University of New York Press, 2005.

———. *The Early Chinese Empires: Qin and Han*. Cambridge, MA: Belknap Press, 2007.

Li Guangdi 李光地. *Lixiang benyao* 曆象本要 (Essentials of cosmology and calendrics). In *Ronchun quanshu* 榕村全書 (Complete collection of Li Guangdi), volume 26. 1829 edition. Taipei: Academia Sinica, Fu Sinian Library collection.

———. *Rongcun yulu; Rongcun xuyulu* 榕村語錄； 榕村續語錄 (Li Guangdi's oral record and its sequel). 2 vols. Beijing: Zhonghua shuju, 1995.

———. "Xu" 序 (Prologue) to Mei Wending's *Lixue yiwen* 歷算疑問 (Questions in mathematical astronomy). It is included in *Meishi congshu jiyao* 梅氏叢書輯要. (Essential collactanea of the Mei family). 1761. UCLA East Asian Library Collection.

———. *Yuzuan Zhou yi zhezhong* 御纂周易折中 (Imperially authored and eclectic compiltation of "The Book of Changes"). *Siku quanshu*.

Li Kai 李開. *Dai Zhen pingzhuan* 戴震評傳 (Evalutation and biography of Dai Zhen). Nanjing: Nanjing Daxue Chubanshe, 1992.

Li Ling 李零. *Zhongguo fangshukao* 中國方術考 (An investigation of divination, alchemy, astrology, and magic in ancient China). Beijing: Dongfang, 2001.

Li Yan and Qian Baozong. *Li Yan Qian Baozong kexueshi quanji* 李儼錢寶琮科學史全集 (*Complete collection of Li Yan's and Qian Baozong's essays on the history of science*). Edited by Du Shiran 杜石然. Shenyang: Liaoning Jiaoyu Chubanshe: Liaoning sheng xinhua shudian faxing, 1998.

Li Zubai 李祖白. *Tianxue chuangai* 天學傳概 (*An outline of the dissemination of heavenly learning.* In *Tianzhujiao Dongchuan wenxianxubian* 天主教東傳文獻續編 (The archives of the Eastern transmission of Catholicism, the sequel, 1055–62. Taipei: Taiwan xuesheng Bookstore, 1966.

Liang Qichao 梁啟超. *Jindai xuefeng zhi dili de fenbu* 近代學風之地理分布 (The geographical distribution of scholarship in early modern China). In *Yinbingshi zhuanji* 飲冰室專集, vol. 9, Taipei: Taiwan zhonghua bookstore, 1972.

———. *Qingdai xueshu gailun* 清代學術概論 (The intellectual history of the Qing dynasty). Taipei: Commercial Press, 1921.

———. "Qingdai xuezhe zhengli jiuxue zhi zongchengji" 清代學者整理舊學之總成績 (The cumulative achievement of the study of traditional scholarship in the Qing dynasty). In *Zhongguo jinsanbainian lai xueshushi* 中國近三百年來學術史 (Intellectual history of China during the last 300 years), 249-511. Taipei: Liren, 1995.

———. *Zhongguo jinsanbainian lai xueshushi* 中國近三百年來學術史 (The intellectual history of China during the last 300 years). Taipei: Zhonghua Shuju, 1955.

Libbrecht, Ulrich. "What Kind of Science Did the Jesuits Bring to China?" In *Western Humanistic Culture Presented to China by Jesuit Missionaries (XVII-XVIII centuries)*, edited by Federico Massini and Marina Battaglini, 221-34. Rome: Bibliotheca Instituti Historici, 1996.

Lin, Yutang. *The Gay Genius: The Life and Times of Su Tungpo*. New York: J. Day Co., 1947.

Lin Cunyang 林存陽. "Lu Jianzeng yu Hanxue de jueqi" 盧見曾與漢學的崛起 (Lu Jianzeng and the rise of Han learning). *Qingshi luncong* 清史論叢 (2007): 580–601.

———. *San Li Guan: Qingchu xueshu yu zhengzhi hudong de lianhuan* 三禮館： 清代學術與政治互動的鏈環 (The Institute of the Three Ritual Classics: The interaction between knowledge and politics in the Qing dynasty). Beijing: Shehui Kexue Wenxian, 2008.

Lin Li-yueh 林麗月. "Yishang yu fengjiao: Wan Ming de fushi fengshang yu 'fuyao' yilun" 衣裳與風教—晚明的服飾風尚與服妖議論 (Clothes and fashion: Fashion in the late Ming and its critics). *Xin shixue* 10, no. 3 (1999): 111–57.

Lin Qingzhang 林慶彰. *Qingchu de qunjing bianweixue* 清初的群經辨偽學 (The scholarship on authenticating classics in the early Qing). Taipei: Wenjin Chubanshe, 1990.

Lin Tingkan 凌廷堪. *Jiaolitang wenji* 校禮堂文集 (Collection in the Hall of Verifying Rituals). *Xuxiu Siku quanshu*, vol. 1480.

Lindberg, David, and Ronald Numbers, eds. *God and Nature: Historical Essays on the Encounter between Christianity and Science*. Berkeley: University of California Press, 1986.

Liu, Shu-hsien. *Essentials of Contemporary Neo-Confucian Philosophy*. Westport, CT: Praeger, 2003.

———. *Understanding Confucian Philosophy: Classical and Sung-Ming*. Westport, CT: Greenwood Press, 1998.

Liu Dakui 劉大櫆. *Liu dakui ji* 劉大櫆集 (Collection of Liu Dakui). Shanghai: Shanghai Guji Chubanshe, 1990.

Liu Dun 劉鈍. "'Shuli jingwen' zhong 'jiheyuanben' de diben wenti" 數理精蘊中幾何原本的底本問題 (The problem of manuscripts of the "Fundamentals of Geometry" in "The Essence of Mathematical Principles"), *Zhonguo keji shilao* 中國科技史料 (China historical materials of science and technology) 12, no. 3 (1991): 88-96.

Liu Fengyun 劉風雲. *Qingdai sanfan yanjiu* 清代三藩研究. Beijing: Renmin Daxue Chubanshe, 1994.

Liu Lu 劉潞 (ed). *Qinggong xiyang yiqi qing* 清宮西洋儀器 (Scientific and technological instruments of the Qing dynasty). Hong Kong: Commercial Press, 1998.

Liu Ning 劉凝. *Jue si lu* 覺斯錄 (Records of true awareness). Chinois 7172, Bibliothèque

Nationale de France. Reprinted in *Yesu hui Luoma dang an guan Ming Qing tian zhu jiao wen xian* 耶穌會羅馬檔案館明清天主教文獻 (Chinese Christian texts fom the Roman archives of the Society of Jesus), edited by Nicolas Standaert and Adrian Dudink, 9:530-89. Taipei: Taipei Ricci Institute 2002.

Liu Shipei 劉師培. *Liu Shipei shixue lunzhu xuanji* 劉師培史學論著選集 (A selection of Liu Shipei's essays on historiograpy). Shanghai: Shanghai Guji Chubanshe, 2006.

———. *Liu Shipei xinhai qian wenxuan* 劉師培辛亥前文選 (A selection of Liu Shipei's essays published before 1911). Beijing: Shenghuo, Dushu, Xinzhi Sanlian Shudian, 1998.

Liu Zhi 劉智. *Tianfang xingli* 天方性理 (Nature and principles of Islam). *Xuxiu Siku quanshu*, vol. 1296.

Lloyd, G. E. R. *Adversaries and Authorities: Investigations into Ancient Greek and Chinese Science*. Cambridge: Cambridge University Press, 1996.

———. *The Ambitions of Curiosity: Understanding the World in Ancient Greece and China*. Cambridge: Cambridge University Press, 2002.

Lloyd, G. E. R., and Nathan Sivin. *The Way and the Word: Science and Medicine in Early China and Greece*. New Haven, CT: Yale University Press, 2002.

Loewe, Michael. *Divination, Mythology, and Monarchy in Han China*. Cambridge: Cambridge University Press, 1994.

———, ed. *Early Chinese Texts: A Bibliographical Guide*. Berkeley: SSEC and IEAS, 1993.

Loewe, Michael, and Edward L. Shaughnessy. *The Cambridge History of Ancient China: From the Origins of Civilization to 221 B.C.* Cambridge: Cambridge University Press, 1999.

Lu Jiuyuan 陸九淵. *Xiangshan ji* 象山集 (Works by Lu Jiuyuan). *Siku quanshu*.

Lu Longqi 陸隴其. *Lu Longqi nianpu* 陸隴其年譜 (The chronicle of Lu Longqi). Beijing: Zhonghua Shuju, 1993.

———. *Sanyutang riji* 三魚堂日記 (Daily records from Three Fish Hall). *Xuxiu siku quanshu*, vol. 559.

———. *Sishu jiangyi kunmianlu* 四書講義困勉錄 (Record of diligently overcoming obstacles in discussing the meanings of the Four Books). *Siku quanshu*.

Lu Miaw-fen 呂妙芬. *Xiaozhi tianxia: "Xiaojing" yu jinshi Zhongguo de zhengzhi yu wenhua* 孝治天下：《孝經》與近世中國的政治與文化 (Governing the world with filial piety: "The Classic of Filial Piety," politics, and culture in early modern China). Taipei: Lianjing, 2011.

———. *Yangmingxue shiren shequn: Lishi, sixiang yu shijian* 陽明學士人社群：歷史，思想與實踐 (Literati communities of the Wang Yangming school: History, thought, and practice). Taipei: Zhongyang Yanjiuyuan Jindaishi Yanjiusuo, 2010.

Lu Wenzhao 盧文昭. *Baojingtang wenji* 抱經堂文集 (A collection from the Hall of Embracing the Classics). Beijing: Zhonghua Shuju, 1990.

Lu Xianwen 盧仙文 and Jiang Xiaoyuan 江曉原. "Lulun qingdai xuezhe dui gudai lifa de zhengli yanjiu" 略論清代學者對古代曆法的整理研究 (A sketch of Qing scholars' research on ancient calendars). *Zhongguo keji shiliao* 中國科技史料 20, no. 1 (1999): 85-86.

Lundbaek, Knud. *Joseph de Prémare (1666–1736): Chinese Philology and Figurism*. Denmark: Aarhus University Press, 1991.

———. *The Traditional History of the Chinese Script: From a Seventeenth-Century Jesuit Manuscript*. Denmark: Aarhus University Press, 1988.

Luo Lida 羅麗達. "Baijin yanjiu Yi jing shishi jikao" 白晉研究易經史事稽考 (The analysis of Bouvet's research on "The Book of Changes"). *Hanxue yanjiu* 漢學研究 15, no. 1 (1997): 173–85.

Lynn, Richard John. *The Classic of Changes: A New Translation of the "I Ching" as Interpreted by Wang Bi*. New York: Columbia University, 1994.

Ma Duanlin 馬端臨. *Wenxian tongkao* 文獻通考 (Comprehensive studies of documents). *Siku quanshu*, vol. 612.

Macauley, Melissa. *Social Power and Legal Culture: Litigation Masters in Late Imperial China*. Stanford, CA: Stanford University Press, 1998.

Man-Cheong, Iona. *The Class of 1761: Examinations, State, and Elites in Eighteenth-Century China*. Stanford, CA: Stanford University Press, 2004.

Mann, Michael. *The Sources of Social Power*. 2 vols. Cambridge: Cambridge University Press, 1986, 1993.

Mann, Susan. "The Education of Daughters in the Mid-Ch'ing Period." *Education and Society in Late Imperial China, 1600–1900*. Edited by Benjamin Elman and Alexander Woodside, 19–49. Berkeley: University of California Press, 1994.

———. *Precious Records: Women in China's Long Eighteenth Century*. Stanford, CA: Stanford University Press, 1997.

———. "Widows in the Kinship, Class, and Community Structures of Qing Dynasty China." *Journal of Asian Studies* 46, no. 1 (1987): 37-56.

Mao Qiling 毛奇齡. *Ming Tangwen* 明堂問 (An inquiry into the Palace of Light). *Siku quanshu cunmu congshu*, vol. 108.

———. *Sishu gaichuo* 四書改錯 (Correcting mistakes in the Four Books). *Xuxiu Siku quanshu*, vol. 165.

Mao Yuesheng 毛嶽生. "Buxu" 補序 (Supplementary preface). In *Xiuhailou zazhu* 袖海樓雜著 (A miscellany from Sleeve Sea Hall), by Huang Rucheng 黃汝成. Reprint. Beiping: Yanjing Daxue Tushuguan.

Martzloff, Jean-Claude. *A History of Chinese Mathematics*. Berlin: Springer-Verlag, 1997.

———. "Space and Time in Chinese Texts of Astronomy and of Matheamatical Astronomy in the Seventeenth and Eighteenth Centuries." *Chinese Science* 11 (1993-94): 66-92.

McDermott, Joseph, ed. *State and Court Ritual in China*. Cambridge: Cambridge University Press, 1999.

Mei Wending 梅文鼎. *Jianjitang zuanke Mei Wu'an xiansheng lisuan quanshu* 兼濟堂纂刻梅勿菴先生曆算全書 (Essential collectanea of Mei Lineage's work). 1859. Harvard University Rare Book Collection (digitized by Google Books).

———. *Jixuetang shichao* 績學堂詩鈔 (Poems from the Hall of Accumulating Learning). *Xuxiu Siku quanshu*, vol. 1413.

———. *Jixuetang wenchao* 績學堂文鈔 (Collection of essays in the Hall of Accumulating Learning). *Xuxiu Siku quanshu*, vol. 1413.

———. *Lisuan quanshu* 歷算全書 (Complete works of mathematical astronomy). *Siku quanshu*.

———. *Lixue yiwen* 歷算疑問 (Questions in mathematical astronomy). *Meishi congshu jiyao* 梅氏叢書輯要.

———. *Lixue yiwen bu* 歷算疑問補 (Sequel to questions in mathematical astronomy). In *Meishi congshu jiyao* 梅氏叢書輯要.

———. *Meishi congshu jiyao* 梅氏叢書輯要 (Essential collactanea of the Mei family). 1761. UCLA East Asian Library Collection.

———. *Wuan lisuan shumu* 勿菴曆算書目 (Annotated bibliography of Mei Weiding's works). Repring of Zhibuzuzhai congshu 知不足齋叢書. Taipei: Xingzhong Bookstore, 1964.

Menegon, Eugenio. *Ancestors, Virgins, and Friars: Christianity as a Local Religion in Late Imperial China*. Cambridge, MA: Harvard University Asia Center, 2009.

Merton, Robert. *Social Theory and Social Structure*. New York: Free Press, 1968.

Minamiki, George. *The Chinese Rites Controversy: From Its Beginning to Modern Times*. Chicago: Loyola University Press, 1985.

Mingshi 明史 (History of the Ming dynasty). *Shiku quanshu*.

Miura Shūichi 三浦修一. "Hō Shōshō no shisō: Kenryūki no shitaifu to bukkyō ni kansuru ikkousatsu" 彭紹升の思想—乾隆期の士大夫と佛教に關する一考察 (The thought of Peng Shaosheng). Tōhō gakuhō 60 (1988): 439-79.

———. "Hō Shōshō to Tai Shin no shisōken" 彭紹升と戴震の思想圈 (Peng Shaosheng and Dai Zhen's intellectual circle) *Tōhō gakuhō* 62 (1988): 447-81.

Miyazaki, Ichisada. *China's Examination Hell: The Civil Service Examinations of Imperial China*. Translated by Conrad Shirokauer. New York: Weatherhill, 1976.

Mizoguchi Yūzō 溝口雄三. *Chūgoku no kō to shi* 中国の公と私 (The idea of public and private in China). Tokyo: Kenbun Shuppan, 1995.

———. *Chūgoku zenkindai shisō no kussetsu to tenkai* 中国前近代思想の屈折と展開 (The refraction and unfolding of premodern Chinese thought). Tokyo: Tokyo Daigaku Shuppankai, 1980.

———. *Hōhō toshite no Chūgoku* 方法としての中国 (China as method). Tokyo: Tokyo Daigaku Shuppankai, 1989.

———. "Iwayuru Tōrinha jinshi no shisō—Zenkindaiki ni okeru Chūgoku shisō no tenkai (jō)" いわゆる東林派人士の思想: 前近代期における中国思想の展開(上) (The thought of the members of the so-called Donglin school: The development of Chinese thought in the premodern period). *Tōyō bunka kenkyūjo kiyō* 75 (1978): 111–341.

———. "Mōshi jigi soshō no rekishi teki kōsatsu" 孟子字義疏証の歷史的考察 (A historical study of the evidential analysis of "The Meaning of Terms in Mencius"). *Tōyō bunka kenkyūjo kiyō* 48 (1969): 144–65.

———. "Zhongguo sixiang yu sixiangshi yanjiu de shijiao" 中國思想與思想史研究的視角 (Chinese intellectual history and perspectives on intellectual history). *Weshizhe* 270, no. 3 (2002): 5–10.

Mosshammer, Alden. *The Easter Computus and the Origins of the Christian Era*. Oxford: Oxford University Press, 2008.

Mungello, David, ed. *The Chinese Rites Controversy: Its History and Meaning*. Nettetal: Seyler Verlag, 1994.

Mungello, D. E. *Curious Land: Jesuit Accommodation and the Origins of Sinology*. Honolulu: University of Hawai'i Press, 1989.

———. *The Forgotten Christians of Hangzhou*. Honolulu: University of Hawai'i Press, 1994.

———. *Leibniz and Confucianism: The Search for Accord*. Honolulu: University of Hawai'i Press, 1977.

———. "Unearthing the Manuscripts of Bouvet's 'Gujin' after Nearly Three Centuries." *Sino-Western Cultural Relations Journal* 10 (1988): 34–61.

Murase Hiroya 村瀬裕也. *Tai Shin no tetsugaku: Yuibutsuron to dōtokuteki kachi* 戴震の哲学—唯物論と道徳的価値 (Philosophy of Dai Zhen: Materialism and moral values). Tokyo: Nitchū Shuppan, 1984.

Murata, Sachiko, William Chittick, and Weiming Tu. *The Sage Learning of Liu Zhi:*

Islamic Thought in Confucian Terms. Cambridge, MA: Harvard University Press, 2009.

Nakayama, Shigeru. *A History of Japanese Astronomy: Chinese Background and Western Impact*. Cambridge, MA: Harvard University Press, 1969.

Nalan Chengde 納蘭成德 (1655-85) and Xu Qianxue 徐乾學 (1631-94), comps. *Tongzhitang jingjie* 通志堂經解 (Exegeses of Tongzhi Hall). Reprint of 1680 edition. Taipei: Datong shuju, 1969.

Naquin, Susan. *Peking: Temples and City Life, 1400–1900*. Berkeley: University of California Press, 2000.

Needham, Joseph. *The Grand Titration: Science and Society in East and West*. Toronto: University of Toronto Press, 1969.

———. *Science and Civilisation in China*. 7 vols. Cambridge: Cambridge University Press, 1954-2004.

Ng, On-cho. *Cheng-Zhu Confucianism in the Early Qing: Li Guangdi (1642-1718) and Qing Learning*. Albany: State University of New York Press, 2001.

Nishi Junzō 西順蔵. *Chūgoku shisō* ronshū 中国思想論集 (Essays on Chinese thoughts). Tokyo: Chikuma Shobō, 1969.

Nivision, David. *The Life and Thought of Chang Hsueh-ch'eng (1738–1801)*. Stanford, CA: Stanford University Press, 1966.

———. *The Ways of Confucianism: Investigations in Chinese Philosophy*. Chicago: Open Court, 1996.

Nōda Chūryō 能田忠亮. *Kansho ritsurekishi no kenkyū* 漢書律暦志の研究 (The study of calandrical astronomy section in the standard Han history). Kyoto: Zenkoku Shobō, 1947.

———. *Tōyō tenmongakushi ronsō* 東洋天文學史論叢 (Collective essays on the history of Eastern astronomy) Tokyo: Kōseisha, 1943.

Nylan, Michael. "The Many Dukes of Zhou in Early Sources." In *Statecraft and Classical Learning: The Rituals of Zhou in East Asian History*, edited by Benjamin Elman and Martin Kern, 94-128. Leiden: Brill, 2010.

Okubo Eiko 小久保英子. *Minshin jidai no shoin no kenkyū* 明清時代の書院の研究 (Research on academies in the Ming-Qing period). Tokyo: Kokusho Kankōkao, 1976.

Okuzaki Hiroshi 奥崎裕司, ed. *Minshinwa ikanaru jidaide attaka: Shisōshi ronshū* 明清はいかなる時代であったか： 思想史論集 (What exactly was the Ming-Qing period? Essays of intellectual history). Tokyo: Kyūko Shoin, 2006.

O'Malley, John W. *The First Jesuits*. Cambridge, MA: Harvard University Press, 1993.

Ono Kazuko 小野和子. *Minki Tōsha kō: Tōrintō to Fukusha* 明季党社考—東林党と復社 (An investigation of political parties and association in the Ming Dynasty: Donglin Party and Returning [to Antiquity] Association). Kyoto: Dōhōsha Shuppan, 1996.

Onogawa Hidemi 小野川秀美. *Shinmatsu seiji shisō kenkyū* 清末政治思想研究 (A study of political thought in late Qing China). 2 vols. Tokyo: Heibonsha, 2009.

Ōtani Toshio 大谷敏夫. *Shindai seiji shisōshi kenkyū* 清代政治思想史研究 (Historical studies of Qing political thought). Tokyo: Kyūko, 1991.

Oxnam, Robert. *Ruling from Horseback: Manchu Politics in the Oboi Regency*. Chicago: University of Chicago Press, 1975.

Pan Xinghui 潘星輝. "Dai Zhen lingyiguan fawei" 戴震靈異觀發微 (A subtle interpretation of Dai Zhen's view of the supernatural). *Qingshi luncong* 清史論叢 (2009): 314-24.

Perdue, Peter C. *China Marches West: The Qing Conquest of Central Eurasia*. Cambridge, MA: Belknap Press, 2005.

Perkins, Franklin. *Leibniz and China: A Commerce of Light*. Cambridge: Cambridge University Press, 2004.

Peterson, Willard. "Learning from Heaven: The Introduction of Christianity and Other Western Ideas into Late Ming China." In *The Cambridge History of China*. Vol. 8, *The Ming Dynasty, 1368-1644, part 2*, edited by Denis C. Twitchett and Frederick W. Mote, 789-839. Cambridge: Cambridge University Press, 1998.

———. "The Life of Ku Yen-wu (1613-1682)." Pts. 1 and 2. *Harvard Journal of Asiatic Studies* 28 (1968): 114-56; 29 (1969): 201-47.

———. "Squares and Circles: Mapping the History of Chinese Thought." *Journal of the History of Ideas* 49, no. 1 (1988): 47-60.

Pfister, Louis. *Notices biographiques et bibliographiques sur les Jésuites de l'ancienne mission de Chine, 1552–1773*. Shanghai: Imprimerie de la Mission Catholique, 1932.

Pocock, J. G. A. *Barbarism and Religion*. 4 vols. Cambridge: Cambridge University Press, 1999.

———. *Virtue, Commerce, and History*. Cambridge: Cambridge University Press, [1985] 1995.

Pollock, Sheldon. *The Language of the Gods in the World of Men: Sanskrit, Culture, and Power in Premodern India*. Berkeley: University of California Press, 2006.

Qi Yongxiang 漆永祥, comp. *Dong Wu san Hui shiwenji* 東吳三惠詩文集 (A collection of essays and poems by Hui Zhouxi, Hui Shiqi, and Hui Dong). Taipei: Zhongyang Yanjiuyuan, Zhongguo Wenzhe Yanjiusuo, 2006.

———. *Jiang Fan yu "Hanxue shicheng ji" yanjiu* 江藩與《漢學師承記》研究 (An investigation of Jiang Fan and his "Records of the Genealogy of Han Learning"). Shanghai: Shanghai Guji Chubanshe, 2006.

———. *Qianjia Kaojuxue yanjiu* 乾嘉考據學研究 (An investigation of evidential scholarship in the Qianlong and Jiaqing reigns). Beijing: Zhongguo Shehui Kexue Chubanshe, 1998.

———. "Xinfaxian Dai Zhen 'Jiang Shenxiu xiansheng qishi shouxu' yiwen yipian" 新發現戴震《江慎修先生七十壽序》佚文一篇 (Dai Zhen's lost piece "Dedication to the Seventieth Birthday of Mr. Jiang Yong," recently discovered). *Zhongguo dianji wenhua* 中國典籍與文化 (Chinese documents and culture), no. 1 (2005): 122–23.

Qian Baozong 錢寶琮. "Mei Wu'an xiansheng nianpu" 梅勿庵先生年譜 (Chronological biography of Mei Wending). In *Li Yan Qian Baozong kexueshi quanji* 李儼錢寶琮科學史全集, edited by Du Shiran 杜石然, 9:608–38. Shenyang: Liaoning Jiaoyu Chubanshe: Liaoning Sheng Xinhua Shudian Faxing, 1998.

Qian Daxin 錢大昕. "*Bu de yi tiji*" 不得已題記 (Remark on "I Cannot Do Otherwise"). In *Bu de yi: Fu erzhong* 不得已: 附二种 (I cannot do otherwise, with two [Jesuit] responses), by Yang Guangxian 楊光先 et al., compiled and edited by Chen Zhanshan 陳占山, 195. Hefei: Huangshan Shushe Chubanshe, 2000.

———. *Jiading Qian Daxin quanji* 嘉定錢大昕全集 (The complete works of Qian Daxin). Edited by Chen Wenhe 陳文和. 10 vols. Nanjing: Jiangsu Guji Chubanshe, 1997.

———. *Qian Xinmei xiansheng nianpu* 錢辛楣先生年譜 (The annalistic biography of Qian Xinmei). In *Jiading Qian Daxin quanji* 嘉定錢大昕全集 (The complete works of Qian Daxin), edited by Chen Wenhe 陳文和, 1:1–46. Nanjing: Jiangsu Guji Chubanshe, 1997.

Qian Mu 錢穆. *Zhongguo jinsanbainian lai xueshushi* 中國近三百年來學術史 (The intellectual history of China during the last three hundred years). 2 vols. Shanghai: Shangwu Yinshuguan, 1937.

Qian Xuan 錢玄. *San li tonglun* 三禮通論 (A general discussion of three ritual classics). Nanjing: Shifan Daxue Chubanshe, 1994.

Qian Yong 錢泳. *Luyuan conghua* 履園叢話 (Collective jottings in the Garden of Footsteps). Beijing: Zhonghua shuju, 1997.

Qian-Jia ming ru nianpu 乾嘉名儒年譜 (Chronological biographies of famous literati from the Qianlong and Jiaqing reigns). Compiled by Chen Zuwu 陳祖武. 14 vols. Beijing: Beijing Rushuguan Chubanshe, 2006.

Qinding guozijianzhi 欽定國子監志 (Imperially sanctioned history of the imperial college). Vol. 600. *Siku quanshu*.

Qing bai leichao 清稗類鈔 (Categorical records of the unofficial history of the Qing dynasty). Taipei: Taiwan Shangwu Yinshuguan, 1966.

Qing shilu 清實錄 (The veritable records of the Qing dynasty). Beijing: Zhonghua shuju, 1985.

Qing zhong qian qi xiyang Tianzhujiao zai Hua huodong dang'an shiliao 清中前期西洋天主教在華活動檔案史料 (Archives of Catholic activities in the early and middle periods of Qing China). Beijing: Zhonghua Shuju, 2003.

Qiu Weijun 丘為君. *Dai Zhen xue de xingcheng* 戴震學的形成 (The formation of Dai Zhen studies). Taipei: Lianjing, 2004.

Qu Dajun 屈大均. *Guangdong xinyu* 廣東新語 (New jottings from Guangdong). 2 vols. Beijing: Zhonghua Shuju, 1985.

Qu Zengfa 屈曾發. *Jiushu tongkao* 九數通考 (A comprehensive study of nine chapters). Guangzhou: Xuehaitang, 1871.

Rawski, Evelyn S. *Education and Popular Literacy in Ch'ing China*. Ann Arbor: University of Michigan Press, 1979.

———. *The Last Emperors: A Social History of Qing Imperial Institutions*. Berkeley: University of California Press, 1998.

Rho, Giacomo. *Bili guijie* 比例規解 (Illustration of compass). Borgia Cinese 318 (1), Bibliotheca Apostolica Vaticana.

Ricci, Matteo. *China in the Sixteenth Century: The Journals of Matthew Ricci 1583–1610*. Translated by Louis Gallagher, S. J. New York: Random House, 1953.

Ricoeur, Paul. *Time and Narrative*. 3 vols. Chicago: University of Chicago Press, 1984–1988.

Rijiang liji jieyi 日講禮記解義 (Daily lecture on the explanation and meaning of the *Ritual records*). Vol. 123. *Siku quanshu*.

Roddy, Stephen. *Literati Identity and its Fictional Representations in Late Imperial China*. Stanford, CA: Stanford University Press, 1998.

Romano, Antonella. "Observer, vénérer, servir: Une polémique jésuite autour du tribunal des mathématiques de Pékin." *Annales. Histoire, Sciences Sociales* 59 (2004): 729-56.

Rorty, Richard. *Philosophy and the Mirror of Nature*. Princeton, NJ: Princeton University Press, 1979.

Rowbotham, Arnold. *Missionary and Mandarin: The Jesuits at the Court of China*. Berkeley: University of California Press, 1942.

Rowe, William T. *China's Last Empire: The Great Qing*. Cambridge, MA: Harvard University Press, 2009.

———. *Saving the World: Chen Hongmou and Elite Consciousness in Eighteenth-Century China*. Stanford, CA: Stanford University Press, 2001.

Ruan Kuisheng 阮葵生. *Chayu kehua* 茶餘客話 (Chatting with guests over tea). *Congshu jicheng chubian*, vol. 2826.

Ruan Yuan 阮元, comp. and ed. *Chouren zhuan* 疇人傳 (Biographies of mathematical astronomers). 揚州阮氏琅環仙館刻本. 1840 Edition. *Xuxiu Siku quanshu*, vol. 516.

———. *Shisanjing zhushu* 十三經注疏 (Texts and commentaries of the Thirteen Classics). Taipei: Dahua, 1982.

———. "Xu" 序 (Prologue). In *Chouren zhuan* 疇人傳, 1a-2a.

———. *Yan jing shi ji* 揅經室集 (Collection of the Studio of Studying Classics). Beijing: Zhonghua Shuju, 1993.

Rule, Paul. *K'ung-tzu or Confucius? The Jesuit Interpretation of Confucianism*. Boston: Allen & Unwin, 1986.

———. "Moses or China." *Images de la Chine: Le contexte occidental de la sinologie naissante*. Actes du VIIe Colloque international de sinologie de Chantilly, 11–14, septembre 1989 (*Varietes sinologiques-Nouvelle série* 78), 303–31. Taipei: Ricci Institute, 1995.

Rusk, Bruce. "The Rogue Classicist: Feng Fang (1493–1566) and His Forgeries." PhD diss., UCLA, 2004.

Sakade Yoshinobu 坂出祥伸. *Chūgoku kindai no shisō* to kagaku 中国近代の思想と科学 (History of thought and science in modern China). Kyoto: Dōhōsha, 1983.

Saliba, George. *Islamic Science and the Making of the European Renaissance*. Cambridge, MA: MIT Press, 2007.

Sasaki Megumi 佐々木愛. "Mō kilei no sisō henreki" 毛奇齡の思想遍歷 (The intellectual journey of Mao Qiling). *Tōyō shi kenkyū* 東洋史研究 56, no. 2 (1997): 191-234.

Satō, Masayuki. *The Confucian Quest for Order: The Origin and Formation of the Political Thought of Xun Zi*. Leiden: Brill, 2003.

Schaab-Hanke, Dorothee. "The Power of an Alleged Tradition: A Prophecy Flattering Han Emperor Wu and Its Relation to the Sima Clan." *Bulletin of the Museum of Far Eastern Antiquities* (*BMFEA*) 74 (2002): 243-90.

Schäfer, Dagmar. *The Crafting of the 10,000 Things: Knowledge and Technology in Seventeenth-Century China*. Chicago: University of Chicago Press, 2011.

Schneewind, Sarah. *Community Schools and the State in Ming China*. Stanford, CA: Stanford University Press, 2006.

Schneider, Lawrence. *Ku Chieh-Kang and China's New History: Nationalism and the Quest for Alternative Traditions*. Berkeley: University of California Press, 1971.

Schwartz, Benjamin. *The World of Thought in Ancient China*. Cambridge, MA: Belknap Press of Harvard University Press, 1985.

Science and Civilisation in China. Joseph Needham et al. 7 vols. Cambridge: Cambridge University Press, 1954–.

Scott, James C. *Seeing Like a State: How Certain Schemes to Improve the Human Condition Have Failed*. New Haven, CT: Yale University Press, 1998.

Sela, Ori. "Qian Daxin (1728–1804): Knowledge, Identity, and Reception of History in China, 1750–1930." PhD diss., Princeton University, 2011.

Shang, Wei. *Rulin waishi and Cultural Transformation in Late Imperial China*. Cambridge, MA: Harvard University Press, 2003.

Shelford, April. *Transforming the Republic of Letters: Pierre-Daniel Huet and European Intellectual Life, 1650-1720*. Rochester, NY: University of Rochester Press, 2007.

Shen Congwen 沈從文. *Zhongguo gudai fushi yanjiu* 中國古代服飾研究 (A study of clothing in traditional China). Taipei: Taiwan shangwu yinshuguan, 1992.

Shen Zhiqi 沈之奇 (comp). *Daqinglü jizhu* 大清律輯註 (Commentary on the Great Qing Code). First published in 1715. Beijing: Falu Chubanshe, 1998.

Shengzu renhuangdi yuzhi wenji 聖祖仁皇帝御製文集 (The literary collection of the Kangxi emperor). *Siku quanshu*, v. 1299.

Shi, Yunli. "Reforming Astronomy and Compiling Imperial Science in the Post-Kangxi Era: The Social Dimension of the Yuzhi lixiang kaocheng houbian," *East Asian Science, Technology, and Medicine* 28 (2008): 36–81.

Shields, Philip R. *Logic and Sin in the Writings of Ludwig Wittgenstein.* Chicago: University of Chicago Press, 1993.

Shigeta Atsushi 重田德. *Shindai shakai keizaishi kenkyū* 清代社会経済史研究 (The study of the social and economic history of Qing period). Tokyo: Iwanami Shoten, 1975.

Shimada Kenji 島田虔次. *Chūgoku ni okeru kindai shii no zasetsu* 中國における近代思惟の挫折 (The setback of modern Chinese thought). Commentary by Inoue Susumu 井上進. 2 vols. Tokyo: Heibonsha, 2003.

———. *Chūgoku shisōshi no kenkyū* 中国思想史の研究 (A study for the history of Confucianism in China). Kyoto: Kyoto Daigaku gakujutsu Shuppan kai, 2002.

———. *Pioneer of the Chinese Revolution: Zhang Binglin and Confucianism.* Stanford, CA: Stanford University Press, 1990.

Shinjō Shinzō 新城新蔵. *Tōyō tenmongakushi kenkyū* 東洋天文學史研究 (The study of the history of East Asian astronomy). Kyoto: Kōbundō, 1928.

Shunshan Shi Zhongming xiansheng nianpu 舜山是仲明先生年譜 (Chronological biography of Mr. Shi Jing from Shunshan). Vol. 2, *Qianjia mingru nianpu* 乾嘉名儒年譜 (Chronolgoical biographies of famous scholars in the Qianlong and Jiaqing reigns). Beijing: Beijing Tushu Chubanshe, 2006.

Shustermna, Noah. *Religion and the Politics of Time: Holidays in France from Louis XIV through Napoleon.* Washington, D.C.: The Catholic University of America Press, 2010.

Siku jinhuishu congka 四庫禁毀書叢刊 (The books censored in the complete collection of the four treasuries). Beijing: Beijing Chubanshe, 2000.

Siku quanshu 四庫全書 (The complete collection of the four treasuries). The electronic version of the *Siku quanshu* (Wenyuange Edition).

Siku quanshu cunmu congshu 四庫全書存目叢刊 (The books listed in the catalogue, but not included in the complete collection of the four treasuries). Jinan: Qilu Shushe, 1997.

Siku quanshu zongmu 四庫全書總目 (Catalogue of the complete collection of the four treasuries). I Yun et al., comp. Taipei: Yiwen, 1974.

Siku weishoushu jika 四庫未收書輯刊 (The books not included in the complete collection of the four treasuries). Beijing: Beijing Chubanshe, 1997.

Sima Zhen 司馬貞. *Shiji suoyin* 史記索隱 (Search for subtle meanings in "Records of the Grand Historian"). *Siku quanshu.*

Sivin, Nathan. "Copernicus in China." In *Science in Ancient China: Researches and Reflections*, 4: 1–52. Aldershot: Variorum, 1995.

———. "Cosmos and Computation in Early Chinese Mathematical Astronomy." *T'oung Pao* 55 (1969): 1–73.

———. *Granting the Seasons: The Chinese Astronomical Reform of 1280, with a Study of Its Many Dimensions and an Annotated Translation of Its Records*. New York: Springer, 2009.

———. "State, Cosmos, and Body in the Last Three Centuries B.C.," *Harvard Journal of Asiatic Studies* 55, no. 1 (1995): 5–37.

Skinner, G. William "Marketing and Social Structure in Rural China." *Journal of Asian Studies* 24, nos. 1, 2, 3 (1964): 3–44, 195–228, 363–99.

Skinner, G. William, and Hugh D. R. Baker. *The City in Late Imperial China*. Stanford, CA: Stanford University Press, 1977.

Skinner, Quentin. *The Foundations of Modern Political Thought*. 2 vols. Cambridge: Cambridge University Press, [1978] 2006.

Smart, W. M. *Textbook on Spherical Astronomy*. Cambridge: Cambridge University Press, 1977.

Smith, Richard. *Fortune-tellers and Philosophers: Divination in Traditional Chinese Society*. Boulder, CO: Westview Press, 1993.

———. "A Note on Qing Dynasty Calendars." *Late Imperial China* 9, no. 1 (1998): 123–45.

Soothill, William Edward. *The Hall of Light: A Study of Early Chinese Kingship*. Edited by Lady Hosie and G. F. Hudson. London: Lutterworth, 1951.

Spence, Jonathan D. "Ambivalence and Action: Some Frustrated Scholars of the K'ang-Hsi Period." In *From Ming to Ch'ing: Conquest, Region, and Continuity in Seventeenth-Century China*, 321–65. Edited by Jonathan Spence and John Wills. New Haven, CT: Yale University Press, 1979.

———. *Chinese Roundabout: Essays in History and Culture*. New York: Norton, 1992.

———. *Emperor of China: Self-Protrait of K'ang Hsi*. New York: Vintage, 1988.

———. *Treason by the Book*. New York: Penguin, 2001.

Standaert, Nicolas (ed). *Handbook of Christianity in China: 635–1800*. Leiden: Brill, 2001.

———. *The Interweaving of Rituals: Funerals in the Cultural Exchange between China and Europe*. Seattle: University of Washington Press, 2008.

———. *Yang Tingyun, Confucian and Christian in Late Ming China*. Leiden: Brill, 1988.

Struve, Lynn A. "The Hsü Brothers and Semiofficial Patronage of Scholars in the K'ang-hsi Period." *Harvard Journal of Asiatic Studies* 42 (1982): 231–66.

———. "Huang Zongxi in Context: A Reappraisal of His Major Writings." *Journal of Asian Studies* 47, no. 3 (1988): 474–502.

———. *The Ming-Qing Conflict, 1619–1683: A Historiography and Source Guide*. Ann Arbor, MI: Association for Asian Studies, 1998.

———. *The Qing Formation in World-Historical Time*. Cambridge, MA: Harvard University Asia Center, 2004.

Sun Xidan 孫希旦. *Liji jijie* 禮記集解 (An explanation of "The Ritual Records"). Beijing: Zhonghua shuju, 1989.

Sun Xingyan 孫星衍. "Sunshi chongkan Songben shuowen jiezi xu" 孫氏重刊宋本說文解字序 (Prologue to the Sun family reprint of Song dynasty edition of "The explication of simple and compound characters"). In *Shuowen jiezi* 說文解字 (The explication of simple and compound characters), complied by Xu Shen 許慎 [CE 58–147]. Reprint. Hong Kong: Taiping Bookstore, 1966.

Sun Yirang 孫詒讓. *Zhouli zhengyi* 周禮正義 (Orthodox meanings of "The Rituals of Zhou"). 14 vols. Beijing: Zhonghua Shuju, 1987.

Tan Tai 談泰. "Chouren jie" 疇人解 (An exegesis of Chouren). *Chouren zhuan*, 1a–5b.

Tasaka, Kodo. "An Aspect of Islamic Culture Introduced into China," *Memoirs of the Research Department of the Toyo Bunko* 16 (1957): 75–160.

Tian Miao 田淼. *Zhongguo shuxue de Xihua licheng* 中國數學的西化歷程 (The Westernization of mathematics in China). Jinan: Shandong Jiaoyu Chubanshe, 2005.

Tiwald, Justin. "Aquiring 'Feelings That Do Not Err': Moral Deliberation and the Sympathetic Point of View." PhD diss., University of Chicago, 2006.

Toulmin, Stephen. *Cosmopolis: The Hidden Agenda of Modernity.* New York: Free Press, 1990.

Usui Sachiko 臼井佐知子. *Kishū shounin no kenkyū* 徽州商人の研究 (The study of Huizhou merchants). Tokyo: Kyūko Shoin, 2005.

Van Brummelen, Glen. *The Mathematics of the Heavens and the Earth: The Early History of Trigonometry.* Princeton, NJ: Princeton University Press, 2009.

Van Kley, Edward. "Europe's 'Discovery' of China and the Writing of World History." *American Historical Review* 76 (1971): 358–85.

Väth, Alfons. *Johann Adam Schall von Bell SJ: Missionar in China, kaiserlicher Astronom und Ratgeber am Hofe von Peking, 1592–1666.* Monumenta Sinica monograph series 25. Nettetal: Steyler, 1991.

Verbiest, Ferdinand. *Lifa bu de yi bian* 曆法不得已辯 (A rebuttal to the astronomical part of "I Cannot Do Otherwise"). In *Bu de yi: Fu erzhong* 不得已： 附二種 (I cannot do otherwise, with two comments), by Yang Guangxian 楊光先 et al., compiled by Chen Zhanshan 陳占山, 137–93. Hefei: Huangshan, 2000.

von Collani, Claudia. "Claudio Filippo Grimaldi S. J. zur Ankenft des Päpstilichen Legaten Charles-Thomas Maillard de Tournon in China," *Monumenta Serica* 42 (1994): 329–59.

———. "Introduction: The Author of the 'Journal.'" In *Joachim Bouvet, S. J. Journal des Voyages*, edited by Claudia von Collani, 1-60. Taipei: Taipei Ricci Insitute, 2005.

———. *P. Joachim Bouvet S. J. Sein Leben und sein Werk.* Nettetal: Steyler Verlag, 1985.

Wagner, Rudolf. "The Zhouli as the Late Qing Path to the Future." In *Statecraft and Classical Learning: The Rituals of Zhou in East Asian History*, edited by Benjamin Elman and Martin Kern, 359-87. Leiden: Brill, 2010.

Wakeman, Frederic E. *Telling Chinese History: A Selection of Essays.* Berkeley: University of California Press, 2009.

Waley, Arthur. *Yuan Mei: Eighteenth Century Chinese Poet.* Stanford, CA: Stanford University Press, 1956.

Waley, Authur, and Joseph Allen, trans. *The Book of Songs: The Ancient Chinese Classic of Poetry.* New York: Grove, 1996.

Walker, D. P. *The Ancient Theology: Studies in Christian Platonism from the Fifteenth to the Eighteenth Century.* London: Duckworth, 1972.

Wan Sida 萬斯大. *Jingxue wushu* 經學五書 (Five studies of classical texts). Reprint. Taipei: Guangwen, 1977.

Wan Sitong 萬斯同. *Miaozhi tukao* 廟製圖考 (An analysis and illustration of the configuration of imperial palaces). Taipei: Shangwu yinshuguan, 1983.

Wang, Jing. *High Culture Fever: Politics, Aesthetics, and Ideology in Deng's China.* Berkeley: University of California Press, 1996.

Wang Biao 王標. *Chengshi zhishifenzi de shehui xingtai—Yuan Mei ji qi jiaoyou wangluo de yanjiu* 城市知識分子的社會形態—袁枚及其交游網絡的研究 (The social patterns of urban intellectuals: A study of Yuan Mei and his social networks). Shanghai: Shanghai Sanlian, 2008.

Wang Chang 王昶. *Chunrongtang ji* 春融堂集 (Collection from the Hall of Early Spring). *Xuxiu Siku quanshu*, vol. 1437.

———. "Dai Dongyuan xiansheng muzhi ming" 戴東原先生墓志銘 (The funerary writing for Dai Zhen). In *Dongyuan xiansheng quanji* 戴東原先生全集 (The complete work of Dai Zhen), vol. 6 of Anhui Collectanea (Anhui congshu, 安徽叢書), compiled by Anhui Collectanea, 64. Reprint, Dahua bookstore, 1987.

Wang Damin 王達敏. *Yao Nai yu Qian-Jia xuepai* 姚鼐與乾嘉學派 (Yao Nai and the schools of the Qianlong and Jiaqing reigns). Beijing: Xueyuan, 2007.

Wang Fansen 王汎森. "Mingmo Qingchu de renpu yu shengguohui" 明末清初的人譜與省過會 (Genealogies and associations for moral cultivation in the late Ming and early Qing). *Zhongyang yanjiuyuan lishi yuyan yanjiusuo jikan* 63, no. 3 (1993): 679–712.

———. *Zhongguo jindaisixiang yu xueshudexipu* 中國近代思想與學術的系譜 (Genealogy of scholarship and modern thought in China). Taipei: Lianjing, 2003.

Wang Guowei 王國維. *Wang Guowei yi shu* 王國維遺書 (A collection of Wang Guowei's posthumous works). Shanghai: Shanghai Guji Shudian, 1983.

Wang Hui 汪暉. *Xiandai Zhongguo sixiang de xingqi* 現代中國思想的興起 (The rise of modern Chinese thought). Beijing: Shenghuo, Dushu, Xingzhi Sanlian, 2004.

Wang Jingqi 汪景祺. "Gao Wenque yishi" 高文恪遺事 (Gao Shiqi's anecdotes). In *Dushutang xizheng suibi* 讀書堂西征隨筆 (Jottings on a westbound military campaign). Reprint. Hong Kong: Longmen, 1967.

Wang Li 王力. *Hanyu yinyunxue* 漢語音韻學 (Chinese phonology). Beijing: Zhonghua, 1956.

Wang Mingsheng 王鳴盛. "Eshu pian" 蛾術篇 (Notes from diligent learning). In *Wang mingsheng dushu biji shiqi zhong* 王鳴盛讀書筆記十七種 (Seventeen of Wang Mingsheng's notebooks). In *Jin sanbainian dushu biji huibian* 近三百年讀書筆記彙編 (A collection of academic jottings and miscellanies in the past three hundred years), edited by Yang Jialuo 楊家駱. Taipei: Dingwen, 1979.

———. *Xi zhuang shi cun gao* 西莊始存稿 (Wang Mingsheng's preserved writings). *Xuxiu Siku quanshu*, vol. 1434.

Wang Shiduo 汪士鐸. "Xu" 序 (Prologue). In *Chuhan zhuhou jianyuzhi* 楚漢諸侯疆域志 (Historical geography of Chu and Han feudal lords), Liu Wenqi 劉文淇, 1a-1b. Reprint. Taipei: Yiwen, 1964.

Wang Xichan 王錫闡. *Xiao'an xinfa* 曉庵新法 (The new methods of Wang Xichan). Vol. 793. *Siku quanshu*.

Wang Yangzong 王揚宗. "'Xixue zhongyuan' shuo zai Ming Qing zhi ji de youlai ji qi yanbian" 「西學中源」說在明清之際的由來及其演變 (The origin and evolution of the doctrine of the "Chinese origins of Western learning" during the Ming-Qing transition). *Dalu zazhi* 90, no. 6 (June 1995): 39–45.

Wang Yingdian 王應電. *Zhouli tushuo* 周禮圖說 (Illustrations and explanations of "The Rituals of Zhou"). *Siku quanshu*, vol. 96.

Wang Yinzhi 王引之. *Jing zhuan shi ci* 經傳釋詞 (An etymology of the classics and the commentaries). Changsha: Yuelu Shushe, 1984.

———. *Jingyi shuwen* 經義述聞 (Statement and hearsay of the meanings of the classics). Taipei: Taiwan Zhonghua, 1965.

Wang Zhenzhong 王振忠. *Huizhou shehui wenhuashi tanwei: Xin faxian de 16-20 shiji minjian wenshu yanjiu* 徽州社會文化史探微：新發現的16-20世紀民間檔案文書研究 (A subtle inquiry into the social and cultural history of Huizhou: Studies of newly discovered archives and contracts in rural Huizhou). Shanghai: Shanghai shehui kexueyuan chubanshe, 2002.

Wang Zhong 汪中. *Wang Zhong ji* 汪中集 (A collection by Wang Zhong). Edited by Wang Qingxin 王清信 and Ye Chunfang 葉純芳. Taipei: Zhongyang Yanjiu Yuan, 1990.

Wang Zhonghan 王鍾翰, comp. *Qing shi liezhuan* 清史列傳 (Official biographies of the Qing History). Beijing: Zhonghua shuju, 2005.

Wantong xiangshan daishi zongpu 皖桐香山戴氏宗譜 (Geneaology of the Dai Lineage in Xiangshan, Tongcheng County, Anhui Province). Taipei: Academia Sinica, Fusinian Library, Rare Book Collection, 1773.

Watson, Burton, trans. *Early Chinese Literature*. New York: Columbia University Press, 1962.

———. *Records of the Grand Historian of China*. 2 vols. New York: Columbia University Press, 1961.

———. *Ssu-Ma Ch'ien, Grand Historian of China*. New York: Columbia University Press, 1958.

Weber, Max. *The Protestant Ethic and the Spirit of Capitalism*. Translated by Talcott Parsons. New York: Charles Scribner's Sons, 1958.

Wechsler, Howard J. *Offerings of Jade and Silk: Ritual and Symbol in the Legitimation of the T'ang Dynasty*. New Haven, CT: Yale University Press, 1985.

Wei Jiangong 魏建功. *Wei Jiangong wenji* 魏建功文集 (A collection of writings by Wei Jiangong). 5 vols. Nanjing: Jiangsu Jiaoyu Chubanshe, 2001.

Wei Liaozi 尉繚子. *Siku quanshu*.

Wei Yuan 魏源. *Wei yuan ji* 魏源集 (Collection of Wei Yuan). Beijing: Zhonghua Shuju, 1976.

Weng Fanggang 翁方綱. *Fuchuzhai wenji* 復初齋文集 (Collection in the Studio of Restoring the Beginning). In *Xuxiu Siku quanshu* 續修四庫全書 (The sequel to the complete collection of the four treasuries), vols. 1454–55. Reprint. Shanghai: Shanghai Guji Chubanshe, 2002.

Wenxian congbian quanbian 文獻叢編全編 (A complete collection of archival sources), edited by the National Palace Musuem. Beijing: Beijing Tushuguan Chubanshe, 2008.

Wenxian tongkao 文獻通考 (Comprehensive historical studies of the long-term development of government institutions based on authoritative sources and later interpretation). Vol. 612. *Siku quanshu*.

Westman, Robert. *The Copernican Question: Prognostication, Skepticism, and Celestial Order*. Berkeley: University of California Press, 2011.

Wilkinson, Endymion. *Chinese History: A New Manual*. Cambridge, MA: Harvard University Press, 2012.

Will, Pierre-Etienne. *Bureaucracy and Famine in Eighteenth-Century China*. Translated by Elborg Forster. Stanford, CA: Stanford University Press, 1990.

Wilson, Thomas. *Genealogy of the Way: The Construction and Uses of the Confucian Tradition in Late Imperial China*. Stanford, CA: Stanford University Press, 1995.

Witek, John W. *Controversial Ideas in China and in Europe: A Biography of Jean-François Foucquet, S. J., (1665–1741)*. Rome: Institutum Historicum S. I., 1982.

Wolf, Arthur, ed. *Studies in Chinese Society*. Stanford, CA: Stanford University Press, 1978.

Woodside, Alexander. "State, Scholars, and Orthodoxy: The Ch'ing Academies, 1736–1839," In *Orthodoxy in Late Imperial China*, edited by Kwang-ching Liu, 158–84. Berkeley: University of California Press, 1990.

Woodside, Alexsander, and Benjamin Elman. "The Expansion of Education in Ch'ing China." In *Education and Society in Late Imperial China, 1600–1900*, edited by Benjamin Elman and Alexander Woodside, 525-61. Berkeley: University of California Press, 1994.

Wright, Arthur. *The Sui Dynasty*. New York: Knopf, 1978.

Wu, Genyou. "A Preliminary Discussion of Dai Zhen's Philosophy of Language." *Frontiers of Philosophy in China* 5, no. 4 (2010): 523–42.

Wu, Silas H. L. *Communication and Imperial Control in China: Evolution of the Palace Memorial System, 1693–1735*. Cambridge, MA: Harvard University Press, 1970.

———. *Passage to Power: K'ang-Hsi and His Heir Apparent, 1661–1722*. Cambridge, MA: Harvard University Press, 1979.

Wu Boya 吳伯婭. "Guanyu yongzheng jinjiao de jige wenti" 關於雍正禁教的幾個問題 (A few questions regarding the persecution of Catholicism during the Yongzhen reign). *Qingshi luncong* 清史論叢 (Academic colleactanea of Qing history; 2003-4): 160–82.

———. "Yesu huishi Baijin dui Yi jing de yanjiu" 耶穌會士白晉對易經的研究 (The Jesuit Joachim Bouvet's research on "The Book of Changes"). In *Zhong-Xi chushi er bian* 中西初識二編 (The first encounter between China and the West, second series), edited by Zhong wai guan xishi xuehui, 57–58. Zhengzhou: Daxiang Chubanshe, 2002.

Wu Hung. *Monumentality in Early Chinese Art and Architecture*. Stanford, CA: Stanford University Press, 1995.

———. "Picturing or Diagramming the Universe." In *Graphics and Text in the Production of Technical Knowledge in China: The Warp and the Weft*, edited by Francesca Bray, Vera Dorofeeva-Lichtmann, and Georges Metailie, 191-214. Leiden: Brill, 2007.

———. *Remaking Beijing: Tiananmen Square and the Creation of a Political Space*. Chicago: University of Chicago Press, 2005.

Wu Jen-shu 巫仁恕. "Mingdai pingmin fushi de liuxing fengshang yu shidafu de fanying" 明代平民服飾的流行風尚與士大夫的反應 (Changing fashions in plebeian clothes and the literati response during the Ming dynasty). *Xin shixue* 10, no. 3 (1999): 55–109.

Wu Jingzi 吳敬梓. *Rulin Waishi* 儒林外史 (An unofficial history of scholars). Taipei: Wenhua Tushu Gongsi, 1994.

Wu Min 吴旻 and Han Qi 韓琦 (eds.). *Ouzhou suo zang Yongzheng Qianlong chao Tianzhujiao wenxian huibian* 歐洲所藏雍正乾隆朝天主教文獻匯編 (Catholicism during the Yongzheng and Qianlong reigns: Documents from European libraries and archives). Shanghai: Shanghai Renmin Chubanshe, 2008.

Wu Sixiao 吳思孝. "Xu" (Prologue). In *Dongyuan xiansheng quanji* 戴東原先生全集 (The complete work of Dai Zhen), vol. 6 of Anhui Collectanea (Anhui congshu, 安徽叢書), compiled by Anhui Collectanea, 832. Taipei: Dahua, 1936.

Wu Xiu 吳修, ed. *Zhaodai mingren chidu xiaozhuan* 昭代名人尺牘小傳 (Short biographies, correspondence, and manuscripts of renowned Qing literati). 2 vols. Taipei: Mingwen, 1985.

Wu Zhefu 吳哲夫. *Siku quanshu zuanxiu zhi yanjiu* 四庫全書纂修之研究 (A study of the compilation of the complete collection of the four treasuries). Taipei: National Palace Musuem, 1990.

Wu Zhenyu 吳振棫. *Yangjizhai conglu* 養吉齋叢錄 (A collection of records from Auspiciousness Fostering Studio). Beijing: Zhonghua, 2005.

Wuli tongkao 五禮通考 (A comprehensive investigation of five rites). Reprint. Taipei: Xinxing, 1970.

Xi Zezong 席澤宗. "Lun Kangxi kexue zhengce de shiwu" 論康熙科學政策的失誤 (On the shortcomings of the Kangxi emperor's scientific policies). *Ziran kexueshi yanjiu* 19, no. 1 (2000): 18–29.

Xia Shang Zhou duandai gongcheng zhuanjiazu 夏商周斷代工程傳家組. *Xia-Shang-Zhou duandai gongcheng 1996–2000 nian jieduan chengguo baogao* (Preliminary report of the Xia-Shang-Shou periodization project, 1996–2000). Beijing: Shijie, 2000.

Xie Guozhen 謝國楨. "Qingchu Shunzhi-Kangxi jian zhi dangzheng" 清初順治康熙間之黨爭 (The factional struggle in Shunzhi's and Kangxi's reigns in early Qing). In *Ming-Qing zhiji dangshe yundong kao* 明清之際黨社運動考 (Evidential analysis of factional movements during the Ming-Qing transition), 96–118. Beijing: Zhonghua, 1982.

Xiyang xinfa lishu 西洋新法曆書 (The new method of astronomy from the Western Ocean). Heiko: Heinan, 2000.

Xu Chengyao 許承堯 et al., comp. *Shexian zhi* 歙縣志 (She County gazetteer). Reprint: vol. 246 of *Zhongguo fangzhi congshu*. Taipei: Chengwen Chubanshe, [1937] 1975.

Xu Fuguan 徐復觀. *Lianghan sixiang shi* 兩漢思想史 (Intellectual history of the former and later Han dynasties). 3 vols. Taipei: Xuesheng, 1979.

Xu Yuanzheng 徐元正. "Xu" 序 (Prologue). In *Tianfang xingli* 天方性理 (Nature and principles of Islam), Liu Zhi 劉智, 5-7. *Xuxiu Siku quanshu*, vol. 1296.

Xuxiu siku quanshu 續修四庫全書 (Sequel to the complete collection of the four treasuries). Shanghai: Shanghai Guji Chubanshe, 2002.

Yabuuchi Kiyoshi 薮内清. *Chūgoku no sūgaku* 中国の数学 (Chinese mathematics). Tokyo: Iwanami, 1974.

———. *Chūgoku no tenmon rekihō* 中国の天文暦法 (Chinese mathematical astronomy). Tokyo: Heibonsha, 1969.

———. "Taishin no rekisan gaku" 戴震の暦算学 (Dai Zhen's calendrical mathematics). In *Min-Shin jidai no kagaku gijutsu shi* 明清時代の科学技術史 (History of science and technology in the Ming and Qing periods), edited by Yabuuchi Kiyoshi and Yoshida Mitsukuni 吉田光邦, 27–34. Kyoto: Kyoto daigaku jinbun kagaku kenkyūjo, 1970.

Yamada Keiji 山田慶児. *Jujireki no michi: Chūgoku chūsei no kagaku to kokka* 授時歴の道—中国中世の科学と国家 (The way to the season-granting calendar: Science and state in medieval China). Tokyo: Misuzu, 1980.

———. *Shushi no shizengaku* 朱子の自然学 (The nature studies of Zhu Xi). Tokyo: Iwanami, 1978.

Yamaguchi Hisakazu 山口久和. *Shō Gakusei no chishikiron: Kōshogaku hihan o chūshin to shite* 章學誠の知識論 : 考證学批判を中心として (Epistemology of Zhang Xuecheng: Centered on his critique of evidential scholarship). Tokyo: Sōbunsha, 1998.

Yamamoto Eishi 山本英史. *Dentō Chūgoku no chiikizō* 伝統中国の地域像 (Local society in traditional China). Tokyo: Keiō Gijuku Daigaku Shuppankai, 2000.

Yamanoi Yū 山井湧. *Min-Shin shisōshi no kenkyū* 明清思想史の研究 (A study of Ming and Qing intellectual history). Tokyo: Tokyo Daigaku Shuppankai, 1980.

Yan Dunjie 嚴敦傑. "Xinbian zhizhi suanfa tongzong yishu de liuchuan" 新編直指算法統宗一書的流傳 (The circulation of the book"Principal and General Methods in

Abacus Mathematics" [suanfa tongzong]). In *Suanfa tongzong jiaoshi* 算法統宗校釋, attributed to Cheng Dawei 程大位 (1533-1606), 1–5. This version is collated and commented on by Mei Rongzhao 梅榮照 and Li Zhaohua 李兆華. Taipei: Jiuzhang Chubanshe, 1992.

Yan Wenyu 嚴文郁, ed. *Qing ru zhuanlue* 清儒傳略 (Biographical sketches of Qing literati). Taipei: Shanwu Press, 1990.

Yang Guangxian 楊光先. *Bu de yi: Fu erzhong* 不得已: 附二种 (I cannot do otherwise, with two [Jesuit] responses). Compiled and edited by Chen Zhanshan 陳占山. Hefei: Huangshan Shushe Chubanshe, 2000.

Yang Hongxun 楊鴻勛. *Gongdian kaogu tonglun* 宮殿考古通論 (A general discussion on architectural archaeology of ancient palaces). Beijing: Zijingcheng chubanshe, 2001.

———. "Mingtang fanlun—Mingtang de kaoguxue yanjiu" 明堂泛論—明堂的考古學研究 (A general discussion of Bright Hall: An archaeological study of Bright Hall). *Tōhō gakuhō* 70 (1998): 1–194.

———. *Yang Hongxun jianzhu kaoguxue lunwenji* 楊鴻勛建築考古學論文集 (A collection of Yang Hongxu's essays on architectural archaeology). Beijing: Qinghua daxue chubanshe, 2008.

Yang Liansheng 楊聯陞. *Yang Liansheng lunwenji* 楊聯陞論文集 (A collection of Yang Liansheng's essays). Beijing: Zhongguo Shehui Kexue Chubanshe, 1992.

Yang Nianqun 楊念群. *Rujia diyuhua de jindai xingtai: Sanda zhishi qinti hudong de bijiao yanjiu* 儒家地域化的近代形態： 三大知識群體的比較研究 (Modern patterns of regional Confucianism: A comparative study of three major intellectual groups). Beijing: Shenghuo, Dushu, Xinzhi Sanlian Bookstore, 1997.

Yao Nai 姚鼐. *Xibao xiansheng chi du* 惜抱先生尺牘 (Collection of Yao Nai's correspondences). Beijing: Beijingshi Zhongguo Shudian, 1986.

———. *Xibaoxuan shiwen ji* 惜抱軒詩文集 (A collection of Yao Nai's poetry and prose). Shanghai: Shanghai Guji Chubanshe, 1992.

Yasuda Jirō 安田二郎 and Kondō Mitsuo 近藤光男. *Tai Shin shū* 戴震集 (A collection by Dai Zhen). Tokyo: Asahi Shinbunsha, 1971.

Ye Dehui 葉德輝. *Shulin qinghua* 書林清話 (Plain talks on books). Taipei: Wenshizhe Chubanshe, 1998.

Ye Gongchuo 葉恭綽, ed. *Qingdai xuezhe xiangzhua*n 清代學者象傳 (The portraits of Qing scholars). Shanghai: Shangwu yinshuguan, 1930.

Ye Shuxian 葉舒憲. *Zhongguo shenhua zhexue* 中國神話哲學 (Mythology and philosophy in ancient China). Beijing: Zhongguo Shehui Kexue Chubanshe, 1992.

Yeh, Wen-hsin. *Provincial Passages: Culture, Space, and the Origins of Chinese Communism*. Berkeley: University of California Press, 1996.

Yongzhengchao manwen zhupi zouzhe quanyi 雍正朝滿文朱批奏折全譯 (Complete translation of Manchu memorials with vermillion comments in the Yongzheng emperor's reign). Hefei: Huangshan shushe, 1998.

Yoshida Jun 吉田純. *Shinchō kōshogalu no gunzō* 清朝考処学の群像 (The collective image of evidential scholars in the Qing dynasty). Tokyo: Sobunsha, 2006.

Yü Ying-shih 余英時. *Lun Dai Zhen yu Zhang Xuecheng: Qingdai zhongqi xueshu sixiang shi yanjiu* 論戴震與章學誠: 清代中期學術史研究 (On Dai Zhen and Zhang Xuecheng: A study of mid-Qing intellectual history). Taipei: Dongda, 1996.

———. "Tai Chen's Choice between Philosophy and Philology." *Asia Major, Third Series*, 2, no. 1 (1989): 79–108.

———. *Zhongguo jinshi zongjiao lunli yu shangren jingshen* 中國近世宗教倫理與商人精神 (The religious ethics and merchant spirit in modern China). Taipei: Lianjing Shuju, 1987.

———. *Zhu Xi de lishi shijie: Songdai shidaifu zhengzhi wenhua de yanjiu* 朱熹的歷史世界：宋代士大夫政治文化的研究 (The historical world of Zhu Xi: A study of political culture of Song literati). 2 vols. Taipei: Yuncheng Wenhua, 2003.

Yuan Mei 袁枚. *Yuan Mei quanji* 袁枚全集 (The complete works of Yuan Mei). Edited by Wang Yingzhi 王英志. 8 vols. Nanjing: Jiangsu Guji Chubanshe, 1993.

Yuzhi lixiang kaocheng houbian 御製曆象考成後編 (Supplement to "The Compendium of Observational and Computational Astronomy"). "Zouyi"奏議 (Report and discussion), 1a–13b. *Siku quanshu.*

Zhang, Qiong. "About God, Demons, and Miracles: The Jesuit Discourse on the Supernatural in Late Ming China." *Early Science and Medicine*, 4, no. 1 (1999): 1–36.

Zhang Bochun 張柏春. *Ming Qing cetian yiqi zhi Ouhua* 明清測天儀器之歐化 (The Europeanization of astronomical instruments in Ming-Qing China). Shenyang: Liaoning Jiaoyu Chubanshe, 2000.

Zhang Shilu 張世祿. *Zhongguo guyinxue* 中國古音學 (Phonology of old Chinese). Shanghai: Shangwu Yinshuguan, 1933.

Zhang Shunhui 張舜徽. *Qingdai Yangzhou xueji* 清代揚州學記 (A record of learning at Yangzhou during the Qing dynasty). Shanghai: Shanghai Rennin Chubanshe, 1962.

Zhang Taiyan 章太炎. *Zhang Taiyan quanji* 章太炎全集 (The complete collection of Zhang Taiyan). Shanghai: Shanghai Renmin Chubanshe, 1985.

Zhang Xingyao 張星曜. *Tianru tongyi kao* 天儒同異考 (Analysis on the similarity and difference between Catholicism and Confucianism). Chinois 7171, Bibliothèque Nationale de France. Reprinted in *Faguo guojia tushuguan mingqing tianzhujiao wenxian* 法國國家圖書館明清天主教文獻 (Chinese Christian texts from the Bibliothèque Nationale de France), edited by Nicolas Standaert, Ad Dudink, and Nathalie Monnet, 8:429–558. Taipei: Taipei Ricci Institute, 2009.

Zhang Xuecheng 章學誠. *Wenshi tongyi jiaozhu* 文史通義校注 (General and comprehensive meanings of culture and history, a punctuated version). Beijing: Zhonghua, 1985.

Zhang Yibing 張一兵. *Mingtang zhidu yanjiu* 明堂制度研究 (An investigation of the institutions of the Palace of Light). Beijing: Zhonghua, 2005.

Zhao Erxun 趙爾巽 et al., comp. *Qing shi gao* 清史稿 (Draft history of the Qing dynasty). Beijing: Zhonghua shuju, 1997.

Zheng Jixiong 鄭吉雄. "Ping Qiu Weijun 'Dai Zhen xue de xingcheng'" 評丘為君《戴震學的形成》 (Review of Qiu Weijung's "The Form of Dai Zhen Studies"). *Taiwan Dongya wenming yanjiu xuekan* 2, no. 1 (June 2005): 197–213.

Zhongguo kexue jishu dianji tonghui: Shuxue juan 中國科學技術典籍通匯：數學卷 (General introduction to comprehensive compendium of materials and sources on Chinese science and technology: Mathematics volume). Edited by Guo Shuchun 郭書. Shanghai: Henan Jiaoyu Chubanshe, 1993.

Zhongguo kexue jishu dianji tonghui: Wuli juan 中國科學技術典籍通匯：物理卷 (General introduction to comprehensive compendium of materials and sources on Chinese science and technology: Physics volume). Edited by Dai Nianzu 戴念祖. Shanghai: Henan Jiaoyu Chubanshe, 1995.

Zhongguo kexue jishu dianji tonghui: Zonghe juan 中國科學技術典籍通匯：綜合卷 (Comprehensive compendium of materials and sources on Chinese science and technology: General introduction volume). Edited by Lin Wenzhao 林文照. Shanghai: Henan Jiaoyu Chubanshe, 1993.

Zhu Weizheng 朱維錚. *Coming Out of the Middle Ages: Comparative Reflections on China and the West*. Translated by Ruth Hayhoe. Armonk, NY: M. E. Sharpe, 1990.

———. *Zouchu zhongshiji* 走出中世纪 (Coming out of the Middle Ages). Shanghai: Fudan Daxue Chubanshe, 2007.

Zhu Xi 朱熹. *Sishu zhangju jizhu* 四書章句集註 (Commentary on chapters, sections, sentences and phrases of the Four Books). *Siku quanshu*.

———. *Zhouyi benyi* 周易本義 (Fundamental meaning of "The Book of Changes"). *Siku quanshu*.

———. *Zhuzi daquan* 朱子大全 (The complete collection of Master Zhu Xi). 2 vols. Sibu beiyao 四部備要 edition. Shanghai: Zhonghua, 1936.

———. *Zhuzi quanshu* 朱子全書 (The complete works of Master Zhu Xi). Shanghai: Shanghai Guji Chubanshe, 2002.

Zito, Angela. *Of Body and Brush: Grand Sacrifice as Text/Performance in Eighteenth-Century China*. Chicago: University of Chicago Press, 1997.

Zurndorfer, Harriet T. *Change and Continuity in Chinese Local History: The Development of Huizhou Prefecture, 800 to 1800*. Leiden: Brill, 1989.

———. "'One Adam Having Driven Us Out of Paradise, Another Has Driven Us Out of China' Yang Kuang-hsien's Challenge of Adam Schall von Bell." In *Conflict and Accommodation in Early Modern East Asia: Essays in Honour of Erik* Zürcher, edited by Leonard Blusse and Harriet Zurndorfer, 141–68. Leiden: Brill, 1993.

INDEX